Ionic Liquids in the Biorefinery Concept
Challenges and Perspectives

RSC Green Chemistry

Editor-in-Chief:
Professor James Clark, *Department of Chemistry, University of York, UK*

Series Editors:
Professor George A. Kraus, *Department of Chemistry, Iowa State University, Ames, Iowa, USA*
Professor Andrzej Stankiewicz, *Delft University of Technology, The Netherlands*
Professor Peter Siedl, *Federal University of Rio de Janeiro, Brazil*

How to obtain future titles on publication:
A standing order plan is available for this series. A standing order will bring
delivery of each new volume immediately on publication.

For further information please contact:
Book Sales Department, Royal Society of Chemistry, Thomas Graham
House, Science Park, Milton Road, Cambridge, CB4 0WF, UK
Telephone: +44 (0)1223 420066, Fax: +44 (0)1223 420247
Email: booksales@rsc.org
Visit our website at www.rsc.org/books

Ionic Liquids in the Biorefinery Concept

Challenges and Perspectives

Edited by

Rafal Bogel-Lukasik
National Laboratory of Energy and Geology, Lisbon, Portugal
Email: rafal.lukasik@lneg.pt

RSC Green Chemistry No. 36

Print ISBN: 978-1-84973-976-4
PDF eISBN: 978-1-78262-259-8
ISSN: 1757-7039

A catalogue record for this book is available from the British Library

Published by The Royal Society of Chemistry,
Thomas Graham House, Science Park, Milton Road,
Cambridge CB4 0WF, UK

Registered Charity Number 207890

For further information see our web site at www.rsc.org

Printed and bound by CPI Group (UK) Ltd, Croydon, CR0 4Y

Foreword

In 1983 Chuck Hussey wrote a review, *Room Temperature Molten Salt Systems*,[1] that reported everything that was known at that time about ionic liquids by referencing only 82 articles. A relatively small band of devotees continued to make steady progress for the next 15 years until, at the turn of this century, interest in ionic liquids began to explode. By the end of 2014, Web of Science™ listed 58 771 articles with the phrase 'ionic liquid' (or 'ionic liquids') in the title. If one wanted to produce a text that covered all that is known about ionic liquids today, it would undoubtedly be a multivolume collection and, like the never-ending task of painting the Forth Bridge, no sooner would it be finished than it would be time to start again. Consequently, in recent years we have been provided with a number of monographs and collections that tell us about one aspect or another of ionic liquids and their applications. *Ionic Liquids in the Biorefinery Concept* complements some other excellent texts, notably: *Electrodeposition in Ionic Liquids*, edited by Frank Endres, Andy Abbott and Doug MacFarlane; *Electrochemical Aspects of Ionic Liquids*, edited by Hiro Ohno; *Ionic Liquids in Synthesis*, edited by Peter Wasserscheid and myself; *Handbook of Green Chemistry, Vol. 6: Ionic Liquids*, edited by Peter Wasserscheid and Annegret Stark, and (recently) *Topics in Current Chemistry: Ionic Liquids*, edited by Barbara Kirchner.

Over a very similar timeframe the concept of the biorefinery has also been being actively developed to a very similar scale, with 74 091 articles with 'biorefinery' or 'biofuels' recorded as topics in Web of Science™ by the end of 2014. *Ionic Liquids in the Biorefinery Concept* complements some other excellent texts, notably: *Biorefineries – Industrial Processes and Products*, edited by Birgit Kamm, Patrick R. Gruber and Michael Kamm, *Biorefinery Co-Products*, edited by Chantal Bergeron, Danielle J. Carrier and Shri Ramaswamy, *Integrated Forest Biorefineries: Challenges and Opportunities*,

RSC Green Chemistry No. 36
Ionic Liquids in the Biorefinery Concept: Challenges and Perspectives
Edited by Rafal Bogel-Lukasik
© The Royal Society of Chemistry 2016
Published by the Royal Society of Chemistry, www.rsc.org

edited by Lew Christopher, and *Renewable Resources for Biorefineries*, edited by Carol Lin and Rafael Luque.

It was perhaps inevitable that these two fields of study would have been brought together at some point; the first papers bringing the two together appeared in 2001. However, that they should generate so much interest so quickly could not have been so easily predicted. It is this activity that is reviewed so effectively here.

Of course, it is first necessary to understand what is meant by the term 'biorefinery' and how it fits into the history of the utilization of biomass, which is by no means a new activity (Chapter 1). We discover that these older technologies were heavily polluting and far from any modern definition of Green Chemistry, in spite of their use of a bio-renewable starting material. We also discover that there is more than one type of biorefinery – many versions are beginning to spring up around the world, tuned to match a local biomass source. We must not forget the importance of applying rigorous Green analysis to any proposed processes (Chapter 6).

The very first step in the use of lignocellulose biomass is to break it down into its component parts, predominantly the polymers cellulose, hemicellulose and lignin. This is also the most difficult step in the economic exploitation of biomass. Ionic liquid pre-treatments have shown much promise in this area (Chapter 2). Ionic liquids with imidazolium cations and basic (mostly acetate) anions have received the greatest attention, with the intention being the dissolving and reprecipitation of cellulose. A whole variety of process variables have been investigated – biomass type, loading and particle size, experimental conditions, and ionic liquids' physical properties.

Technoeconomic analysis based on the results of such studies have emphasized the importance of ionic liquid recycling in achieving an economically viable ionic liquid-based biomass pre-treatment process (Chapter 3).

To obtain fermentable sugars, it is necessary to depolymerize the carbohydrate polymers by hydrolysis (Chapter 4). Acid-catalysed hydrolysis of either whole biomass (without prior pre-treatment) or cellulose itself has been shown to be a viable process option, either by adding acid to an ionic liquid or using an ionic liquid that has an acidic functionality. The use of extremophile-derived enzymes to catalyse the hydrolysis has also been shown to be possible.

When using plant biomass as a source of chemicals and fuels, it is vital that its hugely variable nature is understood. Not only does the composition of biomass depend upon its type (hardwood, softwood, grass, different species, ...), but variables such as point of harvest in the growing season have the potential to lead to the need for different processing conditions (Chapter 5). One should also not forget that different crops might be selected to produce different potential products.

Although ethanol is often thought of as the primary output of the biorefinery, many other potential platform chemicals have been identified. 5-Hydroxymethylfurfural (HMF) production has received particular attention in ionic liquids (Chapter 7).

A number of ionic liquid-based approaches have also been suggested for the extraction of high-value products from biomass (Chapter 8).

In spite of all of these advances and promising results, no ionic liquid based biorefining process has yet been commercialised. It is, of course, very early days for these approaches. It is necessary to think about what barriers exist to the implementation of ionic liquids in this area (Chapter 10) and to work on technological approaches that can overcome these.

I have no doubt that research into the applications of ionic liquids in the utilization of biomass has a very exciting future.

Tom Welton
Professor of Sustainable Chemistry
Imperial College London

Reference

1. C. L. Hussey, *Adv. Molten Salt Chem.*, 1983, **5**, 185.

Preface

The chapters in this book discuss the unique work concerning the role of ionic liquids in the biorefinery concept. Judging by the extraordinary interest in this field and its rapid forward movement with high-quality works, it can be surely stated that ionic liquids have found their place in the biorefinery concept.

The unique and especially easily tuneable properties of ionic liquids make them ideal candidates for reaction media for biomass dissolution, pre-treatment, bio- and chemo- conversion as well as solvents for downstream processing. It is also very important that ionic liquids are no longer expensive solvents; it is especially important that recycling and reuse of ionic liquids is no longer a challenge. Because of this, ionic liquids may contribute to make the green biorefinery concept more economically sustainable and environmentally benign and especially more realistic. As such, ionic liquids can be used in the production of not only specific, high-value niche products but they can also compete in the more sustainable production of bulk commodities.

Nonetheless there is still a strong requirement to develop more and more green and sustainable processes using ionic liquids in the context of the biorefinery concept, and the chapters of this book draw the guidelines to achieve this. This book shows both the state-of-the-art and the future for ionic liquids and biorefineries. Furthermore, it also depicts the most important aspects of the use of ionic liquids, starting from biomass dissolution, pre-treatment and fractionation and moving towards the main fractions of biomass. The specific chapters tackle aspects of direct extraction of diverse value-added compounds from biomass, the separation and purification of bioproducts and biomolecules with ionic liquids, the integrated pre-treatment and hydrolysis of biomass catalysed by ionic liquids, as well as biomass conversion to pivot chemicals. Looking at the toxicity and bioacceptability

RSC Green Chemistry No. 36
Ionic Liquids in the Biorefinery Concept: Challenges and Perspectives
Edited by Rafal Bogel-Lukasik
© The Royal Society of Chemistry 2016
Published by the Royal Society of Chemistry, www.rsc.org

of ionic liquids in the context of the biological processes also shows how important findings in this field are helping to answer questions about the greenness of a biorefinery that uses ionic liquids. The content of the chapters in this book portrays the challenges and achievements in the field of green biorefinery concept.

I hope that this book will achieve my aim of communicating the excitement, breadth and depth of the application of ionic liquids to an audience from both academia and industry.

This book is made possible only by the invaluable contributions of all the authors of the chapters. I also wish to thank Fundação para a Ciência e a Tecnologia (Portugal) and Coordenação de Aperfeiçoamento de Pessoal de Nível Superior (Brazil) for their financial support.

Rafal Bogel-Lukasik
National Laboratory of Energy and
Geology, Lisbon, Portugal

Contents

RSC Green Chemistry No. 36
Ionic Liquids in the Biorefinery Concept: Challenges and Perspectives
Edited by Rafal Bogel-Lukasik
© The Royal Society of Chemistry 2016
Published by the Royal Society of Chemistry, www.rsc.org

Chapter 8 Ionic Liquids as Efficient Tools for the Purification of Biomolecules and Bioproducts from Natural Sources 227
Matheus M. Pereira, João A. P. Coutinho, and Mara G. Freire

Chapter 9 Ionic Liquids in the Biorefinery: How Green and Sustainable Are They? 258
Roger A. Sheldon

The Biorefinery and Green Chemistry

JYRI-PEKKA MIKKOLA*[a,b], EVANGELOS SKLAVOUNOS[c], ALISTAIR W. T. KING[c], AND PASI VIRTANEN[a]

[a]Johan Gadolin Process Chemistry Centre, c/o Laboratory of Industrial Chemistry and Reaction Engineering, Åbo Akademi University, Biskopsgatan 8, 20500 Åbo/Turku, Finland; [b]Technical Chemistry, Department of Chemistry, Chemical-Biological Center, Umeå University, 90187 Umeå, Sweden; [c]Laboratory of Organic Chemistry, Department of Chemistry, University of Helsinki, AI Virtasen Aukio 1, PO Box 55, FIN-00014 Helsinki, Finland
*E-mail: jpmikkol@abo.fi

1.1 Introduction

1.1.1 Definitions

A biorefinery is a facility where different low-value renewable biomass materials are the feedstock to the processes where they are transformed, in multiple steps including fractionations, separations and conversions, to several higher-value bio-based products. Examples of these products can include fibres, food, feed, fine chemicals, transportation fuels and heat. A biorefinery can be formed by a single unit or can combine several facilities targeted for a single purpose that further process products as well as by-products or wastes of combined facilities. In biorefining one can find similarities to oil refining, with the exception

RSC Green Chemistry No. 36
Ionic Liquids in the Biorefinery Concept: Challenges and Perspectives
Edited by Rafal Bogel-Lukasik

Published by the Royal Society of Chemistry, www.rsc.org

that in oil refining the raw material comes from fossil resources. According to the International Energy Agency 'Biorefineries will contribute significantly to the sustainable and efficient use of biomass resources, by providing a variety of products to different markets and sectors. They also have the potential to reduce conflicts and competition over land and feedstock, but it is necessary to measure and compare the benefits of biorefineries with other possible solutions to define the most sustainable option.'[1] Although it is possible to produce the same products in a biorefinery as in an oil refinery, this is not the target, which instead is to produce products which can replace the products from oil refining.

In the development of biorefinery processes, as well as any industrial processes, it is crucial for the future of the Earth that the new processes follow the principals of sustainable development and green chemistry. It is good to remind what these terms really mean.

The term 'sustainable development' was famously used by the Brundtland Commission in its report to the United Nations. In the report the term 'sustainable development' was defined as, 'development that meets the needs of the present without compromising the ability of future generations to meet their own needs.'[2] In other words, it can be said that we have every right to utilize resources that the Earth provides to us for our needs as long as we make sure that future generations have the same possibility. The United Nations Millennium Declaration identified principles and treaties on sustainable development, including economic development, social development and environmental protection.[3]

'Green Chemistry' is a term which is often applied when chemistry and chemical processes are defined as environmentally benign. Paul Anastas and John Werner developed and introduced widely accepted 12 principles of Green Chemistry. The following list briefly presents the principles which, if followed, would make chemical processes or products greener.[4]

(1) Prevention: it is better to prevent waste than to treat or clean up waste after it has been created.

(2) Atom economy: synthetic methods should be designed to maximize the incorporation of all materials used in the process into the final product.

(3) Less hazardous chemical syntheses: where ever practicable, synthetic methods should be designed to use and generate substances that possess little or no toxicity to human health and the environment.

(4) Designing safer chemicals: chemical products should be designed to affect their desired function while minimizing their toxicity.

(5) Safer solvents and auxiliaries: the use of auxiliary substances should be made unnecessary wherever possible and harmless when used.

(6) Design for energy efficiency: energy requirements of chemical processes should be recognized for their environmental and economic impacts and should be minimized. If possible, synthetic methods should be conducted at ambient temperature and pressure.

(7) Use of renewable feedstock: a raw material or feedstock should be renewable rather than depleting whenever technically and economically practicable.

(8) Reduce derivatives: unnecessary derivatization should be minimized or avoided if possible, because such steps require additional reagent and can generate waste.

(9) Catalysis: catalytic reagents are superior to stoichiometric reagents.

(10) Design for degradation: chemical products should be designed so that at the end of their function they break down into harmless degradation products and do not remain in the environment.

(11) Real-time analysis for pollution prevention: analytical methodologies need to further develop to allow for real-time, in-process monitoring and control prior to the formation of hazardous substances.

(12) Inherently safer chemistry for accident prevention: substances and the form of a substance used in chemical process should be chosen to minimize the potential for chemical accidents, including releases, explosions and fires.

Based on these Green Chemistry principles, Paul Anastas and Julie Zimmermann have also developed 12 principles of Green Engineering which should be kept in mind when developing new processes. The principles are briefly listed here, but more detailed information can be found from their publication.[5]

(1) Inherent rather than circumstantial.
(2) Prevention instead of treatment.
(3) Design for separation.
(4) Maximize efficiency.
(5) Output-pulled *versus* input-pushed.
(6) Converse complexity.
(7) Durability rather than immortality.
(8) Meet need, minimize excess.
(9) Minimize material diversity.
(10) Integrate material and energy flows.
(11) Design for commercial 'afterlife'.
(12) Renewable rather than depleting.

1.1.2 Value of Major Biorefinery Products

1.1.2.1 Fibre Versus Chemicals

A major question concerning the definition of a 'biorefinery' comes when you consider existing fibre-lines and pulping technology. Is the modern kraft or sulfite mill to be considered as a biorefinery? If so, modern kraft mills can already be considered as 'green biorefineries'.

- The more modern mills use relatively harmless chemicals.
- Most of the materials are recycled so the net consumption of sulfur or caustic is very low.
- Importantly, the process is aqueous-based.

- With current totally chlorine free (TCF) bleaching stages, the production of chlorinated compounds in waste waters can be eliminated.
- Release of sulfides can also be minimized.

However, developing a green-field site with the most modern technology is a billion euro investment. The main product from kraft pulping is fibrous pulp. Bleached European softwood kraft pulp has a current value of ~0.8 € per kg. The price has been known to fluctuate within an approximate range of ±0.5 € per kg. Softwood kraft, in particular pine and spruce, are most valued due to the long fibre length. Dissolving pulp, produced through the sulfite pulping process or pre-hydrolysis kraft pulping, is even more valuable but is typically used for chemical production, *e.g.* cellulose acetate, carboxymethyl cellulose (CMC) and other esters or ethers. Therefore, this nicely fits the definition of a biorefinery. The fibrous properties of kraft pulp are the basis of its value, with high-quality graphical papers being the largest market for this pulp.

However, over the last 10 years demand for graphical papers has decreased, in particular in North America and Europe (*i.e.* the 'saturated' markets) and this lowers the value of pulp. This is due to competition mainly from China and Brazil and the use of graphical paper in general has decreased. This makes the use of fibrous biomass for production of energy and chemicals more attractive, despite the lower cost structure. In addition, unstable oil prices, global warming and eventually international commitment by governments to increase biofuels share in the transportation sector (EU 10% by 2020) has resulted in strong demand for bioethanol from biomass.

This is also reflected in the growing market price of bioethanol over the last 10 years, *i.e.* from about 0.3 € per kg to about 0.6 € per kg.[6] This is uncertain to continue in the long term, as crude oil prices are currently low (at the time of writing), which will decrease bioethanol demand, and will also decrease demand from applications other than biofuels, *i.e.* as platform chemical for synthesis of green chemicals (diethyl ether, ethylene). Ethanol is a high volume bioproduct; however, its market value is considered low-to-average *versus* specialty chemicals from biomass, importantly in comparison to chemical pulp.

1.1.2.2 Bulk Chemicals

The production of bulk chemicals (>1000 tonnes per year) from biomass remains rather limited as the majority of organic chemicals and polymers are still derived from fossil-based feedstocks, predominantly oil and gas.[7] Hence, most of the bulk chemicals originating from biomass do not show any dramatic increase in market value. However, a steady increase in demand is reported for lactic acid at 10% annual growth rate.[7] Lactic acid can be converted, *e.g.* to polylactic acid (PLA). PLA is mainly used in production of sustainable biopolymers for use in the packaging industry (thin films) but is also found in applications elsewhere. According to reports, European demand for

PLA is currently 25 000 tonnes per year and could reach 650 000 tonnes per year in 2025.[7] Furans derived from biomass, such as furfural derived from pentoses and 5-hydroxymethylfurfural (HMF) derived from hexoses, are one major platform feedstock of interest. Furfural has already been produced on an industrial scale for almost 100 years. The first industrial process for its production was by the Quaker Oat company where it was discovered that furfural could be obtained by sulfuric acid-catalysed dehydration of their oat hull stockpiles.[8] A multitude of applications ensued. Nowadays, China produces the majority of the capacity, much of it from grass-based waste material. Most of these producers are dedicated towards furfural production as the main product. There are many estimates on the production of furfural, but they typically range between 300 000 and 800 000 tonnes per year and several smaller producers produce furfural as a secondary product. For example, Lenzing AG produces furfural on a 5000 tonnes per year scale during their pulping of beech wood.[9] The main product for this process is their cellulosic pulp destined for textile production and only about 1% of the dry mass of the wood is converted to furfural. The market price of furfural ranges from roughly between 0.5 and 1.5 € per kg and, not surprisingly, the prices are lowest in China. Hydroxymethyl furfural (HMF) by contrast, is not yet produced industrially due to the difficulty in accessing hexoses, difficulties in isolation as well as the instability of it at the process conditions. Most hexoses are bound up in softwood, which is much more recalcitrant than the abundant pentoses in grass species. Advanced techniques and enabling technologies, such as ionic liquids, are now required to allow us to access and selectively convert these saccharides whereby applications of both furans are likely to be wide ranging. Sequential catalytic dehydration, hydrolysis, hydrogenation and hydrogenolysis steps can be applied to convert them into a wide range of commodity chemicals, into potential biofuels and solvents.

One example of biofuel production is the Sylvan process (Figure 1.1).[10] This process involves the hydrogenation of furfural to 2-methylfuran, dimerization or trimerization of 2-methylfuran and hydrogenation/hydrogenolysis to the fully saturated alkane.[11]

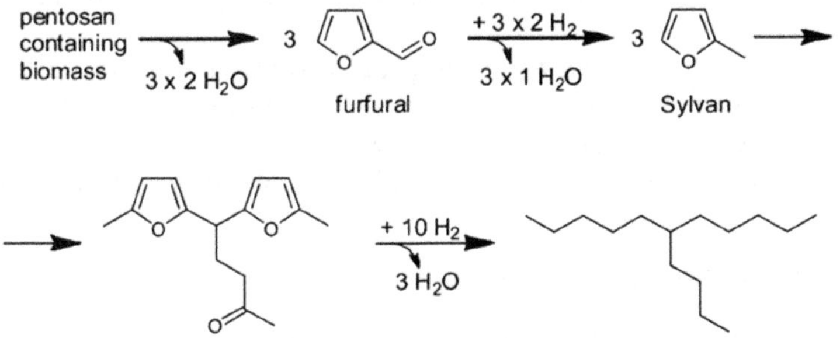

Figure 1.1　　The sylvan process: production of biodiesel from pentose dehydration, furfural hydrogenolysis, 2-methylfuran (sylvan) trimerization and final hydrogenation.[11]

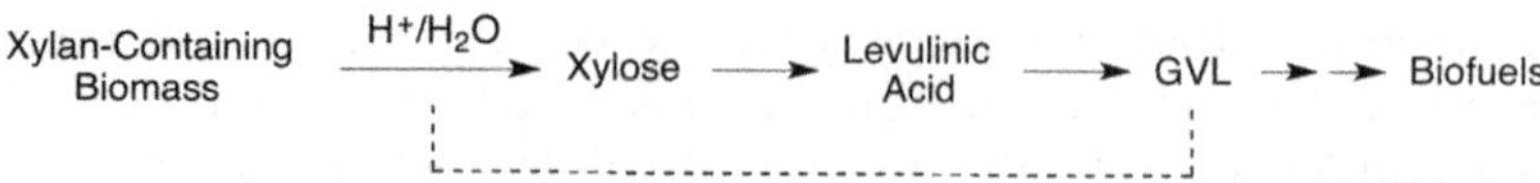

Figure 1.2 Biofuel production by acid-catalysed biomass degradation in GVL:H₂O. GVL itself can be derived in the reaction sequence by fully hydrolysing xylan, dehydrating xylose to furfural, hydrolysing furfural to levulinic acid, cyclizing and then hydrogenating levulinic acid to GVL.[11,12]

This product can be used as a high-quality paraffinic diesel, but the cost is rather high and so it will likely only find immediate access for high-end engines. In general, the cost of furfural is still too high compared to crude oil to have utility as a fuel precursor. Thus, large process improvements need to be made to access fuel markets. Solvents, however, can have a high price. It has been suggested that 2-methyltetrahydrofuran (2-MTHF), accessible from hydrogenation of furfural, has potential to replace THF in certain applications and is a potential biofuel itself. *gamma*-Valerolactone (GVL) is also now being studied intensively as a media for the conversion of polymeric pentoses and hexoses into oligomers and monomers.[12] Thus, the media can be derived from the biomass feedstock. GVL can itself be also be converted into liquid transportation fuels (Figure 1.2).[13]

This media unfortunately cannot access bulk hexoses bound up in soft-wood at lower temperatures. Increased temperatures do allow for almost complete solubilization of hexoses but lower temperature treatments do allow for fractionation of birch sawdust, resulting in high-purity cellulosic pulps.[14]

1.1.2.3 *Specialty Chemicals*

Specialty chemicals from biomass are sold at relatively high prices (>10 € per kg) due to their limited production (<1000 tonnes per year), 'green credentials', unique quality and properties. Demand for specialty chemicals by the industry (pharmaceutical, cosmetics, food sector) is growing stronger and so is their market value. Specialty chemicals fit well to specific, so-called 'niche', markets.

An example is Borregaard's vanillin produced by oxidation of sulfite lignin. The company has established a profitable business from vanillin as it is the exclusive producer and supplier of wood vanillin to the food industry. Vanillin derived from petrochemical sources is sold at much lower market prices.

1.1.3 **Obtaining Pure Bio-Fractions**

One of the major challenges in industrial biorefining is related to the need of fractionation and purification of the heterogeneous biomass streams to be processed; in fact, a majority of industrially and techno-economically feasible routes to chemicals, fuels and bio(composite) materials depend on the availability of pure, non-contaminated raw material fractions, whether those are carbohydrates, their polymers, fats and oils, various fragrances, nutraceuticals, extractives from lignocellulose or lignin. Further, the existence of various ash-elements (inorganic minerals) further complicates the task.

Throughout the years many processes were developed to facilitate the fractionation/separation targets: among others, the early acidic sulfite pulping, the alkaline sulfate pulping, mechanical pulping, various organosolv processes. More recently, ionic liquid or deep eutectic facilitated processes for biomass pre-treatments, fractionations, cellulose as well as hemicellulose separations and manipulations have all shown their potential in providing fractionated biomass components for further use. Indeed, if performing a simple SciFinder® search for the key words 'ionic liquid & biorefinery', this search string gives the very first paper as being published in 2008 with already 10 papers by the year 2014. From the early technologies based on (expensive) alkyl-imidazolium systems, the field has seen the rise of new concepts based on switchable ILs, strong organic bases and acid gases, the application of distillable ILs, as well as the use of bio-based, low-cost ILs, mixtures of 'cooperative' ionic liquids and deep eutectic solvents (DES), also in conjunction to radiofrequency and microwave heating or acoustic cavitation.[15-17] The systems are not easily understood but the hydrogen bond basicity of the ionic liquid–water mixtures apparently relates to cellulose dissolution, lignin depolymerization and even to sugar yields obtained.[18-38]

In addition, liquid CO_2 (sub- or super-critical), gas-expanded liquids or simply water under relatively harsh conditions (often near-critical) offer other alternative processing possibilities.[39,40] Nevertheless, it is estimated that around 60–80% of the processing costs are still related to the separation steps.[41]

1.2   From Historical Milestones to Modern Operations

Early biorefining has its roots deep in the Nordic forests where many small-scale production facilities for tar, a sticky crude oil-like product with a complex chemical composition, were erected. Wood tar was particularly sought after by the rapidly expanding maritime industries of the 18th century to impregnate the (wooden) vessel hulls, manila ropes, *etc.* against rotting. Especially Finland and Sweden capitalized on this lucrative export product produced *via* a kind of slow pyrolysis of pine stumps and fatwood. At the same time, charcoal, ideally suited as a fuel for steam engines, was getting more attractive. Both tar and charcoal were initially produced in primitive tar pits and charcoal kilns. Still, in the early days, a third industrially important product was potash (soluble potassium salts). Potash was produced from birch (*Betula pendula*) ash and used in the production of glass and soap. It was exported in large amounts until sometime in the mid-19th century when the lack of birch trees saw to the downfall of this industry. In fact, timber and sawn wood products took over as more important export products only sometime in the middle of 19th century.[15]

As indicated, the saw-mill industries grew during the first half of the 19th century, soon giving way and allowing the growth of the pulping industry. In the beginning, the pulp mills were a way to increase the value of the saw-mill operations by taking advantage of the waste (saw dust) and under-dimensioned wood fractions.[42] After the invention of the paper machine, the pulping industry grew in relative importance, a trend lasting until the late 20th century. In general, the paper and pulp industry has during the recent years been facing harsh competition and declining profitability. The underlying reasons are obvious: declining home markets in the western developed economies (shrinking consumption of paper); overcapacity in various paper qualities; emergence and development of pulping in Latin America and Asia (particularly from eucalyptus); lack of vision for the industry's future directions; and resistance to change. The once so lucrative business in refining coniferous trees to high-quality paper was until recently considered the most feasible and rational use of Nordic softwood. However, signs of changing attitudes are around: besides sawmills as well as paper and pulp combines, many other new or transformed industries use increasing amounts of lignocellulose from wood.

Here, a short overview is compiled describing the changes and trends occurring today in Fennoscandia. This area was once a 'paper-belt', in future perhaps again it will be known as a 'bio-belt' for its biorefineries – as well as other areas of the world where sustainable and biodegradable chemicals, transportation fuels, and materials are co-produced along with carbon dioxide neutral heat and electric energy.

In particular, managed forests possess a vast resource and potential as sources of biomaterials, chemicals and energy. Also, one of the potential tools enabling increased utilization of woody biomass is presented by careful selective breeding as well as cultivation of new generations of trees with improved genome. Nevertheless, particularly Europe is notoriously frightened by the GMO plants – and not for no reason. For instance, the Nordic boreal forests and rest of the sensitive nature took a long time to evolve, adapt and thrive; if we were to risk that balance overnight (like is already happening due to unpredictable weather chances evidenced by us living in the area), something irreversible might happen. Let us alone consider the currently popular eucalyptus plantations (originally native to Australia and transplanted and gene engineered around the more temperate climate zones of the world) which might end up as deserts in a few decades since the species is well equipped to siphon the aquifers with its deep roots. Indeed, even the imminent threat of climate change might give rise to some unprecedented complications since the domestic fauna has throughout millennia slowly adapted to the current climatic conditions. If the climate changes dramatically and rapidly, the current species are perhaps not any more the optimal ones as a source of raw materials to the industry. Nevertheless, chemical pulping has a long history in Nordic countries but it is difficult to find original literature from earlier times written in any other language than Swedish (such as 'Ur de stora skogarnas historia' by Bertil Boëthius, 1917, or 'Kungliga Vetenskapsakademiens Historia', 1967) which sheds light to the debates of 18th century about forest and forestry.[43]

Later on, we saw the rise of sawmills and chemical pulp mills based on the so-called sulfite pulping. The acidic sulphite mills, nowadays an increasingly rare approach compared to the dominant alkaline sulfate or 'kraft' mills, are characterized by their high flexibility in terms of fine-tuning the cooking procedure (in batches, contrary to the modern kraft mills that operate continuous cookers), although one was not historically able to solve some critical issues related to this process.[44] Pine (Scots pine) was not a suitable raw material, thus limited to spruce (Norway spruce) in the Nordic area. Moreover, the environmental concerns related to the process resulted in serious problems with the air quality (sulfur dioxide), as evidenced by my parents who on a holiday trip through northern Sweden in the year of 1963 entered the town of Örnsköldsvik and still remember it as 'that place where blue-green smoke that made you cough loomed around and caused a throbbing headache' – as well as in the local fjord stretching to the sea: the pollution from the mill (the BOD-load and release of cooking chemicals) rendered the nearby waters of the Gulf of Bothnia nearly lifeless for a long period of time. How a complete change is possible as evidenced by the sustainable operations of the today mill (Aditya Birla Domsjö, see later).

The work horse of today's pulping industry is a kraft pulp mill. The yield is only about 50% because most of the hemicelluloses and almost all the lignin dissolve in the aqueous pulping stream, called black liquor. Typically, black liquor is combusted for steam and electricity generation in the recovery boiler and, at the same time, the dissolved inorganic cooking chemicals are recovered in the reductive environment of the boiler and recycled for pulping. Since the heating value of the carbohydrates is rather low (much less than that of lignin), combustion of dissolved hemicelluloses does not constitute an optimal economical use of this resource. In the older sulfite process, hemicelluloses were frequently recovered and fermented to lignocellulosic bioethanol. Already in 1908, 'diplomingenjör' Gösta Ekström filed a Swedish patent application concerning recovery of sulfite alcohol from a sulfite process, although a rival, 'civilingenjör' Hugo Wallin had filed a competing patent application in 1907. The fight between the two gentlemen and other interest groups went on for years until, in 1917, both of the patents were declared not valid! Finally, in 1919, both patents were again declared valid after vigorous negotiations.

More recently, a boom in bioenergy, partially enforced by the new European Union regulations and political incentives, has given birth to numerous enterprises producing wood pellets (combusted both at small family homes and centrally for district heating for heat). The local energy companies in many towns have built biomass-combusting power stations that supply the grid with electricity, that supply the towns with district heating and, sometimes, even supply nearby industries with steam.

Another example of an industry in transition is presented by the Chemrec company[45] who in collaboration with academia, research institutes,[46] the Smurfit Kappa pulp mill[47] and Volvo Trucks[48] developed 'From Wood-to-Wheel' black liquor gasification technologies for methanol and Bio-DME (bio-dimethyl ether) to be utilized as green, renewable diesel fuel for trucks.

At this site, the Piteå Science Park,[49] another company, Sunpine,[50] in collaboration with a classical petroleum refining company, Preem AB,[51] has launched its crude tall diesel process integrating the tall oil production from the adjacent pulp mills to the petroleum-based transportation fuels.

In other parts of Fennoscandia, the Borregaard company[52] – in its original location in Sarpsborg, Norway, advertising itself as 'the world's most advanced biorefinery' – is a representative of another 'sulfite mill that didn't die' since the focus is on biomaterials, biochemicals and bioethanol, instead of pulp for the paper industry.

In Finland, Varkaus, a demonstration-scale unit was erected and operated for a long period of time for black liquor gasification by NSE Biofuels,[53] producing diesel-type waxes ready to be supplied into the existing refining processes in an oil refinery. Unfortunately, this operation is now discontinued, presumably due to the very large investment costs required to construct full-scale industrial operations.

Chempolis Oy[54] is a small, entrepreneurship-generated company located in a Northern Finnish town of Oulu and primarily marketing a biorefining concept based on chemical pulping taking advantage of formic acid. Nevertheless, the company has also a demonstration-scale and R&D facility in Oulu.

The company ST1[55] has developed concepts utilizing waste materials as the source of bioethanol which is blended with regular gasoline, for example to E85 – a blend with 85% ethanol and 15% gasoline. The production plants are modular and fully automated, thus enabling utilization of local, municipal waste fractions.

A lesser known player is the company Sybimar Oy,[56] who builds and markets technology (besides selling renewable oils) for utilizing various waste fractions, such as fish processing industry wastes to bio-based oils usable either as heating oil or as a raw material for further refining.

In terms of early chemicals production, the utilization of wood hemicelluloses *via* fermentation to ethyl alcohol upon sulfite pulping saw its rise and fall before and after the 2nd World War: in particular at that time on the site of the 'Domsjö' pulp mill in Sweden today,[57] a whole range of chemical production was based on ethanol chemistry. At best, dozens of chemicals (well over 50 of today's bulk chemicals) were produced to compensate for oil that was not available.[44]

Today, other examples of obtainable specialty products produced from side-streams of pulping operations are sodium lignosulfonate (*e.g.* a dispersant), xylose (the raw material of xylitol,[58] a sugar alcohol and alternative sweetener with nutraceutical properties) and vanillin (a flavouring agent).[52] Further, turpentines and tall oil have since long been utilized not only as a source of energy (combustion to energy) but also as components in varnishes, paints and other chemicals,[59,60] including more recent uses as raw materials for renewable diesel synthesis.[61]

Around the same time as the Swedish and Finnish biorefining efforts were starting, in Norway, the Borregaard company also introduced sulfite-based ethyl alcohol production and has since then introduced many other refined products from lignocellulose.

In Switzerland, Attizholts closed its mill in 2008. Georgia-Pacific in the USA closed in 2001 whereas in Canada Tembec is still in operation. Further, several Russian mills used to operate until recently but the current status is unclear.[62]

Biogas – or biomethane – is rapidly gaining momentum as yet another, partial, solution for the oil-deficient future of ours. Besides that, it can be used as a direct energy source (combustion with subsequent co-production of electricity and heat). It is entirely feasible to use it as a transportation fuel, both in gasoline and diesel engines after gas purification and appropriate modifications to the fuel supply systems of the vehicles. Perhaps the biggest challenge is the purification: various technologies aiming at competitive, small-scale economics are under development. Typically, the technologies are based on physical or chemical (or combined) binding of CO_2, the major by-product from anaerobic digestion of waste. Indeed, various lignocellulosic streams as well as municipal (*e.g.* communal waste water treatment plants) or industrial waste (*e.g.* dairy industry) can be used as a feed. The most challenging gas sources are, perhaps, the multiple waste dumps established throughout the years that can contain an astonishing mix of dumped industrial wastes and degradation products of chemicals long ago banned in developed countries. In these gases, besides carbon dioxide, many other toxic and dangerous components detrimental to health and internal combustion engines are found.

In the Jyväskylä area of Finland, a small, entrepreneur-driven facility, Metener Oy,[63] is operational and selling vehicle-grade biogas derived from cattle manure. In addition, co-production of electricity and heat takes place on site. The company also sells biogas digesters and gas purification equipment. However, this is just one example. In particular throughout the European continent many biogas plants are operated.

The textile industry of today is facing enormous challenges since cultivation of the classical raw material, cotton, is associated with a huge water demand and extensive use of pesticides. Moreover, today most of the cotton farming takes place in third-world countries with dubious labour ethics and lack of water.[64,65] Clothing is needed throughout all human societies; however, recently more people are raising the question whether we already have passed 'peak cotton' (analogous to 'peak oil') since the analysis of data shows that the market for the specialty cellulose is booming and the worldwide supply of cotton has fallen. Dissolving cellulose refers to high-quality cellulose with a very low hemicellulose content that is classically processed further by various spinning technologies to form fibres that can be used for textile weaving.[66]

In today's world, many biorefinery operations have been commenced throughout the world. These include Neste Oils NextBtl® renewable diesel and jet fuels,[67] the two operating production lines in Finland and two renewable diesel plants operating in Rotterdam (the Netherlands) and in Singapore. There are also the Brazilian and US efforts towards bioethanol particularly, albeit heavy criticism has been directed towards the use of corn as an ethanol source that merely results in increased greenhouse gas emissions.[68–70] More actors are constantly moving into this environment and one of the latest newcomers is the small BioEndev company[71] commercializing torrefaction technology. Other examples are represented by the Roquette operations in Lestrem, France, albeit the mill suffers from the same ethical dilemma as many operations in the American continent – the use of edible crops for the production of fuels and chemicals.[72] Nevertheless, a recent US opening introduced a cellulosic biorefinery in Hugoton, Kansas, based on the use of corn stover residues advertised to exceed the capacity of the GranBio facility in Alagoas, Brazil, operating on bagasse and straw wastes.[73] In Europe, many companies and countries have operations either aiming at demonstrating or as commercial production in line with the biorefinery philosophy. Examples are the Portucel Soporcel Group or the Respol Group in Portugal that is actively pursuing further development in the area,[74,75] as well as the Austrian focus on mainly grassland and agricultural residues,[76] the German very holistic view[77] or the many Chinese efforts. The reader should turn to the many published reviews, papers and reports to obtain more detailed information and by no means do these examples cover all past and ongoing efforts towards a more sustainable future of tomorrow.

1.3 Modern Competitive Technologies

Biorefineries can be classified into three broad categories on the basis of biomass chemistry: sugar and starch, lignocellulose and triglyceride biorefineries.

1.3.1 Sugar and Starch Biorefineries (SSB)

This type of biorefinery can utilize a wide range of sucrose-containing feedstocks (sugar beet, sugar cane, *etc.*) and starchy biomass (corn, wheat, barley, maize, *etc.*) to produce bioethanol. Bioethanol is a promising biofuel when mixed with gasoline but is also a useful platform chemical for synthesis of diethyl ether, ethylene, *etc.* At present, commercial ethanol production is predominantly based on edible sugar and starchy biomass; for instance, sugar cane in Brazil, corn grains in USA and wheat and sugar beets in European Union countries. The use of edible feedstocks and large areas of arable land for biorefining operations is strongly debated by many as it competes with food production.

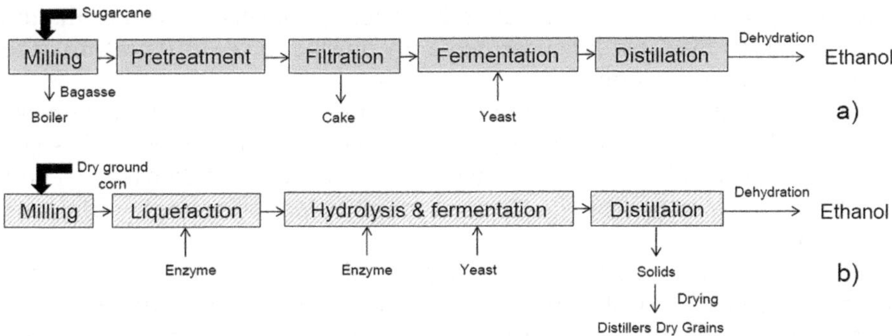

Figure 1.3   Sugar and starchy biorefinery (SSB): process scheme for the production of ethanol from (a) sugar cane (b) dry ground corn.

The main steps involved in the production of ethanol from sugar and starchy feedstocks (SSF) are shown in Figure 1.3.

- Pre-treatment; needed only for lignocellulosic (sucrose containing) feedstocks. Pre-treatment by different methods removes most of the lignin and hemicellulose, increases porosity and disrupts the crystalline structure of cellulose.
- Enzymatic hydrolysis of lignocellulosic and starchy biomass (after milling) by cellulases–hemicellulases and amylases, respectively.
- Fermentation of the hexose-rich hydrolysate from starchy biomass by naturally occurring yeasts (usually Baker's yeast). The hydrolysate obtained from lignocellulosic feedstocks is more complex as it contains both pentose (xylose and arabinose) and hexose sugars (glucose, galactose and mannose). The fermentation of pentose sugars is challenging and costly as only a few yeast strains are available for fermentation to ethanol.[78]

1.3.1.1   Current SSB Biorefinery Examples

A typical sugar and starch-based biorefinery is the 'Les Sohettes' complex located in Pomacle, France.[79] The biorefinery comprises of a sugar beet processing unit, a wheat refinery and a sugar plant, an ethanol distillery (Cristanol), a research centre (ARD), a demo-plant for second-generation ethanol (Futurol), a straw-based paper production pilot unit (CIMV), and a succinic acid pilot plant (BioAmber). Succinic acid is a useful platform chemical for the production of polyurethanes, coatings, adhesives, sealants and personal care ingredients.

This facility is a good example of integration in the product networks as two major crops are being used as feedstocks: sugar beets and wheat. The combination of both crops allows year-long biorefining operations as the harvesting period of sugar beet is rather short (typically a few months in a year) and this would render sugar beet only production of ethanol and other bioproducts uneconomical.

Cargill Inc. has been operating a corn biorefinery since 2002 (Nebraska, USA).[80] This integrated biorefinery processes corn to produce corn oil, sugar, ethanol, lactic acid and polylactic acid (Natureworks LLC). Polylactic acid is a natural biopolymer that can be used to produce biodegradable films and plastic products.

DuPont Tate & Lyle BioProducts is producing 63 000 tonnes per year of 1,3-propanediol (1,3-PDO) from corn in their Loudon plant in Tennessee, USA (2006).[81] In fact, 1,3-PDO is a key building block for producing polypropylene terephthalate which is used as biopolymer film.

1.3.2 Lignocellulosic Biorefinery (LCB)

The lignocellulosic biorefinery uses naturally dry biomass such as cellulosic biomass (wood energy crops) and agriculture waste to produce biofuels and other bioproducts. Woody energy crops are fast growing hardwood trees that are harvested within 5–8 years of planting. These crops include poplar, willow, silver maple, eastern cottonwood, black walnut, sweetgum, *etc.* and they are traditionally used for manufacture of paper and pulp. Agricultural wastes include sugar cane bagasse, corn stover (stalks, leaves, husks and cobs), wheat straw, rice straw, rice hulls, nut hulls, barley straw, olive stones, *etc.* Unused sawdust, bark, branches, and leaves/needles that are produced during processing of wood for bioproducts or pulp are also included in this category of wastes. Animal manure and municipal and industrial wastes are another potential source of biomass for LCB.[82]

In an LCB the raw biomass is first cleaned, pre-treated to improve accessibility of sugars for subsequent processing and then broken down into its primary constituents (cellulose, hemicellulose and lignin) through biochemical (enzymatic hydrolysis, Figure 1.4a) or chemical (acid hydrolysis, Figure 1.4b) routes. The cellulose and hemicellulose are converted to monomeric sugars. The glucose obtained from hydrolysis of cellulose is

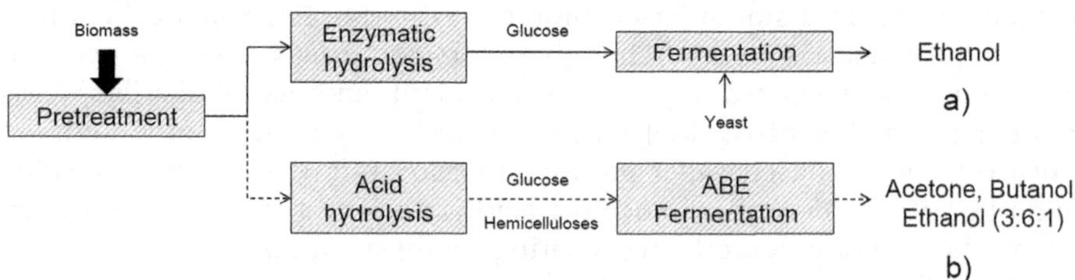

Figure 1.4 Lignocellulosic biorefinery (LCB): basic schematic process flow diagram for the production of ethanol and butanol *via* (a) the biochemical route and (b) the chemical route.

further converted to valuable products such as ethanol, acetic acid, acetone, butanol, succinic acid, *etc. via* fermentation. The fermentation step is typically performed by yeast. Another alternative is to use bacteria to produce biobutanol and other chemicals *via* ABE fermentation. This approach is more advantageous if the hemicellulose hydrolysate is fed simultaneously to fermentation. However, a requirement is that the hemicellulose hydrolysate is free of fermentation inhibitors. Soudham *et al.*[83] and Sklavounos[84] have reported efficient methods to detoxify acidic hydrolysates from wood for fermentation. Except alcohols, the LCB produces furfural from xylose. Lignin is typically used as an adhesive or binder and as a fuel for direct combustion.[82]

The major advantage of LCB is that it does not use edible feedstocks and therefore it eliminates the need of sacrificing arable lands. However, commercialization of the chemical and biochemical LCB is currently limited due to technological constraints and high processing costs (pre-treatment, detoxification and enzymatic hydrolysis steps).

1.3.2.1 Current LCB Biorefinery Examples (Chemical and Biochemical Processing)

Borregaard's integrated biorefinery in Sarpsborg, Norway, is a special case of a pulp mill that has gradually evolved to an LCB. In this biorefinery spruce chips are treated with acidic calcium bisulfite cooking liquor. Hemicellulose is hydrolysed to monomeric sugars during the cooking process. After concentration of the spent sulfite liquor (SSL), the sugars are fermented and ethanol is distilled off in several steps. The Borregaard integrated biorefinery is particularly successful as it can produce cellulosic ethanol but also a whole range of bioproducts.[85] The latter include vanillin from lignin and lignosulfonates. Vanillin is a high value flavouring agent with numerous uses in the food industry whereas lignosulfonates are used in many special applications, *i.e.* as dispersants, in paints, oil drilling agents, *etc.* The company is currently running a demonstration plant that showcases its latest concept – called the 'BALI® biorefinery'.[85,86] This plant can handle sugar cane bagasse, straw, wood, energy crops and other lignocellulosics to produce ethanol and lignin specialty chemicals.

GranBio recently announced (2014) start-up of its 2nd generation lignocellulosic ethanol biorefinery in Alagoas, Brazil.[73] The facility is one of the largest ethanol biorefineries in the world. This biorefinery uses straw and bagasse as feedstocks. GranBio's facility employs feedstock pre-treatment technology from Beta Renewables, enzymatic hydrolysis and fermentation by yeast.

Abengoa's biorefinery in Kansas, USA, also began operation in 2014. The biorefinery is built with a planned capacity of up to 75 000 tonnes per year ethanol. Abengoa's facility utilizes agricultural crop residues (such as stalks and leaves) that do not compete with food or feed grain.[87]

A recent (2012) development in Europe is the Beta Renewables biorefinery in Crescentino, Italy. This biorefinery uses straw and energy crops (giant reed) to produce about 60 000 tonnes per year ethanol.[88]

CIMV built its first pilot plant biorefinery in Pomacle, France in 2007. The facility processes 50–70 kg h^{-1} of hardwood, straw and bagasse.[89] The feedstocks are processed by the organosolv (ethanol–water) method. The produced pulp is then enzymatically hydrolysed and released sugars are fed to fermentation. The plant produces lignin (Biolignin™), C5 sugar syrup, paper grade cellulose pulp, bioethanol and chemicals.[90] The company is planning to commercialize its technology in 2015.

Finnish energy company St1 is planning start-up of a new bioethanol plant in Kajaani, Finland (capacity of 10 million L per year) in 2015. The plant will produce bioethanol transportation fuel from sawdust, which comes as a side-product of the sawmill industry. The company opened its first ethanol plant in 2007 and currently has seven plants in Finland producing ethanol from biowaste and food industry residues using its Etanolix® technology.[91]

An alternative approach to chemical and biochemical processing of lignocellulosic biomass are the thermochemical conversion processes of gasification and fast pyrolysis.

1.3.2.2 Gasification

Gasification of biomass generates synthesis gas (syngas), heat and electricity. In gasification, the biomass is converted to a combustible gas mixture of H_2, CO, CO_2, CH_4, N_2 (for gasification with air) and traces of higher hydrocarbons in the temperatures range of 800–900 °C.[92,93] The gasification is a combination of pyrolysis and partial oxidation. The heat required for endothermic pyrolysis is generated by partial oxidation of biomass using air (most common) or oxygen. However, the former technology suffers from drawback of low heating value (4–7 MJ m^{-3}) of resulting synthesis gas that limits its application for boiler, engine and turbine operation only.[94] Biomass gasification by oxygen has potential to produce syngas with an improved heating value (10–18 MJ m^{-3}); however, the economics favour use of hydrocarbons, *i.e.* natural gas and inexpensive coal, as feedstock.[92]

A key drawback of syngas from biomass gasification is its composition, as it is rich in tars and methane. Tars affect gasification efficiency and foul processing equipment whereas the presence of methane makes syngas unsuitable for Fischer–Tropsch (FTS) synthesis. It is, however, technically possible to overcome these barriers, *i.e.* by operating the gasifier at a slightly lower temperature with use of suitable catalysts. Today there are very few commercial biomass gasifiers operating without government support or subsidies. Most of them are used for power generation only.[82] Downstream catalyst poisoning is also a major issue, requiring multi-step gas purification stages if the syngas is to be used as synthesis gas, *i.e.* for FTS, and not for energy production.

1.3.2.3   Biomass to Liquid Technology

Biomass to liquid (BTL) is the technology that allows the production of synthetic fuels (gasoline, diesel, heating oil, jet fuel, methanol, dimethyl ether, ethanol, *etc.*) from biomass-derived syngas using FTS. The low temperature FTS (200–250 °C) is generally used for the production of jet fuel and diesel whereas the high temperature FTS (300–350 °C) is used to produce gasoline range hydrocarbons.[82] However, with the exception of producing methanol, dimethyl ether and synthetic natural gas, the BTL technology suffers from poor selectivity to fuel products.[95] Moreover, FTS require syngas with a specific H_2/CO ratio, which requires adjustment of biomass-derived syngas composition using the water gas shift reaction. As mentioned earlier, impurities in the gas need to be removed prior to FTS, to avoid catalyst poisoning. These impurities are present in the biomass feedstock and can poison catalysts rapidly even at the ppm level. Other drawbacks include high capital investment costs as the scale of gasification complexes is generally very large, by economic necessity.

1.3.2.4   Current LCB-BTL Biorefinery Examples

It is only a few large companies that have managed to overcome the previously described techno-economic barriers. All of them use natural gas (Sasol, Shell Pearl GTL) or coal (ExonMobil, Linc Energy) as the feedstock for FTS to fuels. A recent effort by NSE biofuels, as already mentioned (a joint venture between Neste Oil and Stora Enso) to build a commercial biorefinery in Porvoo or Imatra, Finland, that would produce Fischer–Tropsch diesel from biomass feedstock (planned capacity of 100 000 tonnes per year) was abandoned due to the massive investment costs needed. Application for funding under EC's NER 300 programme was discontinued in 2012.[96]

Chemrec AB, Sweden, inaugurated the world's first dimethyl ether (DME) plant in Piteå in 2010.[45] The plant produces DME from syngas formed after gasification of black liquor (spent cooking liquor of kraft pulping containing dissolved hemicelluloses and lignin) followed by Fischer–Tropsch synthesis. BioDME is a synthetic 2nd generation biofuel which can be mixed with diesel.

1.3.2.5   Fast Pyrolysis

Fast pyrolysis has better potential over BTL due to its simplicity, lower equipment requirements and hence lower capital investment costs.[97] Fast pyrolysis with high heating rate (500 °C s^{-1}) is performed in the absence of oxygen to produce 'bio-oils' (highly acidic 'oils' with a significant water content) in high yields. Bio-oils are a mixture of more than 300 chemical compounds with considerable variation in physical and chemical composition depending on the type of biomass. They are a potential feedstock for producing chemicals. However, separation of chemical compounds from these mixtures is challenging by fractional distillation or extraction. There are though methods for upgrading bio-oils to get specific types of chemicals in high concentrations (aqueous phase dehydration/hydrogenation or APD/H).[98] Bio-oils could also be used as transportation fuels. However, high water and oxygen contents, immiscibility with petroleum fuels, low heating value, poor storage stability and high corrosiveness currently prohibit their use in engines for vehicles. Upgrading of bio-oils for use as fuels is possible, *i.e.* by steam reforming, hydrodeoxygenation (HDO) and zeolite upgrading. Bio-oils are currently used only for specific applications, *i.e.* in firing boilers, running turbines and heavy duty diesel engines.[82]

It is reported that fast pyrolysis is going to be the leading thermochemical biomass processing technology due to its favourable credentials as one can see from recent technological advancement in this area.

1.3.2.6   *Current LCB-Pyrolysis Biorefinery Examples*

Envergent Technologies (a joint venture between UOP LLC and Ensyn Corp.) started operation of a pyrolysis biorefinery in Les Plains, Illinois, USA, in 2008.[99] The facility converts forest and agricultural waste to bio-oils using RTP® (Rapid Thermal Process); a fast thermal process whereby biomass is heated rapidly (for 2 s) in a fluidized bed reactor. Produced bio-oils can then be used for generation of electricity and for the production of process heat. Development is underway to upgrade RTP bio-oils into green gasoline, green diesel and green jet fuel.

KiOR (a joint venture between Bioecon and Khosla Ventures) has developed a catalytic cracking technology to convert biomass into renewable crude oil that is processed into gasoline, diesel and fuel oil blends. The company built the first commercial scale cellulosic fuel facility in Columbus, MS, USA, which started production in 2012.[100]

1.3.2.7   *The 'Green Biorefinery'*

A special type of lignocellulose-based biorefinery is the 'green biorefinery'. In the context of biorefineries, the term 'green', as used here, does not have any reference to 'green chemistry' but rather the general colour of the ripe feedstock. This biorefinery utilizes naturally 'wet' herbaceous biomass such as switch grass, *Miscanthus*, wheatgrass, reed canary grass, alfalfa hay, *etc.* and can produce a wide range of products, *i.e.* biogas, electricity, novel biomaterials (fibre-reinforced plastics, insulation materials) and fertilizers by employing mechanical means to fractionate biomass into a protein-rich liquid (green juice) and a solid fraction (press cake). The green juice is treated by biotechnological methods (fermentation) towards the production of lactic acid, amino acids, ethanol and proteins. The press cake can be used for the production of green feed pellets, as raw material for the production of chemicals (*i.e.* levulinic acid) and for conversion to syngas and hydrocarbons (synthetic fuels).[76]

Biowert in southern Hessen, Germany, has operated an industrial grass biorefinery to produce high-quality cellulose fibres for many applications since 2005.[101]

1.3.2.8   The Tall Oil Biorefinery

A tall oil biorefinery uses crude tall oil (CTO), which is a residue of the kraft pulping process. CTO is derived from rosin and fatty acids, which occur naturally in wood used for pulping. These acids are converted into corresponding sodium salts by the caustic conditions in kraft cooking. These salts are suspended in the spent black liquor from the kraft cooking and are referred to as 'soap'. The quantity of tall oil soap varies according to wood species, geographical location, season of the year and wood storage practices. For example, the typical CTO yields in Finland are 40–50 kg per tonne of softwood pulp and about 20 kg per tonne of hardwood pulp. Pine and in particular Scots pine (*Pinus sylvestris*) affords the highest yields of CTO. Pine is also highly prized for the long length of the kraft fibre resulting from pulping. CTO consists of around 30 to 50% fatty acids, 15 to 35% rosin acids and 30 to 50% pitch, a bio-liquid that is used for energy generation. These fractions are separated by distillation over wide pressure ranges and they are marketed as wood-based chemicals for use in many applications, such as paper sizing agents, dispersants and surfactants.

1.3.2.9   Current Tall Oil Biorefinery Examples

Forchem Tall Oil biorefinery[59] is located in Rauma, Finland, and its annual distillation capacity is approximately 175 000 tonnes per year of CTO from pine. The biorefinery produces Tall Oil Fatty Acid for use as raw material for many chemical reactions and intermediates. It is also produces Tall Oil Rosin which is used as raw material in adhesives and tackifyers (glues). Distilled Tall Oil, a complex mixture of mainly fatty acids and rosin acids (more than 10% rosin acids), is produced for use in many chemical reactions and blended products. The biorefinery also produces Tall Oil Pitch which is sold as low-sulfur content biofuel to be used in communal and industrial boilers. Forchem's new owner since 2013 is the Portuguese Respol Group which is one of the leading rosin upgraders in Europe.[60]

Arizona Chemical is the largest producer of pine chemicals in the world.[102] The company has been operating a Tall Oil biorefinery in Sandarne, Sweden, since the early 1930s. Today the biorefinery is used to refine and upgrade CTO and Crude Sulfate Turpentine. The latter is also a by-product of the pulp and paper industry. The company sells natural pine-based products to customers in many diverse markets including adhesives, roads and construction, tires and rubber, lubricants, fuel additives, and mining. In 2007, Arizona Chemical was sold by International Paper to the private equity company Rhône Capital to enable the company's further growth. In 2010, Arizona Chemical was acquired by American Securities, LLC, a US-based private equity firm.

As already mentioned, Sunpine has operated a Tall Oil Biorefinery in Piteå, Sweden, since 2010. The facility has a capacity of up to 10 000 m^3 per year crude tall diesel. The raw materials used are CTO, acid vegetable oils and methanol. The latter is used for esterification of CTO to biodiesel. Distillation of the crude fraction gives crude biodiesel (which is purified downstream), rosin acids and bio-oil.[103]

1.3.3   Oil and Fats Biorefinery (OFB)

This biorefinery utilizes vegetable oils and animal fats to produce biodiesel with comparable properties to petrodiesel. The production of biodiesel – as Fatty Acid Methyl Ester (FAME) – is based on transesterification of tri-glycerides with methanol. The reaction is commonly catalysed by acid, alkali or enzymes, depending on the free fatty acids (FFA) content of the feedstock.

The process of making biodiesel produces glycerol as by-product (~10 wt.% of biodiesel).[82] The glycerol is either etherified with alcohols or esterified with acetic acid to produce ether/esters for applications as fuel additives. Alternatively, glycerol can be converted to value-added chemical intermediates such as 1,2-PDO or 1,3-PDO by reduction and acrolein by dehydration or syngas by steam reforming.[82] This versatile feedstock can also be converted to many other chemicals such as epichlorohydrin.

Currently more than 95% of biodiesel is produced from edible oils such as rapeseed and sunflower oil in Europe, soybean oil in USA and palm oil in tropical countries.[104] This is a major issue, as the excessive use of edible oils necessitates that large fractions of arable land – that would otherwise be used for food production – is needed to satisfy the increasing demand for biodiesel. The use of low-cost non-edible feedstocks is thus necessary for economic and sustainable production of biodiesel. In addition, many farmers in tropical Asian countries are deforesting virgin forest to allow for the planting of vegetable oil-producing cash crops, such as oil palm (*Elaeis guineensis*). Therefore the long- and short-term development of vegetable oil sources is hard to justify, from a green perspective.

1.3.3.1   *Current OFB Biorefinery Examples*

UPM has recently announced (2014) the commercial start-up of its Tall Oil biorefinery in Lappeenranta, Finland. This biorefinery uses lower grade CTO fractions to produce biodiesel at a capacity of 100 000 tonnes per year. It is estimated that the produced volumes of biodiesel will cover approximately 25% of Finland's biofuel target. UPM's biodiesel (UPM BioVerno™) is produced by hydrotreatment of CTO to modify its chemical structure. Then a fractionation unit removes hydrogen sulfide and incondensable gases. The remaining liquid fraction is distilled to separate biodiesel.

Likewise, Neste Oil, Finland, operates triglyceride- (or oil-) based industrial biorefineries in Finland (Porvoo), Holland (Rotterdam) and Singapore that produce biodiesel (called NExTBTL®) from vegetable oils and fats. The Singapore and Rotterdam refineries are two of the largest triglyceride biorefineries in the world, with similar capacities of about 800 000 tonnes per year biodiesel, each. Renewable diesel is produced by hydrogenation of vegetable oils (HVO). Currently about 50% of used feedstock is palm oil.[105] In 2011 the company had to face fierce protests by Greenpeace regarding the use of palm oil as a feedstock for its biorefining operations.[106] Neste Oil (among others) was buying palm oil from the IOI Group; a Malaysian company allegedly responsible for illegal deforestation.[107] Since 2013 Neste Oil claims that 100% of the crude palm oil used for its biorefining operations is certified to be sustainably produced.[108]

Research at Neste Oil is currently directed at microbial lipid production by using lignocellulosic biomass as the feedstock. Specifically, a process is being developed that involves hydrolysis of biomass into sugar-rich hydrolysates, which are then used by oleaginous microorganisms as the carbon and energy sources to produce lipids. However, the costs of microbial lipids are currently prohibitively high for commercialization (simultaneous saccharification and enhanced lipid production, SSELP, process).[109] Also research is currently underway on using microbial oil and algae oil to produce renewable diesel (see below). A dedicated pilot plant has been built at Neste Oil's Porvoo refinery to study the opportunities offered by microbial oil and the company is involved in a number of international research projects working on algae oil, in Australia and elsewhere.[108]

BioMCN (Farmsum, Netherlands) uses crude glycerine (residue from biodiesel plants) which is purified, evaporated and cracked to obtain syngas.[110] Syngas is used to synthesize methanol. Methanol is an extremely versatile product, either as a fuel in its own right or as a feedstock for other biofuels. It can be used as a chemical building block for a range of future-oriented products, including MTBE, DME, hydrogen and synthetic biofuels (synthetic hydrocarbons).[82]

1.3.3.2 Algal Biorefinery

A key bottleneck for successful realization of biodiesel is the requirement for large areas of arable for cultivation of oil crops. A promising prospect is the algal biorefinery, which can utilize microalgae as source of lipids for the production of biodiesel. The microalgae grow rapidly (commonly double their biomass within 24 h), they give a high productivity of oils per hectare, have high oil content (up to 80 wt.% of dry biomass with 20–50 wt.% oils being common) and can grow in a variety of aquatic environments.[111,112] Microalgae fix solar energy in the form of biomass and oxygen using CO_2 and inexpensive growth medium containing water and inorganics. Microalgae do not require arable land and need much less water compared to energy crops. They can be cultivated in fresh water, sea water, lakes, rivers and even waste water from municipal, agricultural and industrial activities. Cultivation on a large scale is performed either in open ponds or enclosed tubular photobioreactors.

The latter configuration offers many benefits including higher volumetric productivity.[82] There is a lot of academic research on microalgae currently ongoing. For example, research in SLU Umeå, Sweden, is focused on algal cultivation in waste water and flue gases and transformation to bioethanol and biodiesel.[113] Laboratory tests showed that microalgae (phytoplankton) grew well on a combination of nutrients from untreated sewage and CO_2 dioxide from a thermal (bio) power plant. The algae absorbed CO_2 from flue gases and removed up to 90% of the nitrogen and phosphorus in the sewage. They are also able to fix heavy metals. The selected microalgae were found to have a substantially lower ash concentration than marine algae (seaweed), for example, which is important if dried algae are used directly as biofuel. Research is ongoing and there are plans to use yeast, which is capable of converting the carbohydrates in the biomass into ethanol.

Despite having enormous potential, the commercial scale construction of an algal biorefinery has not been realized yet due to the high costs of oil production.[114] The collection of microalgae from diluted biomass streams, the dewatering and the extraction of oils are energy intensive and therefore expensive processes. The oil extraction step requires addition of organic solvents (*e.g.* hexane or chloroform) and very harsh conditions to allow disruption of the thick cell wall that covers the microalgae. Adoption of an integrated biorefinery approach where a multitude of bioproducts (biofuels, platform chemicals, biogas, fertilizers, animal feed) are produced at a large scale could improve economics of the algal biorefinery and lead to its successful commercialization in the near future.

1.4 Early Industrial Biorefining Examples

1.4.1 Europe

The concept of biomass biorefining is not new, as it has been around for many years in the form of paper mills. Two of the oldest paper mills in Europe that are nowadays operating as biorefineries are the Borregaard biorefinery[115] in Sarpsborg, Norway, and the Domsjö biorefinery[116] in Örnskoldsvik, Sweden.

1.4.1.1 The Borregaard Biorefinery

Production of pulp and paper in Sarpsborg started in 1889, based on the acid sulfite process. In 1938 hemicellulose monosugars (mainly mannose) started to be fermented to ethanol by yeast. Ethanol was intended for use as a solvent and disinfectant. In 1967 when a process line for lignin and vanillin was installed, chemical products were produced from all components of wood (spruce). A wide range of biochemicals were produced during the period 1960–1980. Production of most of them was soon discontinued as they could not compete economically with their petroleum-derived counterparts. Nowadays the company is focused on producing ethanol and specialty cellulose for the manufacture of cellulose derivatives, such as cellulose ethers and esters, acetate cellulose and micro-crystalline cellulose. Borregaard is also one of the leading manufacturers of lignosulfonates and an exclusive producer of vanillin from wood.

1.4.1.2 The Domsjö Biorefinery

The original sulfite mill started operating in 1903 by the then Mo and Domsjö AB in Örnskoldsvik. Manufacturing of viscose pulp began in 1930 and a chemicals plant was developed on the site in the 1940s. Dömsjö is one of the first industrial biorefinery complexes where innovative and environmentally sound measures were implemented in both its processes and waste treatment. It was one of the first facilities in the world to bleach cellulose to the highest degree of brightness in a process totally free from the use of chlorine or chlorine dioxide. Nowadays the Domsjö biorefinery produces dissolving cellulose (255 000 tonnes per year), lignin as sodium lignosulfonate (120 000 tonnes per year) and bioethanol (14 000 tonnes per year). Other products include carbonic acid, biogas and energy. The bioenergy generated by the process is used internally and makes Domsjö virtually independent of fossil energy sources. The raw material consists largely of spruce and pine. Domsjö was acquired by the Indian Aditya Birla Group in April 2011, and is now a member of its pulp and fibre business.[57]

1.4.2 USA

The United States pioneered the corn industry. In 1844, the Wm. Colgate & Company wheat starch plant in Jersey City, NJ, became the first dedicated corn starch plant in world.[117] However, the multi-product corn wet mills appeared in their current form much later – in the 1970s – prompted by the development of commercial technology for the production of high fructose corn syrup for use in the soft drinks industry.[118] Early development of other types of industrial biorefining in the US was rather limited; the only lignocellulosic biorefinery in the United States was operated by Georgia Pacific in Bellingham, Washington. The biorefinery produced ethanol from cellulosic feedstock using the sulfite process. This plant operated from 1976 to 2001.[119]

1.4.3 Brazil

In November 1975, Brazil initiated a program (Pro Alcool) to start the production of ethanol from sugar cane and increase the use of ethanol as a substitute for gasoline. The program was supported by the national petroleum company Petrobras and started partly because of the quadrupling of world oil prices in 1973, and also as a means of assisting the sugar industry in times of low international sugar prices. The program promoted the installation of many 'mini-ethanol plants' capable of producing 20 000 L per day ethanol from sugar cane and other plants such as cassava. Sucrose from sugar cane or hydrolysed starch from cassava were subjected to fermentation by yeast.[120] Bagasse (fibrous residue of sugar cane) was used as boiler fuel and for electricity production to make the ethanol plants energy self-sufficient.

1.4.4   China

China has a long history of biomass biorefining; for instance the ancient Chinese developed soy processing to extract protein for bean curd, to ferment sugars into wine, to produce soybean oil and to make fertilizers for crops (from waste materials).[121] Historical examples of biorefining at an industrial scale are the hydrolysis and ABE batch fermentation plants of the period 1960–2000. These plants aimed mostly at producing acetone and ethanol solvents. The raw materials used were corn, cassava, potato and sweet potato. Strong competition from the petrochemical industry (cheap solvents) and the high costs for product purification led to the gradual demise of these plants at the turn of the last century.[122]

1.4.5   USSR/Russian Wood Hydrolysis Plants

Since the early 1930s, the focus in USSR was on production of ethanol, single cell protein (SCP) yeast and furfural/xylitol. Hydrolysis of softwood/hardwood feedstock was typically performed with weak sulfuric acid (130–150 °C) in one or two steps. Production of ethanol was performed by yeast. A unique facility was the Dokshukino plant, which started operation in 1960. The plant utilized sugars from starch for ABE fermentation by *Clostridia* bacteria. It was a prime example of continuous fermentation efficiency as it yielded a 20% productivity increase *versus* batch fermentation and saved 65 kg of starch per tonne of solvents produced.[123,124] It was the only industrial continuous ABE fermentation plant in the world.

At least seven more ABE plants were constructed; their design was based on the experience acquired from the Dokshukino facility. These ABE plants remained in operation until the late 1980s.[125] All of them were closed down after USSR lost economic power but also because cheaper petrochemical processes became available. A short review on USSR hydrolysis plants by Frölander[85] revealed that over the period of 1935–1985, 18 ethanol, 16 SCP yeast and 15 furfural/xylitol plants were in operation. It was also concluded that none of them was profitable without state subsidies because they lacked the integrated approach (multitude of products from a single feedstock). Today there is only one sulfite ethanol plant that still in operation in Russia (Kirov).

1.5   Future Technologies for Biorefining: Catalysis and Ionic Liquids

Clearly the bioprocessing industry has already adopted or is adopting many of the 12 principals of Green Chemistry. However, economics are still the major deciding factor when it comes to running processes. Legislation often only

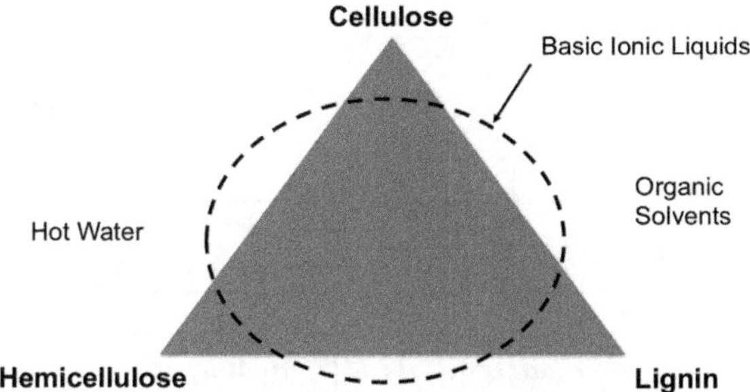

Figure 1.5   Potential for homogenization of biomass using basic ionic liquids.

changes when competitive technologies become available that give market control to a limited number of actors. The major developments in green bioprocessing will be facilitated to a large degree by application of the 9th principal: application of catalysts is superior to the use of non-stoichiometric reagents. The Pimentel report 'Opportunities in Chemistry'[126] in 1985 estimated that 20% of the US gross national product was produced through the use of catalytic processes. Long experience in the understanding and development of catalytic processes has now been gained through petrochemical industrial development. This must now be combined with the existing bioprocessing infrastructure to afford selective reactions to high-value products.

An example of this kind of catalytic process applied in future biorefinery can be upgrading of biomass extractives to fine chemicals over heterogeneous catalysts.[127,128]

Ionic liquids are also seen as an enabling technology. Different biomass components have selective solubility in different solvents. For example, hot water is known to solubilize native (acetylated) hemicelluloses. Dipolar aprotic solvents can also solubilize both lignin and hemicelluloses. However, basic ionic liquids are known to dissolve all three components (Figure 1.5).[129–131]

This therefore allows for the potential for homogeneous processing of biomass, in addition to selective extraction of different components. This opens the door to a wide range of possibilities that were not accessible before.

1.5.1   Recalcitrance Reduction

Recalcitrance reduction is where ionic liquids are used to pre-treat biomass to enhance downstream biofuel and chemical production. Bioethanol production through recalcitrance reduction, polysaccharide enzymatic hydrolysis and fermentation is of major interest. The ionic liquid pre-treatment typically allows for much higher hydrolysis rates and yields, based on the saccharide content of the feedstock. By applying an ionic liquid treatment to lignocellulosic feedstocks the crystallinity of cellulose can be significantly reduced or converted into the more digestible cellulose II. Lignin can also be removed at the same time, which prevents

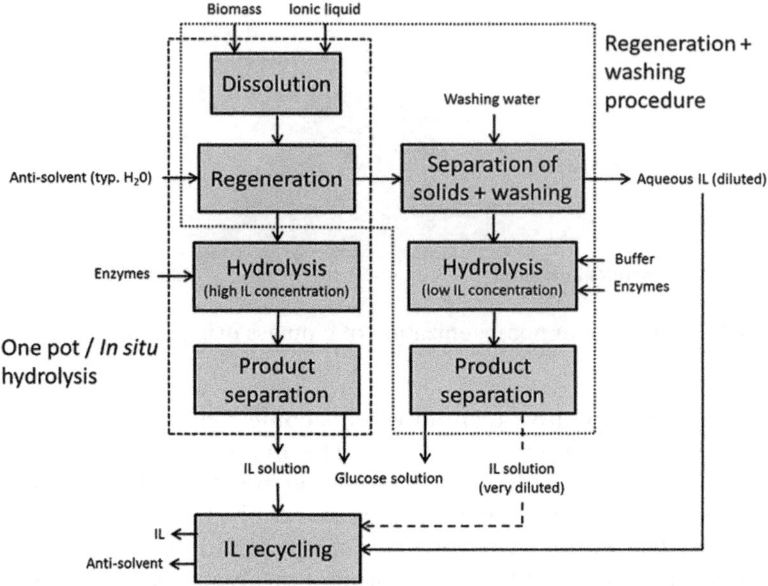

Figure 1.6   One-pot process for ionic liquid-aided biomass recalcitrance reduction and enzymatic saccharification.[133]

enzyme inhibition and allows enzymes or other catalysts to attack the saccharide portions of the biomass. Aside from the high cost of enzymes, the major drawbacks in this process are related to the high cost of the ionic liquid and poor resistance of the downstream enzymes or microorganisms to residual ionic liquid. New strains of microorganism are now being found that can tolerate high doses of ionic liquid.[132] A recent proposal for a one-pot saccharification and enzymatic hydrolysis was also recently made (Figure 1.6).[133]

Recycling the ionic liquid to a very high degree is implicit in all these processes. There are many methods that are under development however, including the use of switchable ionic liquids (SILs),[30–34,38,134,135] distillable ionic liquids (DILs),[136–138] phase-separable ionic liquids (PSILs),[139,140] chromatography[141] and auxiliaries[142] for purification of ionic liquid or isolation of oligomers, post biomass treatment. In the case of SILs and DILs the ionic liquids are protic and are derived from the combination of superbases, such as 1,8-diazabicyclo[5.4.0]undec-7-ene (DBU), 1,5-diazabicyclo[4.3.0]non-5-ene (DBN) or 1,1,3,3-tetramethylguanidine (TMG) with either acid gases (in the case of SILs) or organoacids (in the case of DILs). Therefore, recycling of the components can be performed by vaporization of the neutral starting species (Figure 1.7). These are accessible by increasing the temperature of the mixtures to between 100 and 180 °C.

Major advances are being made, however, to reduce the cost of the ionic liquids designed for biomass pre-treatment and generally improve

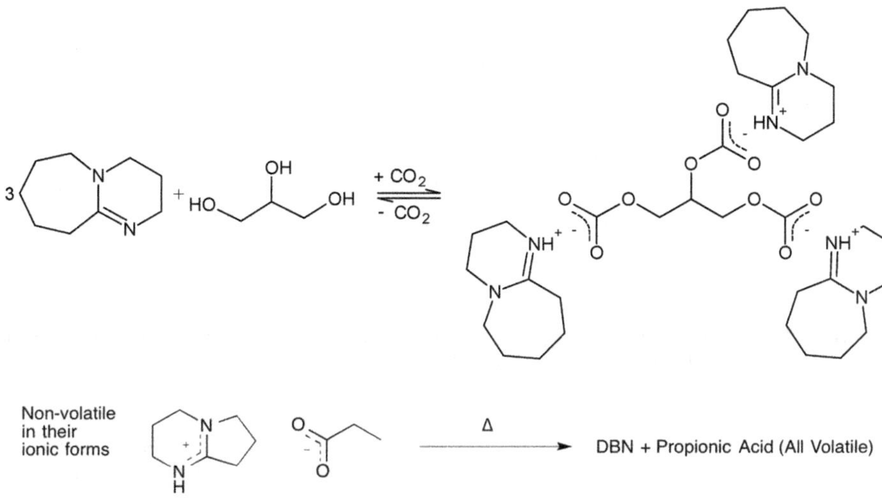

Figure 1.7	Mechanism of distillation for the switchable ionic liquid (SIL) DBU–CO$_2$–Glycerol (top)[38] and the distillable ionic liquid [DBNH][CO$_2$Et] (bottom, adapted from ref. 138 from Wiley-VCH).

performance.[143] Competing recalcitrance reduction technologies are predominantly aqueous acid digestion and ammonia fibre explosion/expansion (AFEX).

1.5.2   Fractionation of Biomass

Fractionation of biomass affords pure or tuneable polymeric fractions. The opportunity to selectively extract components or for complete dissolution and selective precipitation of different components offers the potential for highly tuneable fractions from biomass. Some studies have suggested that fraction purity also depends very much on molecular weight and the degree of linkage between lignin and carbohydrates.[144,145] However there have been some recent success stories with, in particular, SILs[30–34,38,134,135] and pre-treatments being applied to improve the efficiency of separation of components in hard and softwoods.[146] The growing body of work on SIL fractionation in particular offers very high efficiency in separation of lignin from polysaccharide and the production of fibrous pulps. This is in part due to the fact that cellulose is not soluble in the SILs but lignin can be selectively extracted. A typical process scheme for the highly optimized short-time high-temperature (STHT) SIL fractionation is given in Figure 1.8.

In this case, and under the optimum conditions, the SIL was DBU–SO$_2$–monoethanolamine (DBU–SO$_2$–MEA). Wood chips were used (birch or spruce). The water composition in the SIL was 37 wt.%. The tolerance of the process to water is rather important due to the fact that dried chips will not be accessible on an industrial scale, due to the energy required in removal of

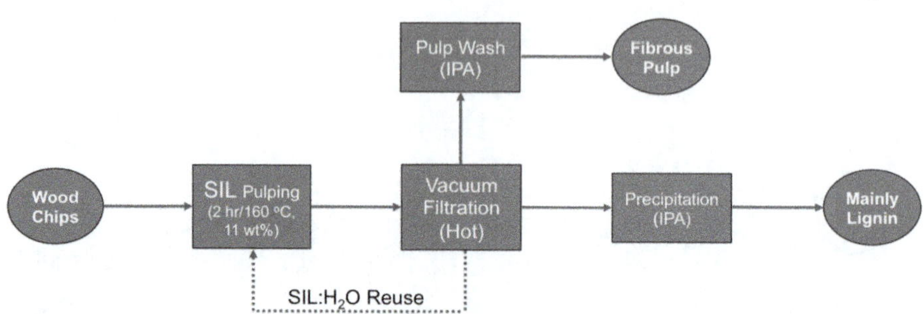

Figure 1.8   Typical simplified process scheme for fractionation of wood chips with switchable ionic liquids (SILs).

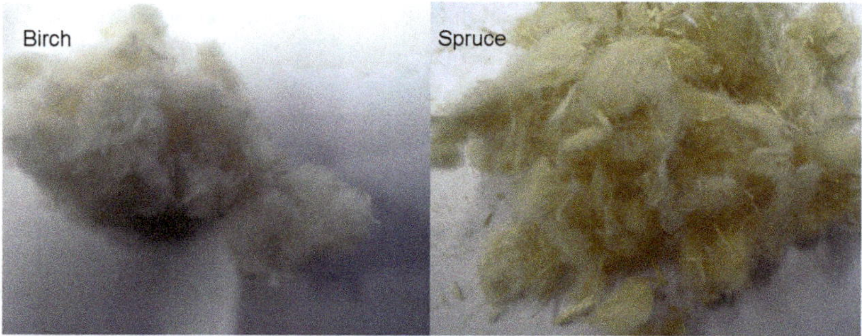

Figure 1.9   Fibrous pulps from birch (left) and spruce (right), resulting from DBU–SO$_2$–MEA fractionation under optimum conditions (adapted from ref. 32 with permission from Wiley-VCH).

water. Water contents can be as high as 50 wt.% for industrial chips. Interestingly, the resulting pulps are actually fibrous (Figure 1.9).

Much of the success of the different fractionation trials is, however, highly dependent on the biomass sources and chemical pre-treatments applied. Particle size is also of major concern. Finely divided sawdust is not a practical feedstock to use due to the high cost of production. Wet wood chips are much more realistic to use, which is where most homogeneous ionic liquid processes will fail, as they cannot dissolve chips, without significant degradation. However, some ionic liquids are known to fibrillate wood chips, potentially avoiding costs associated with biomass pulverization.[31,147,148] Hypothetically, homogeneous processing of wood chips to produce polymers is possible but not achievable in practice, without the development of selective degradation treatments that do not degrade the constituent polymers. The main candidate linkages for selective degradation are suggested to be lignin-carbohydrate complexes[130] but the existence of these entities, as covalent linkages, in native wood has not been rigorously established. At least, maintaining the fibrous properties of the resulting pulps is not undesirable as they may enter more traditional fibre-based value chains. Simpler fractionation concepts are also appearing, such as the extraction of kraft pulp using ionic liquid–water compositions. This allows for quite efficient removal of hemicelluloses, thus affording a higher purity cellulose pulp which may be useful for chemicals or textiles production. This process has been termed as the IONCELL-P process.[149]

1.5.3 Thorough Chemical Modification of Biomass

Thorough chemical modification of biomass to produce high bio-content polymeric materials is facilitated by homogeneous dissolution in ionic liquid. Ionic liquids allow for a wide range of chemistries on wood. The most common of these studied is the acylation of wood.[130,150] Ionic liquids can allow for complete reaction of all hydroxyl groups in wood or wood biopolymers using typical acylating reagents. These include anhydrides, acid chlorides, isocyanates, *etc.* More sustainable methods are also under development, such as alkylcarbonate formation[151] or even acylation by transesterification.[152] Etherification is also possible.[37]

1.5.4 Enhanced Analytics

Enhanced analytics are possible to improve our understanding and analysis of lignocellulosics. As ionic liquids have exquisite biomass dissolving capabilities this has allowed for improved analytics where the native structure of the biomass is largely preserved. Ionic liquids are finding application as media for biomass derivatization for chromatography,[153] as mobile phases for chromatography[154] and as media for high-resolution solution-state NMR.[139,155,156] In the other hand, the analysis of ionic liquid-treated lignocellulosics can become cumbersome, since traces of ionic liquids can cause several problems in traditional chromatographic methods.[157,158] However, these problems can be overcome by developing methods like capillary electrophoresis.[159]

Further applications of ionic liquids in biomass processing are appearing but the major challenges that need to be overcome are:

(1) recycling the ionic liquids
(2) reduction in the cost of ionic liquids
(3) reduction in toxicity of ionic liquids[160]
(4) use of cost-effective and selective pre-treatments to preserve molecular weights upon ionic liquid fractionation and
(5) develop ionic liquid-resistant enzymes and microorganisms for saccharification.

1.6 Conclusions

It is evident that the field of biorefining is experiencing not only a 'renaissance' but also rapid growth in terms of the volume of the research and development. The constant aim towards more green and sustainable processes and products is motivated by the impeding threat of global warming as well as increased awareness about the danger contained into ever increasing chemical loads on biosphere and humankind. To facilitate this, we need to re-think not only the today's extensive use of fossil resources but also what kinds of synthetic molecules are released to our habitat. Future integrated biorefineries need to evolve in order to give solutions in assisting our quest towards these aspirations. In general, neoteric solvents or solvent–catalyst systems will most likely be a part of the solution but only if the principles and the spirit of 'Green Chemistry' are respected. Therefore, appropriate choice of concepts and chemistries, efficient recycling and separation concepts, re-designed benign chemical derivatives as building blocks for our commodities and daily consumables are the trends of today. At the same time, we need to change the way how the global industry, consumer markets, business and societies perceive our consumption: reversing the clock from today's throw away and designed to fail culture to seeing durable, well-designed and long-lasting commodities of the past – only this time with clearly improved energy-efficiency.

References

1. Technology Report – Biofuels for Transport, http://www.iea.org/publications/freepublications/publication/Biofuels_Roadmap_WEB.pdf.
2. *United Nations Report of the World Commission on Environment and Development*, Brundtland Commission, 1987.
3. C. Smith and G. Rees, *Economic Development*, Macmillan, Basingstoke, UK, 1998.
4. P. T. Anastas and J. C. Warner, *Green Chemistry: Theory and Practice*, Oxford University Press, New York, 1998.
5. P. T. Anastas and J. B. Zimmerman, *Environ. Sci. Technol.*, 2003, 37, 94a–101a.
6. J.-L. Dubois, in *Biorefinery: From Biomass to Chemicals and Fuels*, Walter de Gruyter, 2012, ch. 19–48.
7. E. de Jong, *IEA Bioenergy, Task42 Biorefinery*, 2012.
8. R. H. Kottke, in *Kirk-Othmer Encyclopedia of Chemical Technology*, ed. J. I. Kroschwitz and M. Howe-Grant, John Wiley & Sons, New York, 4th edn, 1998.
9. A. G. Lenzing, http://www.lenzing.com.
10. A. Corma, O. de la Torre, M. Renz and N. Villandier, *Angew. Chem., Int. Ed.*, 2011, 50, 2375–2378.
11. A. Corma, O. de la Torre and M. Renz, *Energy Environ. Sci.*, 2012, 5, 6328–6344.
12. J. S. Luterbacher, J. M. Rand, D. M. Alonso, J. Han, J. T. Youngquist, C. T. Maravelias, B. F. Pfleger and J. A. Dumesic, *Science*, 2014, 343, 277–280.
13. J. Q. Bond, D. M. Alonso, D. Wang, R. M. West and J. A. Dumesic, *Science*, 2010, 327, 1110–1114.
14. W. Fang and H. Sixta, *ChemSusChem*, 2015, 8, 73–76.

15. J. Joelsson and T. Tuuttila, The History and current development of forest biorefineries in Finland and Sweden, http://www.biofuelregion. se/UserFiles/file/History%20and%20status%20of%20biorefineries%20 in%20Fi%20and%20Sw_v1_1_20121029.pdf.

16. P. D. de Maria, *J. Chem. Technol. Biotechnol.*, 2014, **89**, 11–18.

17. J. Xu, X. Yao, Q. Zhou, X. Lu and S. Zhang, *RSC Adv.*, 2014, **4**, 27430–27438.

18. W. Song, Y. Deng, Y. Xu and M. Cui, *J. Chem. Pharm. Res.*, 2014, **6**, 260–263.

19. P. Praveen and A. Prasad, WO2014/144588, 2014.

20. X. D. Hou, T. J. Smith, N. Li and M. H. Zong, *Biotechnol. Bioeng.*, 2012, **109**, 2484–2493.

21. A. M. da Costa Lopes and R. Bogel-Lukasik, *ChemSusChem*, 2015, **8**, 947–965.

22. A. M. da Costa Lopes, K. G. João, E. Bogel-Łukasik, L. B. Roseiro and R. Bogel-Lukasik, *J. Agric. Food Chem.*, 2013, **61**, 7874–7882.

23. A. M. da Costa Lopes, K. G. Joao, D. F. Rubik, E. Bogel-Lukasik, L. C. Duarte, J. Andreaus and R. Bogel-Lukasik, *Bioresour. Technol.*, 2013, **142**, 198–208.

24. S. P. Magalhães da Silva, A. M. da Costa Lopes, L. B. Roseiro and R. Bogel-Lukasik, *RSC Adv.*, 2013, **3**, 16040–16050.

25. R. Pezoa, V. Cortinez, S. Hyvarinen, M. Reunanen, J. Hemming, M. E. Lienqueo, O. Salazar, R. Carmona, A. Garcia, D. Y. Murzin and J. P. Mikkola, *Cellul. Chem. Technol.*, 2010, **44**, 165–172.

26. V. P. Soudham, J. Grasvik, B. Alriksson, J. P. Mikkola and L. J. Jonsson, *J. Chem. Technol. Biotechnol.*, 2013, **88**, 2209–2215.

27. F. Cheng, H. Wang, G. Chatel, G. Gurau and R. D. Rogers, *Bioresour. Technol.*, 2014, **164**, 394–401.

28. R. Prado, X. Erdocia and J. Labidi, *J. Chem. Technol. Biotechnol.*, 2013, **88**, 1248–1257.

29. C. Froschauer, M. Hummel, M. Iakovlev, A. Roselli, H. Schottenberger and H. Sixta, *Biomacromolecules*, 2013, **14**, 1741–1750.

30. V. Eta, I. Anugwom, P. Virtanen, P. Mäki-Arvela and J. P. Mikkola, *Ind. Crops Prod.*, 2014, **55**, 109–115.

31. I. Anugwom, V. Eta, P. Virtanen, P. Mäki-Arvela, M. Hedenström, M. Yibo, M. Hummel, H. Sixta and J.-P. Mikkola, *Biomass Bioenergy*, 2014, **70**, 373–381.

32. I. Anugwom, V. Eta, P. Virtanen, P. Mäki-Arvela, M. Hedenstrom, M. Hummel, H. Sixta and J. P. Mikkola, *ChemSusChem*, 2014, 7, 1170–1176.

33. I. Anugwom, P. Mäki-Arvela, P. Virtanen, S. Willför, P. Damlin, M. Hedenstrom and J. P. Mikkola, *Holzforschung*, 2012, **66**, 809–815.

34. I. Anugwom, P. Mäki-Arvela, P. Virtanen, S. Willför, R. Sjoholm and J. P. Mikkola, *Carbohydr. Polym.*, 2012, **87**, 2005–2011.

35. J. Shi, K. Balamurugan, R. Parthasarathi, N. Sathitsuksanoh, S. Zhang, V. Stavila, V. Subramanian, B. A. Simmons and S. Singh, *Green Chem.*, 2014, **16**, 3830–3840.

36. J. X. Long, X. H. Li, B. Guo, F. R. Wang, Y. H. Yu and L. F. Wang, *Green Chem.*, 2012, **14**, 1935–1941.
37. J.-P. Mikkola, A. Kirilin, J.-C. Tuuf, A. Pranovich, B. Holmbom, L. M. Kustov, D. Y. Murzin and T. Salmi, *Green Chem.*, 2007, **9**, 1229–1237.
38. I. Anugwom, P. Mäki-Arvela, P. Virtanen, P. Damlin, R. Sjöholm and J.-P. Mikkola, *RSC Adv.*, 2011, **1**, 452–457.
39. S. A. Nolen, C. L. Liotta, C. A. Eckert and R. Gläser, *Green Chem.*, 2003, **5**, 663–669.
40. C. A. Eckert, C. L. Liotta, D. Bush, J. S. Brown and J. P. Hallett, *J. Phys. Chem. B*, 2004, **108**, 18108–18118.
41. A. J. Ragauskas, C. K. Williams, B. H. Davison, G. Britovsek, J. Cairney, C. A. Eckert, W. J. Frederick, J. P. Hallett, D. J. Leak, C. L. Liotta, J. R. Mielenz, R. Murphy, R. Templer and T. Tschaplinski, *Science*, 2006, **311**, 484–489.
42. C. Valeur, *Papper och massa i Västerbotten och Norrbotten. [från handpappersbruk till processindustri]*, Skogsindustrierna, Stockholm, Sweden, 2003.
43. S. Sörlin and A. Öckerman, *Nordisk pappershistorisk tidskrift*, Föreningen Nordiska Pappershistoriker, Stockholm, Sweden, 2001.
44. B. Persson, *Sulfitsprit. Förhoppningar och besvikelser under 100 år*, DAUS Tryck & Media, Bjästa, Sweden, 2007.
45. Chemrec, http://www.chemrec.se.
46. SP Technical Research Institute of Sweden, http://www.sp.se.
47. S. Kappa, http://www.smurfitkappa.com.
48. Volvo Trucks, http://www.volvotrucks.com.
49. Piteå Science Park, http://www.piteasciencepark.se.
50. SunPine, http://www.sunpine.se.
51. A. B. Preem, http://www.preem.se.
52. Borregaard, http://www.borregaard.com.
53. Neste Oil and Stora Enso inaugurate biofuels demonstration facility at Varkaus in Finland, http://www.nesteoil.com/default.asp?path=1,41,540,1259,1260,11736,12772&output=print.
54. Chempolis Oy, http://www.chempolis.com.
55. ST1, http://www.st1.eu.
56. Sybimar Oy, http://www.sybimar.fi/en/product_categories/bioenergy.
57. Aditya Birla - Domsjö, http://www.domsjo.adityabirla.com.
58. F. Carvalheiro, L. C. Duarte, R. Medeiros and F. M. Girio, *Biotechnol. Lett.*, 2007, **29**, 1887–1891.
59. Forchem, http://www.forchem.com.
60. Arizona Chemicals, http://www.arizonachemical.com.
61. Introducing Renewable Diesel UPM BioVerno, http://www.upmbiofuels.com/renewable-diesel-upm-bioverno.
62. G. Rødsrud, M. Lersch and A. Sjöde, *Biomass Bioenergy*, 2012, **46**, 46–59.
63. Metener Oy, http://www.metener.fi.
64. A. K. Chapagain, A. Y. Hoekstra, H. H. G. Savenije and R. Gautam, *Ecol. Econ.*, 2006, **60**, 186–203.

65. Cotton – Fair Trade, http://www.customerservicefoundation.com/cotton-fair-trade.html.
66. C. J. Biermann, *Handbook of Pulping and Papermaking*, Academic Press, 1996.
67. Neste Oil, http://www.nesteoil.fi.
68. M. L. G. Reno, O. A. del Olmo, J. C. E. Palacio, E. E. S. Lora and O. J. Venturini, *Energy Convers. Manage.*, 2014, **86**, 981–991.
69. A. E. Farrell, R. J. Plevin, B. T. Turner, A. D. Jones, M. O'Hare and D. M. Kammen, *Science*, 2006, **311**, 506–508.
70. T. Searchinger, R. Heimlich, R. A. Houghton, F. Dong, A. Elobeid, J. Fabiosa, S. Tokgoz, D. Hayes and T.-H. Yu, *Science*, 2008, **319**, 1238–1240.
71. BioEndev, http://www.bioendev.se.
72. ROQUETTE, http://www.roquette.com.
73. GranBio starts cellulosic ethanol production at 21 million gallon plant in Alagoas, Brazil, http://www.biofuelsdigest.com/bdigest/2014/09/24/granbio-starts-cellulosic-ethanol-production-at-21-mgy-plant-in-brazil.
74. G. Sousa, Perspective on the Development of Integrated Biorefineries, http://www.lneg.pt/download/1054.
75. RESPOL, http://www.respol.pt.
76. B. Kamm, P. R. Gruber and M. Kamm, *Biorefineries – Industrial Processes and Products. Status Quo and Future Directions*, Wiley-VCH Verlag GmbH & Co. KGaA, Weinheim, 2006.
77. Biorefineries Roadmap as part of the German Federal Government action plans for the material and energetic utilisation of renewable raw materials, http://www.bmbf.de/pub/BMBF_Roadmap-Bioraffinerien_en_bf.pdf.
78. R. C. Kuhad, R. Gupta, Y. P. Khasa, A. Singh and Y.-H. P. Zhang, *Renewable Sustainable Energy Rev.*, 2011, **15**, 4950–4962.
79. J.-L. Dubois, Refinery of the Future: Feedstock, Processes, Products, http://www.eurobioref.org/Summer_School/Lectures_Slides/day2/Lectures/L02_JL%20Dubois.pdf.pdf.
80. Cargill History Timeline, http://www.cargill.com/wcm/groups/public/@ccom/documents/document/doc-cargill-history-timeline.pdf.
81. DuPont Tate & Lyle's Biochemical Plant, Tennessee, United States of America, http://www.chemicals-technology.com/projects/dupont-tate-lyle-biochemical-loudon-tennessee/.
82. S. K. Maity, *Renewable Sustainable Energy Rev.*, 2015, **43**, 1427–1445.
83. V. P. Soudham, T. Brandberg, J.-P. Mikkola and C. Larsson, *Bioresource Technol.*, 2014, **166**, 559–565.
84. E. Sklavounos, Doctoral Thesis, *Conditioning of SO_2-ethanol-water (SEW) spent liquor from lignocellulosics for ABE fermentation to biofuels and chemicals*, Aalto University, 2014.
85. A. Frolander, Conversion of cellulose, hemicellulose and lignin into platform molecules: biotechnological approach http://www.eurobioref.org/Summer_School/Lectures_Slides/day3/Lectures/L06_A.Frolander.pdf.

86. G. Rodsrud, BALI™ demo plant for co-production of bioethanol and green chemicals, http://nobio.no/upload_dir/pics/GudbrandRoedsrud.pdf.

87. Abengoa celebrates grand opening of its first commercial-scale next generation biofuels plant, http://www.abengoa.com/web/en/noticias_y_publicaciones/noticias/historico/2014/10_octubre/abg_20141017.html.

88. Crescentino's Biorefinery Grand Opening, http://www.betarenewables.com/press-release-detail/2/crescentinos-biorefinery-grand-opening.

89. B. Benjelloun, Tomorrow's Biorefineries in Europe – The CIMV organosolv Process, http://www.biocore-europe.org/file/1_4%20BIOCORE%20B%20Benjelloun%20CIMV%20Organasolv.pdf.

90. CIMV, CIMV The Biorefinery Concept, http://www.cimv.fr/cimv-technology/cimv-technology/5-.html.

91. St1 plans new sustainable bioethanol plant in Finland, http://www.cleantechfinland.com/content/st1-plans-new-sustainable-bioethanol-plant-finland.

92. A. V. Bridgwater, *Fuel*, 1995, **74**, 631–653.

93. P. McKendry, *Bioresource Technol.*, 2002, **83**, 47–54.

94. A. V. Bridgwater, *Chem. Eng. J.*, 2003, **91**, 87–102.

95. G. W. Huber, S. Iborra and A. Corma, *Chem. Rev.*, 2006, **106**, 4044–4098.

96. Biomass to Liquids (BtL) - European Biofuels Technology Platform, http://www.biofuelstp.eu/btl.html.

97. J. C. Serrano-Ruiz and J. A. Dumesic, *Energy Environ. Sci.*, 2011, **4**, 83–99.

98. T. P. Vispute and G. W. Huber, *Green Chem.*, 2009, **11**, 1433–1445.

99. UOP and Ensyn to Form Joint Venture to Offer Second Generation Biomass Technology, http://www.uop.com/?press_release=uop-and-ensyn-to-form-joint-venture-to-offer-second-Generation-biomass-technology.

100. KIOR, http://www.kior.com/.

101. BIOWERT Industrie GmbH, http://greennewdeal.eu/industry/successes/biowert-industrie-gmbh.html.

102. SunPine Biorefinery, http://www.energimyndigheten.se/Global/Forskning/Transport/Stigsson,%20SunPine.pdf.

103. O. S. Stamenković, A. V. Veličković and V. B. Veljković, *Fuel*, 2011, **90**, 3141–3155.

104. S. Mannonen, Biofuels from wood-based raw materials – European Biofuels Technology Platform, http://www.biofuelstp.eu/spm6/docs/sari-manonen.pdf.

105. Neste Oil – Renewable raw material procurement, http://2013.nesteoil.com/business/renewable-fuels/Renewable-raw-material-procurement/.

106. Greenpeace, Neste Oil's plans for global leadership in palm oil diesel will drive massive rainforest destruction and climate change, http://www.greenpeace.org/international/en/press/releases/neste-pal-oil-drives-climate-change/.

107. Firends of the Earth Europe – Championed 'green' supplier exposed for illegal expansions, http://www.foeeurope.org/press/2010/Mar15_europes_demand_for_palm_oil_driving_deforestation_and_landgrabbing.html.

108. NESTE OIL – Renewable raw material, http://www.biofuels2050.eu/resource-efficient/renewable-feedstock.
109. Z. Gong, H. Shen, Q. Wang, X. Yang, H. Xie and Z. K. Zhao, *Biotechnol. Biofuels*, 2013, **6**, 1–12.
110. http://www.dutchglycerinrefinery.eu/.
111. Y. Chisti, *Trends Biotechnol.*, 2008, **26**, 126–131.
112. L. Rodolfi, G. Chini Zittelli, N. Bassi, G. Padovani, N. Biondi, G. Bonini and M. R. Tredici, *Biotechnol. Bioeng.*, 2009, **102**, 100–112.
113. N. Adelsköld, Growing algae for fuel and feed, http://www.slu.se/en/collaboration-innovation/knowledge-bank/2013/4/growing-algae-for-fuel-and-feed.
114. P. T. Pienkos and A. Darzins, *Biofuels, Bioprod. Biorefin.*, 2009, **3**, 431–440.
115. G. Rødsrud, A. Frolander, A. Sjode and M. Lersch, in *Biorefinery: From Biomass to Chemicals and Fuels*, ed. M. Aresta, A. Dibenedetto and F. Dumeignil, Walter de Gruyter, 2012, pp. 141–166.
116. http://www.adityabirla.com/businesses/Profile/Domsj%C3%B6-Fabriker.
117. Corn Refiners Association, http://corn.org/about/history/#sthash.D7zdSqtv.dpuf.
118. L. R. Lynd, C. Wyman, M. Laser, D. Johnson and R. Landucci, *Strategic Biorefinery Analysis: Analysis of Biorefineries*, National Renewable Energy Laboratory-NREL, Colorado 80401–3393, 2005.
119. A. Graf and T. Koehler, Oregon Cellulose-Ethanol Study, http://www.oregon.gov/energy/RENEW/Biomass/docs/OCES/OCES.PDF.
120. E. J. Weber, J. H. Cock and A. Chouinard, *Cassava Harvesting and Processing*, The International Development Research Centre, Ottawa, Canada, 1978.
121. L. R. Lynd, C. Wyman, M. Laser, D. Johnson and R. Landucci, *Strategic Biorefinery Analysis: Review of Existing Biorefinery Examples*, National Renewable Energy Laboratory – NREL, http://www.nrel.gov/docs/fy06osti/34895.pdf, Golden, Colorado 80401–3393, 2005.
122. Z. Sun and Z. Shi, The acetone-butanol (ABE) fermentation industries in China, http://dc.engconfintl.org/cgi/viewcontent.cgi?article=1006&context=bioenergy_i.
123. D. Nimcevic and J. R. Gapes, *J. Mol. Microbiol. Biotechnol.*, 2000, **2**, 15–20.
124. J. Hospodka, in *Theoretical and Methodological Basis of Continuous Culture of Microorganisms*, ed. I. Malek and Z. Fencl, Academic Press, New York, USA, 1966, pp. 611–613.
125. V. Zverlov, O. Berezina, G. Velikodvorskaya and W. Schwarz, *Appl. Microbiol. Biotechnol.*, 2006, **71**, 587–597.
126. G. C. Pimentel, *Opportunities in Chemistry*, National Academic Press, Washington, USA, 1985.
127. E. Salminen, P. Mäki-Arvela, P. Virtanen, T. Salmi and J.-P. Mikkola, *Top. Catal.*, 2014, **57**, 1533–1538.
128. E. Salminen, L. Rujana, P. Mäki-Arvela, P. Virtanen, T. Salmi and J.-P. Mikkola, *Catal. Today*, 2015, in press, Corrected Proof, Available online 30 June 2014.

129. I. A. Kilpeläinen, H. Xie, A. King, M. Granstrom, S. Heikkinen and D. S. Argyropoulos, *J. Agric. Food Chem.*, 2007, **55**, 9142–9148.
130. L. Kyllönen, A. Parviainen, S. Deb, M. Lawoko, M. Gorlov, I. Kilpeläinen and A. W. King, *Green Chem.*, 2013, **15**, 2374–2378.
131. D. A. Fort, R. C. Remsing, R. P. Swatloski, P. Moyna, G. Moyna and R. D. Rogers, *Green Chem.*, 2007, **9**, 63–69.
132. J. I. Khudyakov, P. D'haeseleer, S. E. Borglin, K. M. DeAngelis, H. Woo, E. A. Lindquist, T. C. Hazen, B. A. Simmons and M. P. Thelen, *Proc. Natl. Acad. Sci. U. S. A.*, 2012, **109**, E2173–E2182.
133. R. M. Wahlstrom and A. Suurnakki, *Green Chem.*, 2015, **17**, 694–714.
134. V. Eta, I. Anugwom, P. Virtanen, K. Eränen, P. Mäki-Arvela and J.-P. Mikkola, *Chem. Eng. J.*, 2014, **238**, 242–248.
135. I. Anugwom, J. P. Mikkola, P. Mäki-Arvela and A. Virtanen, WO2012/059643, 2012.
136. A. W. King, J. Asikkala, I. Mutikainen, P. Järvi and I. Kilpeläinen, *Angew. Chem.*, 2011, **123**, 6425–6429.
137. A. W. T. King and I. A. Kilpeläinen, WO2011/161326, 2011.
138. A. Parviainen, A. W. King, I. Mutikainen, M. Hummel, C. Selg, L. K. Hauru, H. Sixta and I. Kilpeläinen, *ChemSusChem*, 2013, **6**, 2161–2169.
139. A. J. Holding, M. Heikkilä, I. Kilpeläinen and A. W. King, *ChemSusChem*, 2014, **7**, 1422–1434.
140. A. W. T. King, A. J. Holding and I. A. Kilpeläinen, WO2014/060651, 2014.
141. B. R. Caes, T. R. Van Oosbree, F. Lu, J. Ralph, C. T. Maravelias and R. T. Raines, *ChemSusChem*, 2013, **6**, 2083–2089.
142. T. C. R. Brennan, S. Datta, H. W. Blanch, B. A. Simmons and B. M. Holmes, *BioEnergy Res.*, 2010, **3**, 123–133.
143. A. George, A. Brandt, K. Tran, S. N. S. M. S. Zahari, D. Klein-Marcuscha-mer, N. Sun, N. Sathitsuksanoh, J. Shi, V. Stavila, R. Parthasarathi, S. Singh, B. M. Holmes, T. Welton, B. A. Simmons and J. P. Hallett, *Green Chem.*, 2015, **17**, 1728–1734.
144. T. Leskinen, A. W. King, I. Kilpeläinen and D. S. Argyropoulos, *Ind. Eng. Chem. Res.*, 2013, **52**, 3958–3966.
145. T. Leskinen, A. W. King, I. Kilpeläinen and D. S. Argyropoulos, *Ind. Eng. Chem. Res.*, 2011, **50**, 12349–12357.
146. L. K. Hauru, Y. Ma, M. Hummel, M. Alekhina, A. W. King, I. Kilpeläinen, P. A. Penttilä, R. Serimaa and H. Sixta, *RSC Adv.*, 2013, **3**, 16365–16373.
147. I. A. Kilpeläinen, A. W. T. King, P. Karhunen and J. Matikainen, WO2011/114004, 2011.
148. J. Viell and W. Marquardt, *Holzforschung*, 2011, **65**, 519–525.
149. A. Roselli, M. Hummel, A. Monshizadeh, T. Maloney and H. Sixta, *Cellulose*, 2014, **21**, 3655–3666.
150. H. Xie, A. King, I. Kilpelainen, M. Granstrom and D. S. Argyropoulos, *Biomacromolecules*, 2007, **8**, 3740–3748.
151. S. R. Labafzadeh, K. J. Helminen, I. Kilpeläinen and A. W. King, *ChemSusChem*, 2015, **8**, 77–81.

152. A. Schenzel, A. Hufendiek, C. Barner-Kowollik and M. A. Meier, *Green Chem.*, 2014, **16**, 3266–3271.
153. K. Schlufter, H. P. Schmauder, S. Dorn and T. Heinze, *Macromol. Rapid Commun.*, 2006, **27**, 1670–1676.
154. K. Kuroda, Y. Fukaya and H. Ohno, *Anal. Methods*, 2013, **5**, 3172–3176.
155. K. Kuroda, H. Kunimura, Y. Fukaya, N. Nakamura and H. Ohno, *ACS Sustainable Chem. Eng.*, 2014, **2**, 2204–2210.
156. K. Kuroda, H. Kunimura, Y. Fukaya and H. Ohno, *Cellulose*, 2014, **21**, 2199–2206.
157. S. Hyvärinen, P. Damlin, J. Gräsvik, D. Y. Murzin and J.-P. Mikkola, *Cellul. Chem. Technol.*, 2011, **45**, 483.
158. S. Hyvärinen, P. Virtanen, D. Yu Murzin and J.-P. Mikkola, *Cellul. Chem. Technol.*, 2010, **44**, 187.
159. S. Hyvarinen, J. P. Mikkola, D. Y. Murzin, M. Vaher, M. Kaljurand and M. Koel, *Catal. Today*, 2014, **223**, 18–24.
160. S.-K. Mikkola, A. Robciuc, J. Lokajova, A. J. Holding, M. Lämmerhofer, I. A. Kilpeläinen, J. Holopainen, A. W. T. King and S. K. Wiedmer, *Environ. Sci. Technol.*, 2015, **49**, 1870–1878.

The Dissolution of Biomass in Ionic Liquids Towards Pre-Treatment Approach

ANDREIA A. ROSATELLA[*a] AND CARLOS A. M. AFONSO[*a]

[a]iMed.ULisboa, Faculdade de Farmácia da Universidade de Lisboa, Av. Prof. Gama Pinto, 1649-003 Lisboa, Portugal
*E-mail: rosatella@ff.ulisboa.pt, carlosafonso@ff.ulisboa.pt

2.1 Introduction

The actual world population and income growth have been the key drivers for the growing demand for energy and chemical commodities, and from the Industrial Revolution to the present fossil resources have fulfilled the world's needs as energy and feedstock sources. Although these feedstocks are no longer regarded as sustainable and are questionable from the economic, ecological and environmental point of views. Therefore, the quest for sustainable and environmentally benign sources of energy has become crucial in recent years.

A valuable alternative are biofuels produced from biomass, such as plants. The 'first generation' of biofuels (derived from simple sugar fermentation) is already under commercialization. As their feedstock is derived from the agricultural sector, such as corn, wheat, sugar cane, and oilseeds, they are creating some scepticism to scientists due to the food-*versus*-fuel debate.[1-3]

RSC Green Chemistry No. 36
Ionic Liquids in the Biorefinery Concept: Challenges and Perspectives
Edited by Rafal Bogel-Lukasik

Published by the Royal Society of Chemistry, www.rsc.org

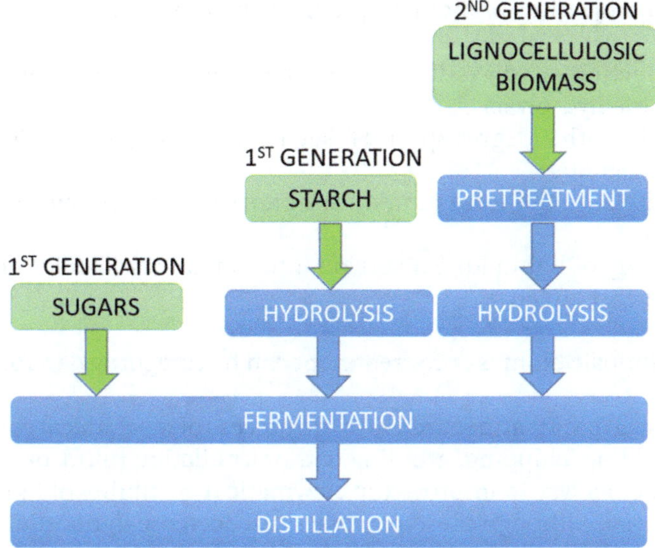

Figure 2.1 Conversion of first-generation and second-generation feedstocks into ethanol *via* the fermentation route.

Second-generation biofuels are produced from non-edible feedstock, mainly raw materials derived from lignocellulosic biomass and crop waste residues, avoiding the direct competition for food crops. However, lignocellulosic feedstock conversion processes are the main challenges for this new biofuels generation, due to their recalcitrant nature to enzymatic or chemical hydrolysis (Figure 2.1).[1,4]

The main component of lignocellulose cell walls is cellulose, a polysaccharide that consists of a linear chain of several glucose units, strongly linked *via* β-1,4-glycosidic linkages. Therefore, a high number of hydroxyl groups are present that form a complex hydrogen bonding network in the same chain or between vicinal chains, resulting in microfibrils with high tensile strength. The degree of structural order of the monomers will establish the cellulose crystallinity, higher the crystallinity more difficult the biodegradation becomes. In plant cell walls these highly organized cellulose microfibrils are linked to each other by hemicellulose (amorphous multicomponent polysaccharide composed manly by glucose, xylose, arabinose, among others) and coated by lignin (polymer of aromatic alcohols).

Therefore there are two major difficulties that prevent the direct access to fermentable sugars present on the plant cell walls: how to access the crystalline cellulose core of cell wall microfibrils and how to decrease the cellulose crystallinity, to improve their enzymatic or chemical hydrolysis. To circumvent this problem several physical and/or chemical pre-treatments have been developed to disrupt the recalcitrant lignocellulosic complex and liberate fermentable sugars for biofuel and chemical production.[5,6]

The main objectives of biomass pre-treatment include:

(i) production of digestible solids that enhances sugar yields during enzyme hydrolysis
(ii) avoiding the degradation of sugars including those derived from hemicellulose
(iii) minimizing the formation of inhibitors for subsequent fermentation steps
(iv) recovery of lignin for conversion into valuable molecules and
(v) to be cost effective.[7]

Lignocellulosic biomass pre-treatment can be categorized in four sections.

(a) Physical pre-treatment refers to the process of mechanical comminution by milling, chipping, grinding and/or irradiation (ultrasound or micro-wave) in a way to improve the enzymatic digestibility of lignocellulosic material. Although these processes can decrease lignocellulose particle size, cellulose crystallinity and degree of polymerization,[8,9] resulting in high yields of biogas,[10,11] they also can be highly energy consuming.

(b) Biological pre-treatment implies the use of microorganisms, usually fungi capable of producing enzymes that degrade lignin and hemicel-lulose present in the biomass.[12,13] This method has gathered increased interest by scientists since it has a high substrate specificity, resulting in high yields of the desired product, there is no toxic by-product gener-ation, and also low energy levels are needed. There are also several dis-advantages associated, such as being time consuming, since it is a slow process, and the need for extreme control of the growing conditions.

(c) Physicochemical pre-treatment involves processes like steam explo-sion, liquid hot water, ammonia fibre explosion (AFEX) or ammonia recycle percolation.[7,14]

(d) Chemical pre-treatments are based on the use of acids, alkalis, more environment friendly solvents, wet oxidation or ozonolysis. The alkaline pre-treatment involves the use of bases, such as sodium, potassium, cal-cium and ammonium hydroxides, and is able to remove lignin, acetyl groups and different uronic acid derivatives which inhibit the cellulose accessibility for enzymatic hydrolysis.[7,15] Wet oxidation uses oxygen as an oxidizer for compounds dissolved in water.[16] Acid pre-treatment can be performed either under low acid concentration and high tempera-ture or under high acid concentration and lower temperatures. This process has the advantage of dissolving hemicellulose, although expen-sive corrosive-resistant equipment is needed.[17,18] More environmental friendly solvents such as ionic liquids (ILs) or supercritical carbon diox-ide $(scCO_2)$[19] have the advantage to dissolve high loadings of different biomass types in mild processing conditions, although, due to the ILs high cost there is a need to recover and reuse the solvent.[14,20] $scCO_2$ can be cost effective, although high pressure is required.

2.2 Ionic Liquids (ILs) Pre-Treatment

IL pre-treatment, when compared with other pre-treatment techniques (ammonia fibre expansion or diluted acid), has shown to decrease biomass crystallinity by breaking inter and intra-chain hydrogen bonds in cellulose fibrils more than other pre-treatment techniques.[21,22] In this chapter we will focus on the use of ionic liquids (ILs) as solvents for biomass pre-treatment. Ionic liquids are salts, usually organic cations in combination with inorganic/organic anions, with the peculiar advantage that physical and chemical properties can be tuned to the desired application, just by changing the cation or the anion structures. ILs have been proposed as volatile organic solvent substituents due to their low volatility, high capability to dissolve inorganic and organic compounds, and also due to their chemical stability and peculiar solubility behaviour that facilitates IL recycling.

In 2002 Rogers *et al.* demonstrated the concept of carbohydrate dissolution in ILs, particularly cellulose.[23] They reported 1-*n*-butyl-3-methylimidazolium chloride ([bmim][Cl]) as the best IL to dissolve cellulose among the ones studied. Furthermore, the dissolution could be improved when microwave irradiation was applied (up to 25 wt.% solubility), although possibly higher polymer degradation could be obtained under microwave radiation when compared with conventional heating conditions.[24] Since then, this new type of solvent has been described for the dissolution of cellulose and biomass lignocellulosic derivatives such as wood (hardwood and softwood) and grass. The ability of ILs to dissolve biomass feedstock depends on the particle size (which can inhibit the diffusion of the IL into the interior of the biomass), the content of lignin, hemicellulose and cellulose present in the biomass sample (Table 2.1),[25] the nature of the native cellulose (cellulose crystallinity, CC, and degree of polymerization, DP), operating conditions (temperature, reaction time, biomass loading) and type of heating (conventional, microwave or sonication). The purity of ILs is also very important,

Table 2.1 Average contents of cellulose, hemicellulose, and lignin from different biomass sources.[25,32,47,76]

Biomass source	Composition (%)		
	Cellulose	Hemicellulose	Lignin
Hardwood	43–47	25–35	16–24
Softwood	40–44	25–29	25–31
Grasses	25–40	35–50	10–30
Bagasse	40	30	20
Coir	32–43	10–20	43–49
Cotton	95	2	1
Wheat straw	30	50	15
Rice straw	35	21	15
Corn cobs	45	35	15
Switch grass	45	30	12
Miscanthus	42	24	27

water being the most common impurity that can drastically change the solubility capability.[24,26]

Table 2.2 presents an overview of the application of ILs as pre-treatment for extraction of biopolymers from different biomass sources.

Moniruzzaman *et al.*[27] studied the recovery of cellulose fibres from wood biomass using 1-ethyl-3-methylimidazolium acetate ([emim][OAc]) for pre-treatment followed by enzymatic delignification, by laccase. The authors observed that wood fibres pre-treated with the IL originate a decrease on the lignin and hemicellulose content, that was justified by the fact that the IL initially dissolves a quantity of these biopolymers. This allowed further enzymatic treatment to be more efficient due to the cellulose being more accessible to the enzyme. They also observed that the treated cellulose fibres presented a higher crystallinity, which was rationalized by the fact that possible amorphous sites in the initial cellulose could be dissolved in the pre-treatment process. It should be noted that the goal of this work was to maintain or increase the cellulose fibres crystallinity for possible applications in textile, composite and other industrial processes.

Five different ILs (1-ethyl-3-methylimidazolium acetate ([emim][OAc]), 1-*n*-butyl-3-methylimidazolium chloride ([bmim][Cl]), 1-ethyl-3-methylimidazolium diethylphosphate ([emim][DEP]), 1-allyl-3-methylimidazolium chloride ([amim][Cl]), and 1-ethyl-3-methylimidazolium hydrogen sulfate ([emim][HSO$_4$])) were tested by Mood *et al.*[20] as pre-treatment solvents of corn stover, followed by enzymatic hydrolysis (cellulase and β-glucosidase). Pre-treatment with [emim][OAc] resulted in higher structural changes than with others ILs tested, for example after 72 h the cellulose digestibility reach 69.7% compared with 21.1% for untreated corn stover. The authors explained this result due to the ability of [emim][OAc] to dissolve cellulose, by breaking intra and intermolecular hydrogen bonding leading to higher amorphous cellulose sites that are consequently more accessible to the enzymes. This result was verified by FTIR, SEM, and XRD analyses that shown [emim][OAc] pre-treated cellulose presented less crystallinity, and higher amorphous sites, when compared with samples pre-treated with the other ILs tested and untreated sample.

Hayes *et al.*[28] also compared different ILs on the pre-treatment performance and further enzymatic saccharification of yellow poplar biomass. The authors have studied different physicochemical parameters in order to identify the best IL ([emim][OAc], [bmim][Cl] or [bmim][OAc]) for a more efficient biomass pre-treatment. The smallest, [emim][OAc], showed once more to be the best choice, since it decreases the biomass crystallinity, cleaves the acetyl groups from its hemicellulosic component, and decreases the decomposition temperature of cellulose and hemicelluloses, when compared with the others ILs.

In order to understand the physicochemical changes in cellulose after IL pre-treatment, Simmons *et al.*[29] reported a detailed study based on differential scanning calorimetry (DSC) and thermogravimetric analysis (TGA). It was previously reported that biomass components, hemicellulose, cellulose

Table 2.2 Overview of the reported application of ILs for biomass pre-treatment.[a]

			Wood			
			Pre-treatment conditions			
Biomass feedstock	Biomass loading, wt.% (particle size)	IL	Temp. (°C)	Time (h)	Observations	Ref.
Yellow poplar	4 (40 mesh)	[emim][OAc]	70	24	[bmim][OAc], [emim][OAc] and [bmim][Cl] were tested as pre-treatment solvents, being the best results obtained with [emim][OAc]. It was observed that a partial dissolution of yellow poplar into the ILs, at 70 °C, resulted in a decrease of the CC and decomposition time of cellulose and hemicellulose. In addition it was observed the cleavage of the acetyl groups from hemicellulose material	28
Poplar (*Populus tomentosa*)	–(10 μm)	[emim][OAc]	120, 25	2, 9	The authors studied the real-time confocal Raman microscopy, and observed that biomass dissolution in the IL processed in two stages, first the slow penetration of IL followed by rapid dissolution of compositions	77
Poplar Maple	4.8–50	[emim][OAc]	125	1	It was observed that for loadings above 33 wt.% lower amounts of lignin were extracted and higher CC was obtained, thus decreasing the enzymatic hydrolysis efficiency. A possible explanation was the related stoichiometry of the hydroxyl protons of glucose unit and the IL acetate ions that allowed the formation of new hydrogen bonds between them disrupting the crystalline structure of cellulose	41

(*continued*)

Table 2.2 (*continued*)

			Wood			
Biomass feedstock	Biomass loading, wt.% (particle size)	IL	Pre-treatment conditions		Observations	Ref.
			Temp. (°C)	Time (h)		
Yellow pine Bagasse	5 (<0.125 mm), 5 (<0.125 mm)	[emim][OAc]	175	0.5	Complete dissolution was obtained for bagasse in only 10 minutes, although 92% of pine was dissolved in 30 minutes. The authors have observed an IL decomposition when it was submitted at high temperatures (185 °C), even if exposure was for short periods of time	78
Pine (*Pinus radiata*)	5 (20–40 mesh)	[emim][OAc]	120 or 150	3	[emim][OAc] pre-treatment increased the enzymatic cellulose digestibility (90%) in spite of no delignification being observed in biomass sample. This result was justified by the authors by the differences in lignin composition and in the anti-solvent used (water), when compared with other reported studies. In addition, the pre-treatment decreased the CC, and also lignin was redistributed allowing a better enzyme accessibility to the cellulose	79
Wood (*acacia dealbata*)	8–20 (1–250 μm)	[emim][OAc]	150	0.5	Different reaction times and temperatures were studied, and the enzymatic hydrolysis showed similar results for pre-treatment samples at 130 °C for 180 minutes, or at 150 °C for 30 minutes. Biomass loading from 8 to 20 wt.% shown similar recovered solid yields and also cellulose conversions (98%, using water as anti-solvent). It was possible to recycle the IL, although the cellulose conversion decreased with the number of cycles	80

Wood chips (*Chamaecyparis obtusa*)	10 (110–550 μm)	[emim][OAc]	80	1	In a way to use cellulose fibres in possible applications such as textile, composite and other industrial processes, the authors have reported a pre-treatment of wood chips at 80 °C. IL pre-treatment removed partially lignin and hemicellulose, allowing further enzymatic delignification. The obtained cellulose fibres presented higher crystallinity and higher thermal stability than the initial one, due to the possible solubilization in the IL of some amorphous sites present in the initial cellulose	27
Wood chips (Norway spruce and pine)	—	[amim][Cl], [bmim][Cl]	80 or 130	8 or 15	Up to 8% wood dissolution in [amim][Cl] or [bmim][Cl] ILs. Biomass particle size (ILs dissolution): wood powder > sawdust > TMP (thermomechanical pulp) fibres > wood chips	35
Southern yellow pine Red oak	5–10	[emim][OAc]	110	25–46	When the loading was increased to 10 wt.% the dissolution decreased from 99.5 to 40% (110 °C, 16 h). Microwave irradiation decreased the required time for total dissolution	37
Wood chips (*Eucalyptus globulus*)	3 (40 mesh)	[emim][OAc]	120	3	It was observed by X-ray that after IL pre-treatment, native cellulose was transformed into cellulose II polymorph, leading to a decrease in CC. The amount of glucose released increased after enzymatic hydrolysis to 7.4% when compared with 2.4% for untreated wood	81

(*continued*)

Table 2.2 *(continued)*

			Wood			
Biomass feedstock	Biomass loading, wt.% (particle size)	IL	Pre-treatment conditions Temp. (°C)	Time (h)	Observations	Ref.
Wood chips (*Pinus radiate* and *Eucalyptus globulus*)	4	[amim][Cl]/DMSO (1/1.3 (w/v))	120 (MW)	1/3	The regenerated celluloses were studied by several methods, such as XRD, FTIR and TGA/DSC that revealed a decrease in the CC and thermal stability when the wood chips were pre-treated with the IL/DMSO solution. It was also possible to separate the lignin and hemicellulose from the cellulose using methanol as anti-solvent. Although MW irradiation was used, the authors refer that no significant degradation was observed in the dissolution process of both woods	53
Spruce Silver fir Beech Chestnut	5 (1–2 mm)	[amim][Cl]	90	12	Over 21 different ILs based on imidazolium, phosphonium and pyridinium cations were tested for the solubilisation of cellulose and wood chips. The best to dissolve cellulose was [emim][OAc], and [amim][Cl] was the best IL to dissolve wood chips	49
Maple wood flour	5 (250 μm)	[emim][OAc]	90	24	[amim][Cl] and [bmim][Cl] could dissolve a higher amount of wood, nevertheless [emim][OAc] was used for this study since it could dissolve a higher amount of lignin, resulting in higher enzymatic cellulose digestibility. It was possible to recycle the IL for at least 4 cycles, where the dissolved lignin was not removed from the IL, resulting in highly concentrated solution of lignin that did not influenced the cellulose enzymatic hydrolysis during the cycles	72

Maple wood flour	5	[emim][OAc], [bmim][OAc], [bmim][MeSO$_4$]	90	24	A correlation was observed between Kamlet–Taft β parameters and pre-treatment efficiency, where ILs with higher β parameter showed higher lignin dissolution, decreased CC and consequently higher enzymatic hydrolysis efficiency	57
Eucalyptus grandis Southern pine Norway spruce	8 (0.1–2 mm)	[amim][Cl]	120	5	It was observed that for woods with higher density, the pre-treatment efficiency decreased (lower glucose released after enzymatic hydrolysis). When compared with methanol, water was a better antisolvent for the regeneration of cellulose. When the IL was recycled, it was observed that enzymatic hydrolysis decreased for number of IL reuse cycles, maybe due to the wood components accumulation in the IL	36
Spruce wood (*Picea abies*)	5 (20–48 mesh)	[emim][OAc]	120	15	[emim][OAc], [bmim][OAc] and NMMO were tested as pre-treatment solvents. The authors refer the IL removal from the treated biomass one of the major disadvantages for this process, since it is necessary a large quantity of hot water to completely remove the IL from cellulosic fibres. After pre-treatment, saccharification and fermentation [emim][OAc] was the best pre-treatment solvent that allowed an ethanol yield of 66.8% and 81.5% from spruce chips and powder, respectively, compared with 2.7% and 9.7% obtained from untreated spruce wood	82

(*continued*)

Table 2.2   (*continued*)

Wood						
			Pre-treatment conditions			
Biomass feedstock	Biomass loading, wt.% (particle size)	IL	Temp. (°C)	Time (h)	Observations	Ref.
Pine Poplar Chinese parasol Catalpa wood	5 (0.45–0.65 mm)	[amim][Cl]/DMSO	100	2	Pine was found to be more suitable for dissolution in IL than the other wood samples studied, with cellulose regeneration up to 78%. For temperatures higher than 120 °C, it was shown that although the dissolution increased, the cellulose degradation also increased, since the regeneration decreased. It was possible to recycle the IL	40

By-products of food production						
			Pre-treatment conditions			
Biomass feedstock	Biomass loading, wt.% (particle size)	IL	Temp. (°C)	Time (h)	Observations	Ref.
Bagasse	5 (<3 mm)	[chol][OAc]	25 (sonicated)	1	[chol][OAc] has shown similar pre-treatment efficiency as [emim][OAc], although the cholinium-based IL showed less of an inhibitory effect on cellulase when compared with usual IL pre-treatment [emim][OAc]	54
Bagasse	5 (<200 µm)	[bmpy][Cl]	120	1/6	The regenerated biomass yield depends on the biomass type, IL nature and also reaction time. Although reduced cellulose crystallinity was not clearly observed, the 8-fold increase of cellulose conversion was due to decrease of cellulose DP	83

Bagasse sugarcane	14 (250–500 μm)	[emim][OAc]	145	1/4	After IL pre-treatment and enzymatic hydrolysis it was possible to recover 69% of reducing sugars yield. The authors have used the central composite design of response surface methodology to predict the reducing sugar yield, and good correlations where obtained with the experimental values	42
Bamboo (*Phyllostachys sulphurea*)	5 (40–60 mesh)	[amim][Cl]	100	12	Cellulosic-rich material was precipitated with water and it was possible to obtain lignin-rich fractions by basic ethanolic solutions. Although the IL could increase the accessibility of cellulose in the biomass sample, it was not observed any significant change in the CC. In addition, lignin and hemicelluloses fractions were slightly degraded during the IL pre-treatment	84
Bamboo	10 (1 mm)	[chol][OAc]	25 (sonicated)	1	Ultrasound irradiation was used for the pre-treatment of bamboo powder, resulting in a decreased crystallinity and higher cellulose saccharification ratio when compared with conventional heating. [chol][OAc] was used as pre-treatment solvent, and 92% of cellulose was hydrolysed to glucose	55
Rice hulls	10 (1.168–1.651 mm)	[emim][OAc]	110	8	[amim][Cl] and [hmim][Cl] were also tested for the rise pre-treatment, although with [emim][OAc] it was possible to dissolve all the lignin present in the biomass. When water was used as anti-solvent it was possible to precipitate cellulose-rich fibres, although hemicellulose was also present. When ethanol was added to the IL, it was possible to precipitate the existing lignin	85

(*continued*)

Table 2.2 (*continued*)

		By-products of food production				
Biomass feedstock	Biomass loading, wt.% (particle size)	IL	Pre-treatment conditions		Observations	Ref.
			Temp. (°C)	Time (h)		
Rice straw	5 (<150 μm)	[chol][Arg]	60	6	Cholinium ILs based on 28 different anions (mono- and dicarboxylates, and amino acids) were tested for the ILs pre-treatment of rice straw. It was observed that the presence of the basic groups in the anion significantly enhanced the IL pre-treatment effectiveness. The best results were obtained with amino acid anions group being arginine the best IL for the pre-treatment of rice straw	86
Rice straw	5 (0.125–0.250 mm)	[emim][OAc]/DMSO (1 : 1 v/v)	130 (MW)	1/60	Co-solvent addition improved the lignin dissolution, decreased the CC of cellulose I polymorph and also increased the enzymatic hydrolysis when compared with neat [emim][OAc] pre-treatment. In addition, [emim][OAc]/DMSO (1 : 1) in combination with MW heating resulted in even better results with the advantage to reduce the IL residual content, although the authors did not referred the possible cellulose degradation due to MW irradiation	32
Corn stover Rice straw	(<2 mm)	[MMIM][DMP]/aq. HCl	110	2	[MMIM][DMP] was chosen over [bmim][Cl] and [emim][OAc], due to the lower inhibition on enzyme activity and also by their biocompatibility. Total sugar conversion was raised to 92.7% with the [MMIM][DMP]/aq. HCl pre-treatment compared to the conversion of only 27.3% obtained with untreated corn stover	87

Straw (*Triticosecale*)	3 (0.5 mm)	[emim][OAc] 50% aq. sol.	150	1,5	Higher concentration of IL in aqueous solution increased the lignin dissolution, improving the cellulose digestibility and also the CC. In addition, the fermentable sugar yield obtained for 50% IL aqueous solution (81%) was higher than for pure IL pre-treatment under the same conditions (67%)	58
Corncob Rice straw	~5 (<150 μm)	[emim][OAc]/DMA (60:40 v/v)	120		[emim][OAc], [bmim][Cl] and co-solvent mixtures (using DMA and ethanolamine) were studied for the pre-treatment of biomass. The addition of co-solvents could decrease viscosity, improving the washout efficiency of the recovered biomass. The selected co-solvents were stable at the pre-treatment temperature, and possess high boiling points, so it was possible the recycling of the IL/co-solvent. The highest sugar yield was obtained with the mixture [emim][OAc]/DMA, although [bmim][Cl]/ethanolamine could decrease the cellulose crystallinity to lower crystallinity index values	66
Corn stover	10 (4 mm)	[emim][OAc]	140	3	A mixture of ketone/alcohol was used for the efficient separation of biomass residues from the IL. The recovered IL was analysed by 1H-NMR and the main impurity observed was water	88

(continued)

Table 2.2 (*continued*)

		By-products of food production				
			Pre-treatment conditions			
Biomass feedstock	Biomass loading, wt.% (particle size)	IL	Temp. (°C)	Time (h)	Observations	Ref.
Corn stover	4 (0.42 mm)	[emim][OAc]	110	1,5	[emim][OAc], [bmim][Cl], [emim][DEP], [amim][Cl], and [emim][HSO$_4$] were tested as pre-treatment solvents, [emim][OAc] the one that presented higher biomass dissolution and higher yields of recovered solid (43.3%). The authors explained the best dissolution efficiency by the high hydrogen basicity of the IL, and being emim cation smaller than bmim resulted into a higher diffusion capacity. In addition [emim][OAc] could decrease the CC to 19% (CrI), resulting in higher enzymatic saccharification yields (69% in 72 h when compared with 21% without pre-treatment)	20

		Grasses				
			Pre-treatment conditions			
Biomass feedstock	Biomass loading, wt.% (particle size)	IL	Temp. (°C)	Time (h)	Observations	Ref.
Switchgrass (*Panicum virgatum*),	5 (0.5 mm)	[bmim][OAc] aq. sol., [bmim][MeSO$_3$] aq. sol.	110	4	Neat [bmim][OAc] and [bmim][MeSO$_3$] could enhance the cellulose hydrolysis rate, although [emim][OAc] aqueous solution pre-treatment was not so effective. Interestingly, aqueous solution of the IL [bmim][MeSO$_3$] resulted in an improved cellulose hydrolysis rate despite no reduction of CC was observed	61

Switchgrass	5 (32–50; 75–100; >200 µm)	[emim][OAc]	120	3	The authors have compared IL pre-treatment with dilute acid and AFEX pre-treatments, concluding that IL pre-treatment resulted in greater cell wall disruption, reduced crystallinity, increased accessible surface area, and higher saccharification efficiencies. The size of biomass particles were also studied, and better pre-treatment efficiency was obtained when higher biomass particles were used	21
Miscanthus	–(4 or 1 mm)	[mmim][OAc], [emim][OAc], [bmim][OAc], [bmim][Cl], [emim][Cl], [mmim][DMP], [emim][DMP], [emim][MeSO$_4$]	130	—	Several hydrophilic ILs were tested for the *Miscanthus* solubility, being chloride, acetate, and phosphate based ILs increased the biomass solubility, due to having high hydrogen-bond-acceptor strength and high polarity. The biomass particle size was important on the dissolution showing that a smaller particle biomass needed shorter times to be solubilized. The temperature and time of biomass pre-treatment was also studied	47
Miscanthus giganteus	~4 (4 mm)	[emim][OAc]	140	1	*Miscanthus* was dissolved in [emim][OAc] and a strongly basic aqueous solution of phosphate was added. The resulting three-phase system has a salt-rich aqueous phase, a solid-phase rich in cellulose, and an IL-rich phase containing most of the lignin. It was possible to recycle the IL for at least 3 times	56

[a]TMP – thermomechanical pulp; NMMO – *N*-methylmorpholine-*N*-oxide; [amim] – 1-allyl-3-methylimidazolium; [mmim] – 1,3-dimethylimidazolium; [emim] – 1-ethyl-3-methylimidazolium; [bmim] – 1-butyl-3-methylimidazolium; [hmim] – 1-hexyl-3-methylimidazolium; [bmpy] – 1-butyl-3-methylpyridinium; [chol] – choline; [OAc] – acetate; [MeSO$_3$] – methanesulfonate; [MeSO$_4$] – methylsulfate; [DMP] – dimethylphosphate; [DEP] – diethylphosphate; [Arg] – arginine; CC – cellulose crystallinity; XRD – X-ray diffraction; FTIR – Fourier transform infrared spectroscopy; TGA – thermogravimetric analysis; DSC – differential scanning calorimetry; SEM – scanning electron microscopy; CrI – crystallinity index.

and lignin, appear in different zones on TGA curves, therefore it is possible to observe the differences on each component when biomass samples are submitted to IL pre-treatment.[30] In addition, pure compounds shown a single peak in these curves, therefore if two peaks appear this indicates the presence of amorphous and crystalline structures of the same compound, as the crystalline structure has a higher decomposition temperature due to amorphous materials being able to break more easily. In this study,[29] it was observed that when cellulose samples (Avicel) were pre-treated with [emim][OAc] a peak appeared at 270 °C which indicates the depolymerization of the polymer or a change from crystalline to amorphous form. Furthermore, the CC decreased from 87 to 12% when Avicel samples were pre-treated at 120 °C, although the CC percentage increased when the pre-treatment temperature increased, showing the reformation of crystalline structures. The authors explained this fact by the formation of cellulose II polymorph that increase the CC, although is easier to be decomposed and breaks down more easily than native cellulose. Three different biomass sources were analysed: pine, eucalyptus and switch grass. It was observed that 120 °C pre-treatment was enough to obtain a high degree of depolymerization for pine and switch grass samples, although for eucalyptus samples the temperature had to be increased to 160 °C. It should be noted that for biomass samples the DSC curves depend on its lignin carbohydrate complex, specifically the nature of the bonds present between the hemicelluloses and lignin.

Using [bmim][OAc] as pre-treatment solvent, Cheng *et al.*[31] also have studied the physicochemical changes of biomass samples (switch grass and corn stover) by TGA. As referred before,[29] the IL pre-treatment increased the cellulose II polymorph by increasing the CC of the recovered cellulose material. Cheng *et al.*[31] showed that the thermal stability of the cellulose II polymorph can be higher than native cellulose, if considerable depolymerization does not occur during the pre-treatment.

Koo *et al.*[32] have studied the solubility of cellulose, hemicellulose and lignin separately, in several [emim][OAc]/co-solvent mixtures, the co-solvents being polar aprotic DMSO, DMF and DMA (dimethyl sulfoxide, dimethylacetamide, dimethylformamide, respectively), known solvents for the dissolution of carbohydrates.[33] The solvent combination [emim][OAc]/DMSO (1 : 1) was the best mixture for the dissolution of these biopolymers and was further studied for the pre-treatment of rice straw. The authors demonstrated that the co-solvent addition improved the lignin dissolution, decreased the CC of the cellulose I polymorph (although an increase was observed of the formation of polymorph cellulose II that has a less crystalline structure), and also increased the enzymatic hydrolysis when compared with [emim][OAc] pre-treatment. In addition, the pre-treatment of rice straw using [emim][OAc]/DMSO (1 : 1) in combination with MW heating resulted in even better results with the advantage that the residual content of the IL in the treated rice straw can be reduced to 0.4%, possibly due to low viscosity of the solvent mixture when compared with the IL alone.[32] It was possible to recycle the [emim][OAc]/DMSO mixture for at least five times without the decrease of the enzymatic conversion.

2.3 Selection of Parameters that Affect Biomass Pre-Treatment with ILs

Over the last years, several studies have been reported for the ILs pre-treatment of lignocellulosic biomass. Different ILs (manly based on the imidazolium cation), have been described using different experimental conditions in a way to destroy the recalcitrance of lignocellulosic biomass to enzymatic hydrolysis. This recalcitrance can be attributed to the accessible surface area, biomass particle size, degree of polymerization, crystallinity and protective lignin. In this section is described the parameters that affect the biomass pre-treatment using ILs.

2.3.1 Type and Composition of Biomass

Three major types of lignocellulosic biomass can be used as renewable feedstock for biodiesel production: softwood (usually fast-growing and tall, such as fir, pine and spruce), hardwood (more complex structures than softwoods, such as willow, oak and poplar), and annual and perennial grasses (usually corn stalks, sugar cane bagasse, straws (*e.g.* wheat, rice or barley) that are by-products of food production).[34] In 2007 Kilpeläinen *et al.*[35] reported that both hardwoods and softwoods are readily dissolved in different ILs, although softwood could be better dissolved since it has lower density than hardwoods. Using [amim][Cl] as solvent, hardwood (*Eucalyptus grandis*) was shown to be less soluble than softwood (southern pine and Norway spruce) in the same pre-treatment conditions.[36] Ford *et al.*[26] have demonstrated that wood dissolution on ILs is highly dependent on the species, with poplar (a hardwood) being more soluble than pine and eucalyptus (softwoods) in [bmim][Cl]/DMSO mixtures, and with oak (hardwood) being the least soluble biomass sample analysed. In 2009 Sun *et al.*[37] reported that hardwood (oak) required less time for dissolution in [emim][OAc] than the softwood pine. In addition Cheng *et al.*[38] reported that eucalyptus (hardwood) required less time to be dissolved in the IL [emim][OAc] than the softwood pine. At current stage is not possible to identify which biomass source can offer better dissolution on ILs, since biomass type and species differs in internal complex ultrastructure, which may affect the subsequent dissolution process. If hardwood is denser, preventing the ILs entrance in the internal structure, it also has a higher lignin and hemicellulose proportion that can lead to better dissolution efficiencies.[37] In general grass biomass is more easily solubilized, while the most difficult substrates are typically softwoods (pine and spruce).[34]

2.3.2 Biomass Loading in the IL

The ratio between biomass and ILs is an important factor for the overall pre-treatment process. The dissolution rate, regeneration efficiency, recyclability and the process cost can be influenced by the biomass loading.[39] Sun *et al.*[37] studied the increase of wood loading from 4 to 10 wt.% in [emim][OAc] and observed a solubility decrease for higher loadings. The same effect was observed by Wang *et al.*[40] where the solubilization of wood in [amim][Cl] decreased from 35 to 26% when biomass loading increased from 1 to 5 wt.%. The authors justified the higher solubility for low loadings by the increased dispersion of the biomass in the IL leading to higher diffusion rate inside the biomass. Wu *et al.*[41] have studied biomass loadings up to 50 wt.% using [emim][OAc] as pre-treatment solvent. They have reported that a maximum loading efficiency was observed at 33 wt.%. For loadings above 33 wt.%, lower amounts of lignin were extracted and higher CC was obtained, thus decreasing the enzymatic hydrolysis efficiency. A possible explanation was the related stoichiometry of the hydroxyl protons of glucose units and the IL acetate ions that allowed the formation of new hydrogen bonds between them, disrupting the crystalline structure of cellulose.[41] Yoon *et al.*[42] has reported that increasing the biomass loading at 120 °C caused a decrease on the pre-treatment efficiency, although for higher temperatures, the final yield could be enhanced. They explained this yield increase by the cellulose solubility at higher temperatures. The same effect of increasing the pre-treatment efficiency for higher biomass loading was observer by Tan and Lee,[43] who justified this by an increase of effective collisions between biomass and IL molecules. Optimization of the biomass loading is not an easy task, especially because it depends on the biomass type and the particle size. To make IL pre-treatments commercially viable, a compromise between the biomass loading and the pre-treatment efficiency needs to be reached.

2.3.3   Biomass Particle Size

One of the most important parameters in biomass pre-treatment is the particle size, since it can increase the biomass dissolution on the IL. Before thermal deconstruction of lignocellulosic biomass, a milling or grinding step is necessary, although this step is usually highly energy consuming.[34] In theory, smaller particle size increases the available biomass surface area and also increases the IL diffusion into the lignocellulosic material, although the perfect size depends on the biomass type, and also on the chosen IL.[34,39,44,45] Bahcegul *et al.*[45] have studied the effect of four different particles sizes of cotton stalks (<0.15 mm, 0.15–0.5 mm, 0.5–1.0 mm and 1.0–2.0 mm) on [emim][OAc] or [emim][Cl] pre-treatment. Using [emim][OAc], higher pre-treatment efficiency (highest glucose recovery) was obtained with the smallest biomass particle size, although with [emim][Cl] the best efficiency was obtained with the largest particle size.[45] Nguyen *et al.*[46] also studied the effect of particle size (<2 mm, 2–5 mm, >10 mm) of rice straw. They observed that cellulose recovery decreased in the smallest particle size samples although it was similar to medium and highest size particle. Furthermore, the glucose conversion of the medium particle sized samples was significantly higher for >10 mm particle sized samples. Sun *et al.*[37] and also Kilpeläinen *et al.*[35] have reached the same conclusion that smaller particles size were more suitable for biomass dissolution. Padmanabhan *et al.*[47] have also shown that particle size of *Miscanthus* of 1 mm can decrease the dissolution time when compared with 4 mm particles. These results confirm that particle size is an important factor, although being biomass type and loading dependent.

2.3.4 Temperature and Heating Source

High temperatures promotes swelling and disintegration of biopolymer matrices. The main reason for these to occur is the decrease in the IL viscosity that allows a better diffusion of the IL in the biomass, resulting in a destabilization of the hydrogen bond network present in the biopolymer.[37,40,48,49] ILs with higher alkyl or aromatic chains need higher temperature processes due to their higher viscosity.[35,49] Wang *et al.*[40] have reported that the dissolution of wood chips and regeneration rates increase when the temperature increases from 90 to 120 °C, using [amim][Cl] as pre-treatment solvent. For temperatures higher than 100 °C, however, it was possible to observe that the cellulose content in the regenerated material started to decrease, indicating the occurrence of degradation (with the degradation rates dependent on the biomass source). Yoon *et al.*[42] observed that for temperatures higher than 140 °C the reducing sugar yield from sugar cane bagasse started to decrease and the effect was emphasized for longer pre-treatment times. The optimum conditions for [emim][OAc] pre-treatment were 145 °C for 15 min, resulting in a reducing sugars yield of 69.7%. Kimon *et al.*[50] also used sugar cane bagasse as the biomass source, although with [bmim][Cl] as solvent. Once more, it was observed that for higher temperatures (160 °C) the dissolution was almost complete; nevertheless, more than half of the material was unrecoverable upon the addition of the anti-solvent. An equilibrium of the dissolution and recovered material was obtained for temperatures between 140 and 150 °C. For switchgrass biomass, using [emim][OAc], the pre-treatment at higher temperatures (from 110 to 160 °C) increased the sugar recovery and cellulose digestibility.[51] The literature has shown that a higher temperature is necessary to increase the biomass dissolution, although caution is needed to avoid biomass decomposition and consequent decrease of the recovered material. On the other hand, for higher temperatures, it is possible that IL decomposition may lead to unwanted cellulose derivatization.[39]

Due to ILs ionic nature, they can readily absorb microwave irradiation (MW) promoting the heating of materials *via* molecular vibrations.[40] This type of heating source has some advantages when compared with conventional heating, such as non-contact, fast heating and better control over the heating process. For these reasons, the use of MW heating for wood dissolution has increased over the years.[52] Swatloski *et al.*[23] have reported that cellulose solubility increased when [bmim][Cl] was used as solvent and heated under MW irradiation. Furthermore in 2009 it was reported that wood dissolution could be accelerated by MW pulses or ultrasound irradiation before heating the biomass by conventional heating.[37] Wang *et al.*[40] used MW heating to enhance the wood in the IL [amim][Cl], and observed that the same dissolution and regeneration results could be achieved with MW irradiation or conventional heating, with the advantage that the dissolution time could be decreased for MW heating. The authors explained this result by the increase in collision frequency between the anions and cations of IL and the wood macromolecules. Casas *et al.*[53] have also used MW heating for wood pre-treatment with [amim] [Cl]/DMSO mixtures, at 120 °C for 20 minutes. The mixture [emim][OAc]/ DMSO in combination with MW heating resulted in even better results than the conventional heating, with the advantage of reducing the IL residual content in the regenerated material.[32] However, the authors did not refer to the possible cellulose degradation due to MW irradiation. Ultrasound irradiation was used for the pre-treatment of bamboo powder, resulting in a decreased crystallinity and higher cellulose saccharification ratio when compared with conventional heating.[54,55] Furthermore, the authors noted that chemical and physical undesirable effects of ultrasound irradiation on cellulose and lignin were negligible under the experimental conditions.

2.3.5   Reaction Time

The reaction time of ILs pre-treatment is directly related to the temperature. Longer periods of reaction at lower temperatures can increase the biomass dissolution and lignin extraction rate by increasing the IL diffusion into the biomass pores,[44] although higher temperatures for longer periods of time can lead to an increase in the biomass degradation, resulting in less regenerated material.[51] Another important factor that can affect the pre-treatment time is the biomass type, since the high density of hardwoods can prevent the diffusion of the IL into the biomass,[40] although for switchgrass samples pre-treatment time has little effect on the enzymatic saccharification.[51]

2.3.6   Effect of Water Content

The ILs water content is an impurity that is not easy to control due to its polar and hygroscopic general nature. In addition, the biomass water content is also a challenging task to control and may have a negative impact on biomass dissolution.[39] Swatlowski *et al.*[23] reported that a water content of more than 1% reduces the cellulose solubility, due to the competition between the water and IL molecules to form new chemical bonds with cellulose. Wood dissolution with chloride- and acetate-containing ILs requires very low water contents, otherwise a decrease on the lignin extraction and sugar yields may be possible. For this reason an almost complete drying of the wood is necessary.[56,57] Padmanabhan *et al.*[47] have added a small amount of water (3–5 wt.%) into chloride-, acetate- and phosphate-based ILs that prevented the dissolution of *Miscanthus* in these solvents. Although these reported studies have demonstrated a negative impact of water on the lignocellulosic biomass dissolution process, several studies have been reported recently on the use of IL aqueous solutions for biomass pre-treatment in a way to decrease the IL pre-treatment cost.[58–61] When switchgrass was pre-treated with neat [bmim][OAc] and [bmim][MeSO$_3$] the cellulose hydrolysis rate was enhanced, although [emim][OAc] aqueous solution pre-treatment was not so effective.[61] Interestingly, an aqueous solution of the IL [bmim][MeSO$_3$] resulted in an improved cellulose hydrolysis rate despite no reduction of CC being observed. After an optimization of pre-treatment conditions, it was possible to obtain a recovery of fermentable sugars of 71.4% using an 49.5 wt.% aqueous solution of [emim][OAc] at 158 °C for 3.6 h.[59] The literature has shown that the impact of water content may depend on the type of IL used for biomass processing.[62]

2.3.7　ILs Physical Properties

ILs physical properties such as cation and/or anion type, viscosity, melting point and hydrogen basicity can strongly influence the biomass dissolution, thus compromising the overall pre-treatment process. Cation and anion type can be important for the viscosity of the IL, where higher alkyl chains in the cation usually results into higher viscosity ILs. There are anions known to decrease the IL viscosity, such as carboxylates, formates and phosphonates.[63] Fendt *et al.*[64] has reported that the viscosity of ILs based on chloride and acetate anions can significantly decrease with the addition of only a small amount of co-solvent, such as acetonitrile, ethylene glycol or dimethylformamide. Several studies have demonstrated that the addition of a co-solvent such as water or DMSO can decrease the IL viscosity, thus increasing the biomass pre-treatment efficiency.[40,53,65–67] Formate-based ILs have shown a good solubility for cellulose under relatively low temperatures due to the strong hydrogen bond basicity, although they showed relatively poor thermal stability because of decarboxylation.[68] Thus the viscosity is not the only determining parameter for a high pre-treatment efficiency. Kilpeläinen *et al.*[35] have shown that although ILs based on dicyanamide anions possess a lower viscosity, they are not good solvents for the dissolution of wood. In addition, the more viscous chloride-based IL had a high hydrogen bond basicity leading to a higher wood dissolution due to the capability to disrupt the hydrogen bonds network of biopolymers.

2.3.8　Enzymatic Saccharification

The parameters mentioned in the above sections have shown that ILs have a high potential to act as solvents for the pre-treatment of lignocellulosic biomass, although the majority of reported ILs are inhibitory to the saccharification enzymes (cellulases or other enzymes capable of hydrolyse cellulosic bonds). Ideally, the IL used for the pre-treatment of biomass not only should dissolve the biomass (increasing their digestibility), but also should be compatible with enzymes such as cellulases. D'Arrigo *et al.*[69] have studied the stability of recombinant monocomponent endocellulase from *Trichoderma reesei* in [bmim][Cl] as an single-batch approach for the pre-treatment, followed by enzymatic hydrolysis. It was observed that the enzyme was more stable in neat

IL than in a solution of IL/phosphate buffer, and it was possible to hydrolyse cellulose without a separated pre-treatment step. However, the authors only explored microcrystalline cellulose rather than real lignocellulosic substrates.

2.4 Identified Current Challenges and Further Optimization

Although the use of IL for the pre-treatment of lignocellulosic biomass has been very well studied, there are some challenges that can be improved in this process.

(a) Recovery and reuse of the ILs are indispensable to enable an economically viable route to produce biofuels, as the high cost of ILs is one huge barrier to achieve commercial viability.[70] Although several efforts have been performed to minimize the overall production costs, the majority of ILs are still expensive.

(b) After cellulose removal by an anti-solvent, the dissolved lignin and hemicellulose remains on the IL. These are economically valuable biopolymers and their recovery and further use have to be developed.[37,71–73]

(c) The water present on the overall process can decrease the pre-treatment efficiency, therefore an additional step of drying the biomass samples prior to the pre-treatment is necessary. For some biomass sources this step can be difficult, especially for high-solid loading where the water content can be difficult to control.[74,75] It is to be noted that for the next saccharification step, water is crucial for the efficiency of enzymatic hydrolysis. In this line, the search for efficient ILs, less dependent on the water content is an important goal.

2.5 Conclusions

Lignocellulosic biomass is an abundant, low cost, sustainable, environmentally benign, non-food feedstock for biofuel production. The main drawback is the recalcitrant nature and low reactivity of the cellulose derived from lignocellulosic biomass. ILs have shown a high potential to process the lignocellulosic biomass into digestible materials for enzymatic hydrolysis and further fermentation. However, for an industrial biorefinery based on ILs to be available, several challenges have to be overcome, such as the high cost of ILs, their toxicity, biodegradability and recyclability, their low water-dependent solubility, and also their enzymatic compatibility.

Acknowledgements

We thank the Fundação para a Ciência e a Tecnologia (SFRH/BPD/75045/2010 and PTDC/QEQ-PRS/2824/2012) for financial support.

References

1. J. van Eijck, B. Batidzirai and A. Faaij, *Appl. Energy*, 2014, **135**, 115–141.
2. S. N. Naik, V. V. Goud, P. K. Rout and A. K. Dalai, *Renewable Sustainable Energy Rev.*, 2010, **14**, 578–597.
3. J. Valentine, J. Clifton-Brown, A. Hastings, P. Robson, G. Allison and P. Smith, *GCB Bioenergy*, 2012, **4**, 1–19.
4. M. E. Himmel, S. Y. Ding, D. K. Johnson, W. S. Adney, M. R. Nimlos, J. W. Brady and T. D. Foust, *Science*, 2007, **315**, 804–807.
5. Y. Zheng, J. Zhao, F. Xu and Y. Li, *Prog. Energy Combust. Sci.*, 2014, **42**, 35–53.
6. S. Behera, R. Arora, N. Nandhagopal and S. Kumar, *Renewable Sustainable Energy Rev.*, 2014, **36**, 91–106.
7. G. Brodeur, E. Yau, K. Badal, J. Collier, K. B. Ramachandran and S. Ramakrishnan, *Enzyme Res.*, 2011, **2011**, 17.
8. J. Y. Zhu, G. S. Wang, X. J. Pan and R. Gleisner, *Chem. Eng. Sci.*, 2009, **64**, 474–485.
9. Y. Yu and H. Wu, *AIChE J.*, 2011, **57**, 793–800.
10. E. Bruni, A. P. Jensen and I. Angelidaki, *Bioresour. Technol.*, 2010, **101**, 8713–8717.
11. H. Hartmann, I. Angelidaki and B. K. Ahring, *Water Sci. Technol.*, 2000, **41**, 145–153.
12. Isroi, R. Millati, S. Syamsiah, C. Niklasson, M. N. Cahyanto, K. Lundquist and M. J. Taherzadeh, *BioResources*, 2011, **6**, 5224–5259.
13. R. Castoldi, A. Bracht, G. R. de Morais, M. L. Baesso, R. C. G. Correa, R. A. Peralta, R. D. P. M. Moreira, M. D. T. D. Polizeli, C. G. M. de Souza and R. M. Peralta, *Chem. Eng. J.*, 2014, **258**, 240–246.
14. S. H. Mood, A. H. Golfeshan, M. Tabatabaei, G. S. Jouzani, G. H. Najafi, M. Gholami and M. Ardjmand, *Renewable Sustainable Energy Rev.*, 2013, **27**, 77–93.
15. S. C. Rabelo, R. Maciel and A. C. Costa, *Appl. Biochem. Biotechnol.*, 2009, **153**, 139–150.
16. C. Martin, H. B. Klinke and A. B. Thomsen, *Enzyme Microb. Technol.*, 2007, **40**, 426–432.
17. J. Xu, M. H. Thomsen and A. B. Thomsen, *J. Microbiol. Biotechnol.*, 2009, **19**, 845–850.
18. M. F. Digman, K. J. Shinners, M. D. Casler, B. S. Dien, R. D. Hatfield, H. J. G. Jung, R. E. Muck and P. J. Weimer, *Bioresour. Technol.*, 2010, **101**, 5305–5314.
19. N. Narayanaswamy, A. Faik, D. J. Goetz and T. Y. Gu, *Bioresour. Technol.*, 2011, **102**, 6995–7000.
20. S. H. Mood, A. H. Golfeshan, M. Tabatabaei, S. Abbasalizadeh, M. Ardjmand and G. S. Jouzani, *Prep. Biochem. Biotechnol.*, 2014, **44**, 451–463.
21. M. J. Dougherty, H. M. Tran, V. Stavila, B. Knierim, A. George, M. Auer, P. D. Adams and M. Z. Hadi, *PLoS One*, 2014, **9**(6), e100836.

22. C. L. Li, B. Knierim, C. Manisseri, R. Arora, H. V. Scheller, M. Auer, K. P. Vogel, B. A. Simmons and S. Singh, *Bioresour. Technol.*, 2010, **101**, 4900–4906.
23. R. P. Swatloski, S. K. Spear, J. D. Holbrey and R. D. Rogers, *J. Am. Chem. Soc.*, 2002, **124**, 4974–4975.
24. J. Vitz, T. Erdmenger, C. Haensch and U. S. Schubert, *Green Chem.*, 2009, **11**, 417–424.
25. D. Klemm, H.-P. Schmauder and T. Heinze, in *Biopolymers Online*, Wiley-VCH Verlag GmbH & Co. KGaA, 2005.
26. D. A. Fort, R. C. Remsing, R. P. Swatloski, P. Moyna, G. Moyna and R. D. Rogers, *Green Chem.*, 2007, **9**, 63–69.
27. M. Moniruzzaman and T. Ono, *Bioresour. Technol.*, 2013, **127**, 132–137.
28. N. Labbe, L. M. Kline, L. Moens, K. Kim, P. C. Kim and D. G. Hayes, *Bioresour. Technol.*, 2012, **104**, 701–707.
29. S. Singh, P. Varanasi, P. Singh, P. D. Adams, M. Auer and B. A. Simmons, *Biomass Bioenergy*, 2013, **54**, 276–283.
30. M. J. Serapiglia, K. D. Cameron, A. J. Stipanovic and L. B. Smart, *Appl. Biochem. Biotechnol.*, 2008, **145**, 3–11.
31. J. Zhang, L. Feng, D. Wang, R. Zhang, G. Liu and G. Cheng, *Bioresour. Technol.*, 2014, **153**, 379–382.
32. N. L. Sun, S. H. Ha and Y.-M. Koo, *Process Biochem.*, 2014, **49**, 1144–1151.
33. A. A. Rosatella, R. F. M. Frade and C. A. M. Afonso, *Curr. Org. Synth.*, 2011, **8**, 840–860.
34. A. Brandt, J. Grasvik, J. P. Hallett and T. Welton, *Green Chem.*, 2013, **15**, 550–583.
35. I. Kilpeläinen, H. Xie, A. King, M. Granstrom, S. Heikkinen and D. S. Argyropoulos, *J. Agric. Food Chem.*, 2007, **55**, 9142–9148.
36. B. Li, J. Asikkala, I. Filpponen and D. S. Argyropoulos, *Ind. Eng. Chem. Res.*, 2010, **49**, 2477–2484.
37. N. Sun, M. Rahman, Y. Qin, M. L. Maxim, H. Rodriguez and R. D. Rogers, *Green Chem.*, 2009, **11**, 646–655.
38. G. Cheng, P. Varanasi, C. L. Li, H. B. Liu, Y. B. Menichenko, B. A. Simmons, M. S. Kent and S. Singh, *Biomacromolecules*, 2011, **12**, 933–941.
39. K. C. Badgujar and B. M. Bhanage, *Bioresour. Technol.*, 2015, **178**, 2–18.
40. X. J. Wang, H. Q. Li, Y. Cao and Q. Tang, *Bioresour. Technol.*, 2011, **102**, 7959–7965.
41. H. Wu, M. Mora-Pale, J. J. Miao, T. V. Doherty, R. J. Linhardt and J. S. Dordick, *Biotechnol. Bioeng.*, 2011, **108**, 2865–2875.
42. L. W. Yoon, T. N. Ang, G. C. Ngoh and A. S. M. Chua, *Biomass Bioenergy*, 2012, **36**, 160–169.
43. H. T. Tan and K. T. Lee, *Chem. Eng. J.*, 2012, **183**, 448–458.
44. Z. Z. Chowdhury, S. M. Zain, S. B. Abd Hamid and K. Khalid, *BioResources*, 2014, **9**, 1787–1823.
45. E. Bahcegul, S. Apaydin, N. I. Haykir, E. Tatli and U. Bakir, *Green Chem.*, 2012, **14**, 1896–1903.

46. T. A. D. Nguyen, K. R. Kim, S. J. Han, H. Y. Cho, J. W. Kim, S. M. Park, J. C. Park and S. J. Sim, *Bioresour. Technol.*, 2010, **101**, 7432–7438.
47. S. Padmanabhan, M. Kim, H. W. Blanch and J. M. Prausnitz, *Fluid Phase Equilib.*, 2011, **309**, 89–96.
48. A. R. Xu, J. J. Wang and H. Y. Wang, *Green Chem.*, 2010, **12**, 268–275.
49. M. Zavrel, D. Bross, M. Funke, J. Buchs and A. C. Spiess, *Bioresour. Technol.*, 2009, **100**, 2580–2587.
50. K. S. Kimon, E. L. Alan and D. W. O. Sinclair, *Bioresour. Technol.*, 2011, **102**, 9325–9329.
51. R. Arora, C. Manisseri, C. L. Li, M. D. Ong, H. V. Scheller, K. Vogel, B. A. Simmons and S. Singh, *BioEnergy Res.*, 2010, **3**, 134–145.
52. G. Torgovnikov and P. Vinden, *For. Prod. J.*, 2010, **60**, 173–182.
53. A. Casas, M. V. Alonso, M. Oliet, T. M. Santos and F. Rodriguez, *Carbohydr. Polym.*, 2013, **92**, 1946–1952.
54. K. Ninomiya, A. Kohori, M. Tatsumi, K. Osawa, T. Endo, R. Kakuchi, C. Ogino, N. Shimizu and K. Takahashi, *Bioresour. Technol.*, 2015, **176**, 169–174.
55. K. Ninomiya, A. Ohta, S. Omote, C. Ogino, K. Takahashi and N. Shimizu, *Chem. Eng. J.*, 2013, **215**, 811–818.
56. K. Shill, S. Padmanabhan, Q. Xin, J. M. Prausnitz, D. S. Clark and H. W. Blanch, *Biotechnol. Bioeng.*, 2011, **108**, 511–520.
57. T. V. Doherty, M. Mora-Pale, S. E. Foley, R. J. Linhardt and J. S. Dordick, *Green Chem.*, 2010, **12**, 1967–1975.
58. D. B. Fu and G. Mazza, *Bioresour. Technol.*, 2011, **102**, 7008–7011.
59. D. B. Fu and G. Mazza, *Bioresour. Technol.*, 2011, **102**, 8003–8010.
60. J. Shi, K. Balamurugan, R. Parthasarathi, N. Sathitsuksanoh, S. Zhang, V. Stavila, V. Subramanian, B. A. Simmons and S. Singh, *Green Chem.*, 2014, **16**, 3830–3840.
61. S. Xia, G. A. Baker, H. Li, S. Ravula and H. Zhao, *RSC Adv.*, 2014, **4**, 10586–10596.
62. A. Brandt, M. J. Ray, T. Q. To, D. J. Leak, R. J. Murphy and T. Welton, *Green Chem.*, 2011, **13**, 2489–2499.
63. P. Mäki-Arvela, I. Anugwom, P. Virtanen, R. Sjoholm and J. P. Mikkola, *Ind. Crops Prod.*, 2010, **32**, 175–201.
64. S. Fendt, S. Padmanabhan, H. W. Blanch and J. M. Prausnitz, *J. Chem. Eng. Data*, 2011, **56**, 31–34.
65. P. Weerachanchai, S. K. Kwak and J. M. Lee, *Bioresour. Technol.*, 2014, **170**, 160–166.
66. P. Weerachanchai and J. M. Lee, *ACS Sustainable Chem. Eng.*, 2013, **1**, 894–902.
67. F. Huo, Z. P. Liu and W. C. Wang, *J. Phys. Chem. B*, 2013, **117**, 11780–11792.
68. Y. Fukaya, A. Sugimoto and H. Ohno, *Biomacromolecules*, 2006, **7**, 3295–3297.
69. P. D'Arrigo, C. Allegretti, S. Tamborini, C. Formantici, Y. Galante, L. Pollegioni and A. Mele, *J. Mol. Catal. B: Enzym.*, 2014, **106**, 76–80.

70. N. M. Konda, J. Shi, S. Singh, H. W. Blanch, B. A. Simmons and D. Klein-Marcuschamer, *Biotechnol. Biofuels*, 2014, **7**, 86.
71. Y. Wang, L. Wei, K. Li, Y. Ma, N. Ma, S. Ding, L. Wang, D. Zhao, B. Yan, W. Wan, Q. Zhang, X. Wang, J. Wang and H. Li, *Bioresour. Technol.*, 2014, **170**, 499–505.
72. S. H. Lee, T. V. Doherty, R. J. Linhardt and J. S. Dordick, *Biotechnol. Bioeng.*, 2009, **102**, 1368–1376.
73. Q. Xin, K. Pfeiffer, J. M. Prausnitz, D. S. Clark and H. W. Blanch, *Biotechnol. Bioeng.*, 2012, **109**, 346–352.
74. A. A. Modenbach and S. E. Nokes, *Biotechnol. Bioeng.*, 2012, **109**, 1430–1442.
75. A. A. Modenbach and S. E. Nokes, *Biomass Bioenergy*, 2013, **56**, 526–544.
76. Y. Sun and J. Y. Cheng, *Bioresour. Technol.*, 2002, **83**, 1–11.
77. X. Zhang, J. Ma, Z. Ji, G. H. Yang, X. Zhou and F. Xu, *Microsc. Res. Tech.*, 2014, **77**, 609–618.
78. W. Li, N. Sun, B. Stoner, X. Jiang, X. Lu and R. D. Rogers, *Green Chem.*, 2011, **13**, 2038–2047.
79. K. M. Torr, K. T. Love, O. P. Cetinkol, L. A. Donaldson, A. George, B. M. Holmes and B. A. Simmons, *Green Chem.*, 2012, **14**, 778–787.
80. R. Yáñez, B. Gómez, M. Martínez, B. Gullón and J. L. Alonso, *J. Chem. Technol. Biotechnol.*, 2014, **89**, 1337–1343.
81. Ö. Çetinkol, D. Dibble, G. Cheng, M. Kent, B. Knierim, M. Auer, D. Wemmer, J. Pelton, Y. Melnichenko, J. Ralph, B. Simmons and B. Holmes, *Biofuels*, 2010, **1**, 33–46.
82. M. Shafiei, H. Zilouei, A. Zamani, M. J. Taherzadeh and K. Karimi, *Appl. Energy*, 2013, **102**, 163–169.
83. Uju, Y. Shoda, A. Nakamoto, M. Goto, W. Tokuhara, Y. Noritake, S. Katahira, N. Ishida, K. Nakashima, C. Ogino and N. Kamiya, *Bioresour. Technol.*, 2012, **103**, 446–452.
84. D. Yang, L. X. Zhong, T. Q. Yuan, X. W. Peng and R. C. Sun, *Ind. Crops Prod.*, 2013, **43**, 141–149.
85. J. G. Lynam, M. T. Reza, V. R. Vasquez and C. J. Coronella, *Bioresour. Technol.*, 2012, **114**, 629–636.
86. X. D. Hou, J. Xu, N. Li and M. H. Zong, *Biotechnol. Bioeng.*, 2015, **112**, 65–73.
87. Q. Qing, R. Hu, Y. He, Y. Zhang and L. Wang, *Appl. Microbiol. Biotechnol.*, 2014, **98**, 5275–5286.
88. D. C. Dibble, C. L. Li, L. Sun, A. George, A. R. L. Cheng, O. P. Cetinkol, P. Benke, B. M. Holmes, S. Singh and B. A. Simmons, *Green Chem.*, 2011, **13**, 3255–3264.

Ionic Liquid Pretreatment of Lignocellulosic Biomass for Biofuels and Chemicals

TANMOY DUTTA[a,b], JIAN SHI[a,b], JIAN SUN[a,b], XIN ZHANG[c], GANG CHENG[c], BLAKE A. SIMMONS[a,b], AND SEEMA SINGH[*a,b]

[a]Deconstruction Division, Joint BioEnergy Institute, Emeryville, USA; [b]Biological and Material Sciences Center, Sandia National Laboratories, Livermore, USA; [c]Beijing Key Laboratory of Bioprocess and College of Life Science and Technology, Beijing University of Chemical Technology, Beijing, China
*E-mail: seesing@sandia.gov

3.1 Introduction

The fluctuating price of crude oil and diminishing fossil fuel reserves are drawing tremendous attention towards biomass as an economical and renewable future energy source and for the development of renewable platform chemicals. According to the American Society for Testing and Materials (ASTM), biomass can be defined as 'any material, excluding fossil fuel, which was a living organism, that can be used as a fuel either directly or after a conversion process'.[1] Fuels such as ethanol produced from the fermentation of sugars, derived from sugar cane or grains (corn, wheat, *etc.*) are known as 'first-generation biofuels'. Bioethanol production has seen tremendous growth in the last few years. In 2006, the corn ethanol industry produced

RSC Green Chemistry No. 36
Ionic Liquids in the Biorefinery Concept: Challenges and Perspectives
Edited by Rafal Bogel-Lukasik

Published by the Royal Society of Chemistry, www.rsc.org

4.9 billion gallons of ethanol. Corn ethanol production is expected to grow rapidly, contributing an estimated 12–20 billion gallons of ethanol by 2016.[2] However, the viability of grain-based ethanol production is often questioned mainly because of the 'food *versus* fuel' debate, as these crops are also major food sources.[3] In addition, the grain-based ethanol process has a poor energy balance, higher greenhouse gas emissions, and corn production affects the water reserve and soil quality.[4-6] For the above-mentioned reasons, interest in the production of biofuel and other industry relevant chemicals from lignocellulosic biomass such as hard and soft woods, agricultural waste (corn stover, wheat straw) and herbaceous materials continues to grow; these fuels are referred to as 'second-generation biofuels'.[7]

Lignocellulosic biomass is mostly composed of cellulose, hemicellulose and lignin. The cell walls of lignocellulosic biomass are made up of crystalline cellulose fibrils bound by non-crystalline hemicellulose and surrounded by hemicellulose and lignin.[8] Cellulose is the main constituent of the lignocellulosic biomass and is a straight-chain polymer comprised of disaccharide cellobiose-repeating units. In cellulose, the β-D-glucopyranose units are linked *via* β-(1,4)-glycosidic linkages with degree of polymerization (DP) up to 15 000. The presence of intra- and inter-molecular hydrogen bonding drives the cellulose chains (up to 300 units) to align to form crystalline microfibrils.[9] This crystalline cellulose hinders enzyme accessibility during the hydrolysis.[10,11]

Hemicellulose is the second largest polysaccharide component present in lignocellulosic biomass. Unlike cellulose, hemicelluloses are branched and random and heterogeneous polymers of pentoses (xylose and arabinose), hexoses (glucose, galactose and mannose), and acetylated sugars. The structure and composition of hemicellulose differs depending on the origin of the biomass. The primary role of hemicellulose is to support the cellulose microfibrils in the plant cell wall. Hemicellulose is covalently linked to lignin, acting as a connection between lignin and cellulose microfibrils and thus providing rigidity to the cellulose–hemicellulose–lignin network.[8] Hemicellulose is amorphous in nature and thus can be easily hydrolysed chemically or enzymatically.[12,13]

Lignin on the other hand, is an amorphous, random, branched heteropolymer comprising of phenylpropane units known as monolignols (*p*-coumaryl, coniferyl and sinapyl alcohol). Lignin forms a protective layer to prevent the microbial degradation of cellulose and hemicellulose. It also provides rigidity and a barrier to oxidative stress. The relative composition and interaction between the constituents of lignocellulosic biomass differ on the basis of origin and environmental conditions. A typical lignocellulosic biomass consists of 35–45% cellulose, 25–30% hemicellulose and 20–30% lignin.[8]

The major challenge in utilizing lignocellulosic biomass for the production of biofuels and industry relevant chemicals is to overcome the inherent recalcitrance. Biomass recalcitrance is an evolved mechanism to protect the plant cell wall from biological, chemical and physical attack.[14] The major factors contributing to biomass recalcitrance are the accessible surface area (or cellulase accessibility), crystallinity and DP of cellulose, lignin content, cellulose sheathing by hemicellulose, heterogeneous nature of biomass particles and fibre strength. In general, it is believed that reducing the crystallinity and DP of the cellulose increases the rate of hydrolysis or sugar yield. Smaller particle size, higher accessible surface area, lower lignin content and removal of hemicellulose increase the extent and rate of hydrolysis.[15-21]

To overcome the problems associated with biomass recalcitrance, a number of technologies are being developed and optimized, including pretreatment of the biomass to alter or deconstruct the structure of lignocellulose, facilitate delignification, and expose the cellulose and hemicellulose to promote efficient hydrolysis into fermentable sugars (Figure 3.1). An effective pretreatment strategy should minimize the loss of cellulose and hemicellulose, limit formation of hydrolysis inhibitors, maximize delignification and provide easy recovery of lignin. It should also be applicable to a wide range of feedstocks (different sources and particle sizes) and be economically viable. Biomass pretreatment can be divided into physical, chemical, biological and hybrid methods. Currently there are a number of effective pretreatment strategies available, each with their merits and drawbacks. Amongst the most important and widely used strategies are acid, alkali, organic solvent, ammonia fibre expansion (AFEX) and ionic liquid pretreatments.[13,19,22]

Dilute acid pretreatment has received extensive attention due to the low cost of acids. Dilute sulfuric acid is most commonly employed and the pretreatment is carried out at temperatures between 140 and 215 °C. Although dilute acid pretreatment is relatively inexpensive and can achieve high hydrolysis rates, subsequent fermentation is impacted due to the presence of inhibitors and the pretreatment is ineffective on woody biomass.[19,23-25] Alkaline pretreatment is also an effective and economic pretreatment strategy that employs sodium hydroxide,[26] calcium hydroxide,[27] potassium hydroxide,[18] aqueous ammonia[28] or sodium hydroxide in combination with hydrogen peroxide.[29] The alkali cleaves the hydrolysable linkages in lignin and glycosidic bonds of polysaccharides causing a reduction in the cellulose DP and crystallinity, swelling of the biomass and disruption of the lignin structure.[24] Alkaline pretreatment also improves the enzymatic accessibility of the polysaccharides. Like the acid pretreatment, alkaline pretreatment cannot be universally applied to a variety of biomasses. It is shown to be more effective on herbaceous biomass with lower lignin content.[19,24,30] Organosolv pretreatment, on the other hand, involves extraction or removal of lignin from the biomass using organic solvents or solvent mixtures in combination with water. The common solvents used for organosolv pretreatment include ethanol, methanol, acetone, ethylene glycol and tetrahydrofurfural alcohol. Often organosolv processes are carried out in the presence of inorganic and organic acid catalysts to break the internal lignin and lignin–hemicellulose bonds. This results in extraction of high-quality lignin as a value-added product from the biomass and thus the cellulose fibres of the delignified biomass become more accessible to hydrolysis. The main drawbacks of the organosolv process are the cost and flammability of the organic solvents along with their enzyme inhibition properties.[24,31] It has been reported that hydrolysis of AFEX pretreated biomass results in high sugar yields at relatively low

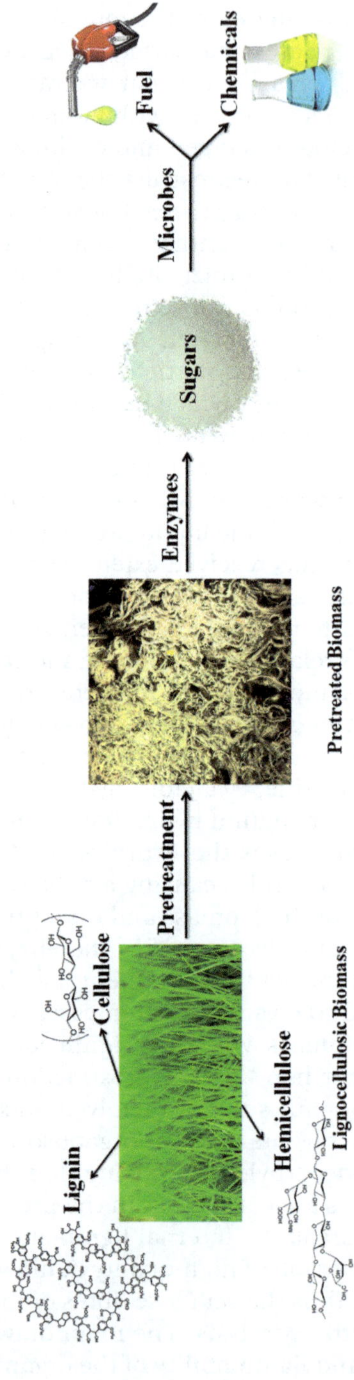

Figure 3.1 Schematic diagram showing the production of fuels and chemicals from lignocellulosic biomass.

enzyme loadings.[32–34] A higher degree of cellulose swelling is believed to be responsible for the enhanced performance. However, whilst this method works well with herbaceous and agricultural residues, it works only moderately for hardwoods and softwoods.[19,25] The success of a lignocellulose-based biorefinery depends largely on the development and implementation of an efficient pretreatment process to overcome the recalcitrance of lignocellulose. The pretreatment process should also facilitate fast enzymatic hydrolysis at high solid loadings with low enzyme concentration and be versatile enough for use with a range of feedstocks. Unfortunately none of the previously described pretreatment processes satisfy all these desired criteria.

Several other pretreatment technologies have been reported in the literature and are beyond the scope of this chapter. Table 3.1 summarizes the comparative effects of different pretreatment technologies.

Ionic liquids (ILs) are molten salts with melting points below 100 °C; they consist of a large organic cation and a small organic or inorganic counter ion. Due to their low vapour pressure at room temperature, non-flammability, wide liquid range and excellent solvation properties, certain ILs can be classified as green solvents. Furthermore, with near infinite possible combinations of anions and cations that can form ILs, they are often referred to as designer solvents.[35–37] Studies have shown that pretreatment using ILs, such as 1-ethyl-3-methylimidazolium acetate, ($[C_2C_1Im][OAc]$) can dramatically reduce biomass recalcitrance and enhance the enzymatic hydrolysis of fermentable sugars.[38,39] Unlike other pretreatment technologies, IL pretreatment can efficiently handle softwoods, hardwoods, herbaceous materials and agricultural residues, both individually and in combination.[40] Moreover, IL pretreatment can handle densified pellets of biomass, which is a unique feature that very few pretreatment technologies can match.[41]

The uniqueness of ILs as tuneable solvents, reaction media and even catalysts holds great potential for biomass pretreatment and fractionation as platform technologies in a biorefinery concept. Figure 3.2 outlines a typical process flow diagram for IL pretreatment under a biorefinery scheme.

(1) The biomass feedstock is mixed with IL and is fed into a pretreatment reactor at operating temperature range from ambient to 160 °C or higher.

(2) Anti-solvent, for example water, alcohol, acetone, or combinations thereof, is applied to the pretreatment slurry to fractionate the cellulose and hemicellulose from the lignin, depending on pretreatment conditions and the type of IL and anti-solvent.

(3) The biomass fractions are then separated by solid/liquid separation or other methods.

(4) In a typical scenario, the sugar stream is taken on to saccharification and fermentation for the production of biofuels or other chemicals while the lignin stream is separated from the IL solution for further valorization.

(5) The IL is recycled for reuse.

Table 3.1 Effect of various pretreatment methods on the composition and structure of lignocellulosic biomass.[a]

Pretreatment method	Increase in accessible surface area	Cellulose decrystallization	Removal of hemicellulose	Delignification	Changes to lignin structure	Ref.
Flow-through acid	■	⊗	■	□	■	150
Liquid hot water	■	⊗	■	⊗	□	151,152
Flow-through liquid hot water	■	⊗	■	□	□	153
Ammonia recycle percolation	■	■	□	■	■	154
Dilute acid	■	⊗	■	⊗	■	19,25,155,156
Lime	■	⊗	□	■	■	19,25,157,158
AFEX	■	■	□	■	■	19,25,159
Uncatalyzed steam explosion	■	⊗	■	⊗	□	160
Wet oxidation	⊗	⊗	■	■	■	161
Organosolv (alcohol and acetone)	■	⊗	■	■	■	7,24,162
IL ([Ch][Lys])	■	□	■	■	⊗	74
IL ([C_2C_1Im][OAc])	■	■	■	■	□	74
Ozonolysis	■	□	■	■	■	163

[a] ■: Major effect; □: minor effect; ⊗: not determined.

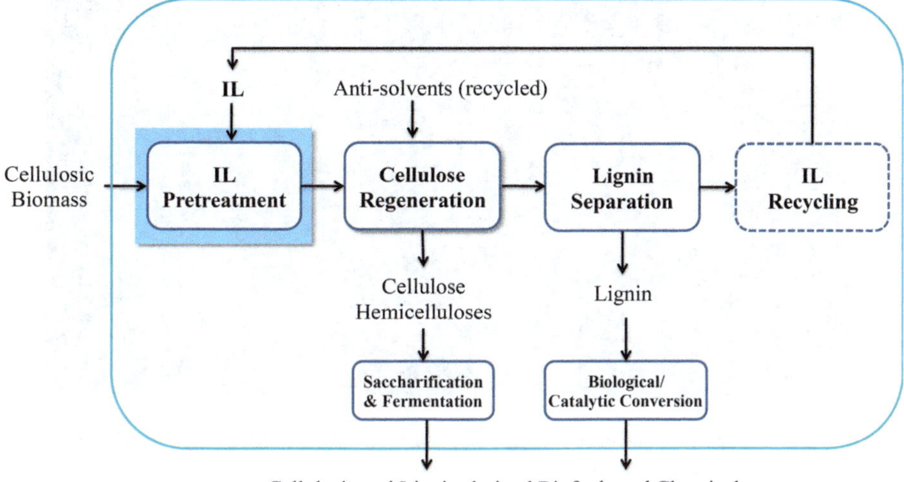

Figure 3.2   A typical process configuration for IL pretreatment.

In this chapter, we provide a brief history of the development of IL pretreatment, an overview of different ILs, pretreatment chemistry and process optimization. We conclude the chapter by identifying the current technical and economic challenges, along with the opportunities and prospects of IL pretreatment for future biorefineries.

3.2   IL Pretreatment of Lignocellulosic Biomass

3.2.1   IL Pretreatment History

Ethylammonium nitrate was the first reported IL, back in 1914 by Walden. It has a melting point of 12 °C.[42] In 1951, Hurley *et al.* reported low-melting salts with chloroaluminate ions for the electroplating of aluminum.[43] In the 1960s, ILs were used primarily for electrochemical applications.[44] In the mid-1980s Pienta *et al.*[45] and Wilkes *et al.*[46] proposed ILs as solvents for organic synthesis. Since then, ILs have shown potential in different applications as diverse as fuel cells,[47] solar cells,[48] electrochemical capacitors,[49] battery systems,[50] catalysts[51] and pharmaceuticals[52] to name a few. The application of ILs for the pretreatment of lignocellulosic biomass is a relatively newer concept. In 2002, Rogers and co-workers demonstrated that 1-butyl-3-methylimidazolium chloride (abbreviated here as [C$_4$C$_1$Im]Cl) can readily solubilize microcrystalline cellulose (MCC).[53] The cellulose can be recovered upon addition of an anti-solvent such as ethanol or water. Varanasi *et al.* and Zhao *et al.* demonstrated that the regenerated cellulose was more susceptible to hydrolysis using cellulases compared to MCC due to a loss in crystallinity during the IL pretreatment and regeneration process.[54–56] In subsequent reports, scientists from the Joint BioEnergy Institute (JBEI), USA, demonstrated the

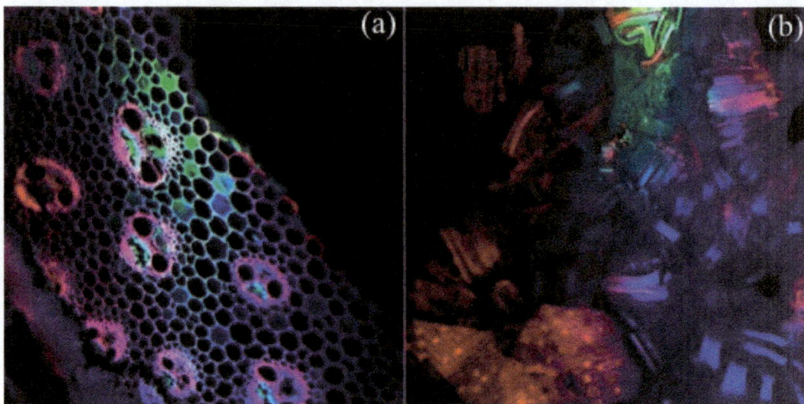

Figure 3.3 Rapid deconstruction of switchgrass before (a) and after (b) IL-pretreatment.

great potential of 1-ethyl-3-methylimidazolium acetate (abbreviated here as $[C_2C_1Im][OAc]$) as an efficient solvent for lignocellulosic biomass with facile regeneration of cellulose and hemicellulose upon anti-solvent addition (Figure 3.3). It also promoted complete delignification, enhanced saccharification kinetics and near theoretical sugar yields.[38,57,58]

3.2.2 Summary of IL-Based Lignocellulosic Biomass Pretreatment

Researchers worldwide have shown many different ILs to be effective for biomass pretreatment. A summary of common ILs, pretreatment conditions and their performances are tabulated in Table 3.2. Currently, no standardized metrics are available to measure the efficiency of IL pretreatments, making it difficult to draw direct comparisons between them. In general, the efficiency of any pretreatment should be a function of biomass feed (type/loading), pretreatment severity (temperature/time), enzyme loading, saccharification efficiency (glucose/xylose yield), and sugar fermentability.

Many different types of ILs have been utilized for biomass pretreatment; however, to date, imidazolium-based ILs have received the most attention. In 2009, Singh *et al.* demonstrated complete and rapid dissolution of switchgrass in $[C_2C_1Im][OAc]$ with subsequent regeneration of cellulose using water as anti-solvent. The accelerated enzymatic hydrolysis of the regenerated cellulose from this process yielded near theoretical amounts of sugar. The authors described a swelling process to precede the dissolution with complete delignification of biomass during cellulose regeneration.[38] The same year, Rogers *et al.* reported that both softwood (southern yellow pine) and hardwood (red oak) can be completely dissolved in the $[C_2C_1Im][OAc]$ by heating at 110 °C for 16 h after mild grinding. Carbohydrate-free lignin and cellulose-rich materials were then able to be regenerated using appropriate reconstitution

Table 3.2 Summary of IL-based lignocellulosic biomass pretreatment.

Entry	Pretreatment system IL/Biomass/ Co-solvent (w/w/w)	Temp./time ($^{\circ}$C h^{-1})	Saccharification enzyme/enzyme loading	Temp./time ($^{\circ}$C h^{-1})	Glucose yield/%	Xylose yield/%	Ref.
1	[C$_4$C$_1$Im]Cl/corn stalk/H$_2$O/HCl (94/5/1/1)	100/0.5	N/A	N/A	66	N/A	68
2	[C$_2$C$_1$Im]DEP/wheat straw (96/4)	130/0.5	30 FPU g^{-1} substrate cellulase	50/12	55	N/A	164
3	[C$_2$C$_1$Im][OAc]/wood (95/5)	110/16	N/A	N/A	N/A	N/A	59
4	[C$_2$C$_1$Im][OAc]/carboxymethyl cellulose (90/10)	130/3	Recombinant enzymes	80/15	N/A	N/A	165
5	(1) 10% Ammonia/rice straw (91/9)	(1)100/6	2 FPU Celluclast 1.5L/0.1 g RS	50/72	97	N/A	166
	(2) [C$_2$C$_1$Im][OAc]/ammonia-treated rice straw (95/5)	(2)130/24	20 CBU Novozyme 188/0.1 g RS				
6	[C$_2$C$_1$Im][OAc]/poplar (95/5)	120/0.5	15 FPU Spezyme CP g^{-1} glucan 30 CBU β-glucosidase g^{-1} glucan	45/24	85	76	61
7	[C$_2$C$_1$Im][OAc]/corn stover (98/2)	70/44	20 FPU g^{-1} of cellulose 62.5 µL β-glucosidase	50/52	80	N/A	147
8	[C$_4$C$_1$im][MeSO$_4$]/*Miscanthus*/ H$_2$O(72/10/18)	120/22	Celluclast (cellulase mix from *Trichoderma reseei*) and Novozyme 188 β-glucosidase	50/96	90	N/A	168
9	[C$_2$C$_1$Im][OAc]/switchgrass (97/3)	120/3	12.5 IU mL^{-1} cellulase, and 5.1 IU/ mL β-glucosidase	50/72	>90	N/A	60
10	50% [C$_2$C$_1$Im][OAc]/triticale straw (97/3)	150/1.5	Cellulase from *Trichoderma reesei*	50/48	>90	>90	139
11	[Ch][Gly]/rice straw (95/5)	90/24	245 U cellulase/20 mg biomass	50/24	N/A	N/A	73
12[a]	[C$_2$C$_1$Im][OAc]/switchgrass (90/10)	160/3	5 mg JTherm g^{-1} switchgrass	70/72	81.2	87.4	65
13	[C$_2$C$_1$Im][OAc]/switchgrass (85/15)	160/3	60 mg (9CTec2 + HTec) g^{-1} glucan	50/72	94.8	62.2	169
14	[C$_2$C$_1$Im][OAc]/corn stover (N/A)	120/3	N/A	N/A	N/A	N/A	170
15	[C$_4$mim]Cl/*A. donax* (95/5) Protic acid resin/(1)	160/1.5	50 µg mL^{-1} kanamycin and 1 µl (0.7 U) Celluclast *Trichoderma* cellulase	37/72	N/A	N/A	171

(*continued*)

Table 3.2 (*continued*)

Entry	Pretreatment system IL/Biomass/Co-solvent (w/w/w)	Temp./time (°C h^{-1})	Saccharification enzyme/enzyme loading	Temp./time (°C h^{-1})	Glucose yield/%	Xylose yield/%	Ref.
16	[C$_2$C$_1$Im][OAc]/switchgrass + lodgepole pine + corn stover + eucalyptus/(90/10)	160/3	20.3 mg (CTec2 + HTec2) g^{-1} glucan	50/24	91.7	85	41
17	[p-AnisEt$_2$NH][H$_2$PO$_4$]/switchgrass (90/10)	160/3	15 mg CTec g^{-1} biomass 1.5 mg HTec g^{-1} glucan	50/72	95	75	75
18	[C$_2$C$_1$Im][OAc]/rice hulls/glycerol (50/10/50)	110/3	N/A	50/72	45	75	172
19	[C$_2$C$_1$Im][OAc]/wood (92/8)	150/0.5	Celluclast 1.5L/Novozym 188	48.5/48	>95	20	173
20	([C$_4$C$_1$Im]Cl/HCl/FeCl$_3$)/softwood/H$_2$O (74.0/1.2/4.8/10/20)	130/0.5	2 FPU CTec2 g^{-1}-starting biomass	50/24	N/A	N/A	69
21	[Ch][OH]/rice straw/H$_2$O (5/2/45)	60/3 (Ultra-sound)	20 mg of treated-rice straw using 0.7 mL of cellulase solution	50/4 (Ultra-sound)	81	N/A	174
22	([Cho][OAc]/bagasse (95/5)	100/24	N/A	N/A	N/A	N/A	175
23	[C$_4$C$_1$Im]Cl/microcrystalline cellulose (96/4)	110/2	Cellulase (17 FPU g^{-1} MCC) and cellobiase (34 IU g^{-1} MCC)	50/4	>90	N/A	114
24	[C$_2$C$_1$Im][OAc]/sugarcane bagasse (97/3)	90/6	35 FPU g^{-1} (cellulase) and 40 CbU (β-glucosidase)	50/100	90	N/A	176
25	[C$_2$C$_1$Im][OAc]/ethanolamine mixture (60/40 vol%)/rice straw (2 mL/100 mg)	120/24	Cellulase from *Trichoderma viride* (crude powder of fungus, 4 U mg^{-1} solid) at a concentration of 34 U L^{-1}	50/24	85.2	N/A	177
26	[C$_4$C$_1$Im]Cl/sugarcane bagasse/polyethylene glycol 4000 (100/10/3)	130/1.5	50 FPU Celluclast 1.5L g^{-1} substrate and 40 CBU Novozyme 188 g^{-1} substrate	50/12	96.2	N/A	71
27	[C$_2$C$_1$Im][OAc]/switchgrass (90/10) or [Ch][Lys]/switchgrass (90/10)	140/1	20 mg (CTec2 + HTec2) g^{-1} glucan	50/72	95.8 or 96.5	93.3 or 95.7	74
28	(1) [C$_4$C$_1$Im]Cl/*A. donax* (95/5) (2) protic acid resin (1)	(1) 160/1.5 (2) 100/1	20 FPU (Celluclast 1.5L) g^{-1} substrate	48/72	92.8	N/A	70
29	[C$_2$C$_1$Im][OAc]/switchgrass (95/5)	120/3	50 mg cellulase g^{-1} glucan, 5 mg β-glucosidase g^{-1} glucan	50/20	80	N/A	178
30	[C$_2$C$_1$Im][OAc]/corn stover (90/10)	140/3	30 mg (CTec2 + HTec2) g^{-1} glucan	50/72	91.6	77	67
31	[C$_2$C$_1$Im][OAc]/wheat straw (85/15)	160/1.5	20 mg protein (CTec 2) g^{-1} glucan	50/72	>95	N/A	126

^{a}No washing requirement after pretreatment of biomass. N/A: Not available.

solvents. The authors compared dissolutions of the hardwood and softwood in [C$_2$C$_1$Im][OAc] and [C$_4$C$_1$Im]Cl.[59] Later in 2010, Singh *et al.* carried out detailed parametric studies of IL pretreatment of switchgrass to shed light on the disposition of cellulose, hemicellulose and lignin in order to optimize the [C$_2$C$_1$Im][OAc] pretreatment process; a pretreatment temperature of 140–160 °C and incubation time of 1–3 h were established as efficient conditions for lignocellulosic biomass processing. The authors observed a correlation between hydrolysis kinetics and delignification efficiency.[60] In 2010, Schall *et al.* used [C$_2$C$_1$Im][OAc] to pretreat poplar and switchgrass with 5% biomass loading at 120 °C for 30 min. Using a commercial cellulase system, 85% glucose and 76% xylose yield were obtained from the pretreated poplar. The authors also observed that IL pretreatment improved glucose and xylose yields compared to untreated substrates.[61] Since 2010, numerous reports of biomass pretreatment using imidazolium-based ILs have appeared in the literature, showcasing their effectiveness in overcoming the recalcitrance of lignocellulosic biomass. The majority of the reported pretreatments have been achieved using relatively low biomass loadings (3–10 wt.%). Klein-Marcuschamer *et al.* conducted techno-economic studies demonstrating that, increasing the biomass loading to 40–50 wt.% during IL pretreatment has a prominent effect on the overall economics of biorefinery operation.[62] Wu *et al.* showed that pretreatment with imidazolium acetate-based ILs dramatically reduced the recalcitrance of corn stover towards enzymatic hydrolysis at biomass loadings as high as 50 wt.%.[63] In 2013, Cruz *et al.* studied in detail the impact of higher biomass loadings of switchgrass on IL pretreatment in terms of viscosity, cellulose crystallinity, chemical composition, saccharification kinetics and sugar yield. The authors show IL pretreatment using [C$_2$C$_1$Im][OAc] can effectively reduce biomass recalcitrance at loadings up to 50 wt.%. Although high biomass loadings result in increased viscosity, the authors observed an increasing enhancement of shear thinning leading to slurries with a lower complex viscosity at higher biomass loadings.[64]

The common enzyme cocktails produced from filamentous fungi and developed for dilute acid pretreatment are inhibited by most of the ILs used for pretreatment and require large amounts of water to remove the ILs from the regenerated biomass after pretreatment. The costs associated with subsequent IL recycling, waste disposal and water treatment pose potential economic hurdles for the commercial scale-up of IL pretreatment-based technologies.[40,62] In 2013, Shi *et al.* addressed this challenge, demonstrating a one-pot, wash-free process that combines IL pretreatment and saccharification into a single vessel. The authors used [C$_2$C$_1$Im][OAc] to pretreat switchgrass (10% w/w) at 160 °C for 3 h. After pretreatment, the IL concentration was reduced to 10–20% by dilution with water, the pretreatment slurry was then directly hydrolysed using the thermostable, IL-tolerant enzyme cocktail JTherm developed at JBEI, yielding 81.2% glucose and 87.4% xylose after 72 h at 70 °C with an enzyme loading of 5.75 mg per gram of biomass. The glucose and xylose were selectively separated by liquid–liquid extraction with over 90% efficiency.[65]

As discussed earlier, the origin (type) of biomass is a very important factor in biomass pretreatment and many pretreatment technologies are not tolerant of a wide range of biomasses. Li *et al.* compared softwood (pine), hardwood (eucalyptus) and switchgrass for the production of fermentable sugars *via* [C$_2$C$_1$Im][OAc] pretreatment. Of the three feedstocks, highest sugar yield and fastest hydrolysis rate was observed for switchgrass. Of the two wood species, eucalyptus gave higher and faster sugar recovery than pine under similar pretreatment conditions (160 °C for 3 h).[66] Shi *et al.* in a subsequent report in 2013, demonstrated pretreatment of a mixed feedstock containing switchgrass, lodgepole pine, corn stover and eucalyptus in flour and pellet form using [C$_2$C$_1$Im][OAc] at 160 °C for 3 h with 91.7% glucose and 85% xylose yield. This result demonstrates the robustness and effectiveness of the IL-pretreatment process.[41]

In 2014, Gao *et al.* compared enzymatic reactivity of corn stover solids from pretreatment by dilute acid, AFEX and IL. The authors found no single factor that dominates early and longer-term glucose yields from enzymatic hydrolysis of solids from these three pretreatment techniques. The initial hydrolysis yield at low enzyme loading was found to be highest from the IL pretreated corn stover, which can be correlated with a higher degree of lignin removal, a large reduction in cellulose crystallinity and high enzyme adsorption.[67]

Additives such as dilute acids, Lewis acids, protic acid resins and surfactants allow a reduction in the severity of the IL-pretreatment conditions. Zhao *et al.* reported the use of acid in IL as an efficient system for hydrolysis of lignocellulosic biomass with an improved yield of total reducing sugars (TRS) under mild conditions (100 °C for 30 min).[68] Ogino *et al.* utilized IL in combination with acid and metal ions for pretreatment with higher solid loading (>10% w/w) under relatively milder conditions (130 °C for 30 min).[69] Xu *et al.* employed protic acid resin along with [C$_4$C$_1$Im]Cl. In this unique pretreatment process, the pretreatment was performed at 160 °C with [C$_4$C$_1$Im]Cl for 1.5 h, followed by treatment with Amberlyst 35DRY (protic acid resin) at 90 °C for 1 h. The authors observed a reduction in cellulose crystallinity and increased porosity caused by extensive swelling of the undissolved biomass caused by Amberlyst, producing a higher glucose yield (92.8%) than for the single [C$_4$C$_1$Im]Cl pretreatment (42.8%).[70] Mousavi *et al.* used 3% w/w polyethylene glycol along with [C$_4$C$_1$Im]Cl for the pretreatment of sugar cane bagasse at 130 °C for 1.5 h yielding 96.2% glucose.[71]

Apart from imidazolium-based ILs, ILs containing cholinium cations and amino acid anions ([Ch][AA]), known as 'bionic liquids', were shown to efficiently pretreat rice straw by selectively removing lignin. These ILs are prepared from naturally occurring, renewable starting materials, are expected to be more biocompatible with enzymes compared to acetate-based ILs and more economic than imidazolium-based ILs.[72,73] In 2014, Sun *et al.* reported pretreatment of switchgrass (10% w/w) using [Ch][Lys] at 140 °C for 1 h, yielding 96.5% glucose and 95.7% xylose. In the same report, under identical pretreatment conditions, [C$_2$C$_1$Im][OAc] yielded 95.8% and 93.3% glucose and xylose respectively.[74] Later in 2014, in a pioneering report Socha *et al.*

demonstrated efficient biomass pretreatment using ILs derived from lignin and hemicellulose. Pretreatment of switchgrass (10% w/w) using lignin-derived *N*-ethyl-*N*-(4-methoxybenzyl)ethanamine dihydrogenphosphate salt ([*p*-AnisEt$_2$NH][H$_2$PO$_4$]) at 160 °C for 3 h yielded 95% and 75% glucose and xylose respectively. The concept of deriving ILs from lignocellulosic biomass shows significant potential for the realization of a 'closed loop' process for lignocellulosic biorefineries.[75]

3.3   Ionic Liquid Pretreatment Chemistry

3.3.1   Dissolution of Cellulose in ILs

Although cellulose is the most abundant polysaccharide on earth, it has poor solubility in common organic solvents, which makes cellulose processing difficult.[76] A great deal of energy is needed to solubilize cellulose by breaking many intra- and inter-molecular hydrogen bonds.

In the 1930s, 1-ethylpyridinium chloride was noted to dissolve cellulose.[76,77] However, this discovery did not draw much attention in either the scientific community or in industry. Until recently, cellulose dissolution was only possible in very few solvents *via* either derivatisation or direct dissolution under relatively harsh conditions, for example, in *N*-methylmorpholine *N*-oxide monohydrate (NMMO), LiCl/*N,N*-dimethylacetamide or ammonium fluorides/dimethyl sulfoxide mixtures, amongst which only NMMO is used on an industrial scale in the Lyocell process to make cellulose fibres.[76]

As stated earlier in this chapter, ILs are able to solubilize cellulose under relatively mild conditions, providing an excellent opportunity for cellulose processing as well as biomass pretreatment. A few years after the pioneering study in 2002,[53] Rogers *et al.* used ^{13}C and $^{35/37}$Cl NMR relaxation measurements to reveal the dissolution mechanism of cellobiose and glucose in [C$_4$C$_1$Im]Cl. They proposed that dissolution involved the formation of hydrogen bonds between the hydroxyl protons of the sugars and the chloride ions of the IL. The NMR relaxation data suggested there was no specific interaction between the IL cation and the sugar.[78] This was supported by another study employing nuclear Overhauser enhancement spectroscopy (NOESY), which showed that there was no direct interaction between sugar hydroxyl groups and acidic hydrogens on the IL cation.[79]

In another investigation, ^{1}H and ^{13}C NMR spectroscopy were used to study the interactions between cellobiose and [C$_2$C$_1$Im][OAc] in dimethyl sulfoxide (DMSO)-d_6. In a thorough analysis of the concentration dependence of chemical shift, it was concluded that hydrogen bonds were formed between the hydroxyl protons of cellobiose and both the cation and anion of the IL.[80] A more recent experimental study on the effect of IL cation structure on the dissolution of cellulose proposed that the acidic protons on the heterocyclic rings may form hydrogen bonds with cellulose and this was essential for dissolution. The solubility data suggested that IL cations can decrease cellulose solubility if the interaction between cation and anion was strong.

The authors also suggested that the effect of steric hindrance of alkyl chains in the cations of the ILs might also decrease the cellulose solubility.[81]

The effect of alkyl chains of the dialkyl imidazolium cation on cellulose solubility has been systematically studied. Erdmenger *et al.* investigated cellulose solubility within a series of 1-alkyl-3-methylimidazolium chloride ILs ([C_nC_1Im]Cl; n = C atoms in the linear alkyl chain = 1–10) and observed higher cellulose solubility with ILs having n = even number than the ILs with n = odd number, for relatively short alkyl chains ($n \leq 5$).[82] However, this 'odd-even' effect was not properly understood. In a subsequent report Vitz *et al.* corroborated the result and also demonstrated that this pattern was no longer observed when the chloride anions were replaced with bromide.[83] In another work, Zhang *et al.* reported 1-allyl-3-methylimidazolium chloride ([AC_1Im]Cl) is an efficient nonderivatizing solvent for cellulose.[84] Subsequently, [AC_1Im]Cl was found to be more efficient at dissolving cellulose than [C_4C_1Im]Cl.[84,85] The increased solubility in ([AC_1Im]Cl can be attributed to the decreased viscosity of the IL as a result of the terminal double bond in the side chain, which could augment the cellulose dissolution. On the other hand, the smaller size and more polarizable character of the [AC_1Im]$^+$ cation increase the interaction with cellulose.[84]

In parallel with experimental studies, computer simulation work has contributed significantly to the understanding of the mechanism of cellulose dissolution in ILs.[86–96] Early molecular dynamics simulations of glucose solvation in [C_1C_1Im]Cl demonstrated the important role of hydrogen bonds between chloride ions and glucose.[86,87] The study also highlighted the contribution of the interaction between the acidic hydrogen at the 2-position of the imidazolium ring and the oxygen atoms of the glucose.[87] Later the concept of cation's influence on the dissolution of cellulose by IL was emphasized in another simulation work.[88] To further understand the synergistic action of anion and cation towards dissolution of cellulose, a 36-chain microfibril was constructed and the interaction between IL and the microfibril was studied.[90,96] It was discovered that both the anion and cation interacted with the glucose residues involved with intersheet formation in the microfibril; this was identified as an important ability of cellulose solvents.[90] In another work, the results indicated that the anion predominantly interacted with the microfibril surface with hydroxyl groups and the cation stack preferentially on the hydrophobic surface. It was believed that the stacking interaction between the cation and glucose ring can stabilize the detached cellulose chains in ILs.[96]

3.3.2 Regeneration of Cellulose and Hydrolysis

Cellulose is extruded from terminal enzyme complexes located in the cell wall whose configuration dictates the resulting microfibril architecture. A microfibril consists of a number of extended parallel chains assembled into nanometre thick crystallites that are millimetres in length.[97,98] Natural microcrystalline cellulose has the cellulose I structure and is recalcitrant to enzymatic hydrolysis. Recent experimental and theoretical studies reveal that other forms of cellulose, cellulose II and cellulose III, are less recalcitrant.[90,99–101] In the cellulose II form, chains with opposite polarity are stacked to form corrugated sheets; hydrogen bonding exists within the sheets as well as between them. Studies have shown that biomass recalcitrance to enzymatic hydrolysis is mitigated by the IL dissolution of the biomass followed by recovery as a solid through the addition of an anti-solvent. The regenerated cellulose has either a cellulose II crystalline structure, an amorphous structure or a mixture of the two. Cellulose II is most often obtained from cellulose I *via* either of two processes: regeneration and mercerization.[57,102]

Cellulose amorphization can also be achieved by swelling and this is a suitable alternative to the dissolution–precipitation pretreatment method to facilitate enzymatic hydrolysis. Swelling is commonly conducted in an IL–water mixture. In one typical example, soaking 10 wt.% Avicel PH-101 in $[C_2C_1Im][OAc]$ containing 20 wt.% water at room temperature, was found to reduce the crystallinity of the cellulose by 35%, while the dissolution was measured to be only 1 wt.%.[103]

It has been reported that along with the degree of crystallinity, the DP of cellulose was also altered during pretreatment process and this was found to affect the biomass recalcitrance.[104,105] Liu *et al.* reported that the DP of the regenerated cellulose from cotton pulp decreased with increasing reaction time and temperature. The authors observed a sharp decrease in DP of regenerated cellulose from 510 to 180 after pretreatment with $[C_4C_1Im]Cl$ at 90 °C for 7 h.[106]

A viscometric study in 2012 showed that the maximum water content allowing dissolution of 1 wt.% cellulose in $[C_2C_1Im][OAc]$ was 15 wt.%; above this the cellulose formed large aggregates.[107] A molecular simulation study demonstrated that the interaction between the cellulose and acetate ion diminished with increasing water content. The addition of water led to disruption of the hydrogen bonds between cellulose/acetate ion and subsequent formation of cellulose/cellulose and acetate ion/water hydrogen bonding interactions.[108] In comparison to acetone and ethanol, water was found to be the most effective solvent at breaking cellulose/acetate ion hydrogen bonds.[109] However, very recent experimental work showed that the use of ethanol afforded regenerated cellulose with a high surface area and low crystallinity, which was ideal for enzymatic hydrolysis.[110] It is noted that carbon dioxide has also been used as an anti-solvent to precipitate cellulose from IL solution.[111]

The potential importance of cellulose hydrolysis in the context of conversion of plant biomass to fuels and chemicals is widely recognized, and also represents one of the largest material flows in the global carbon cycle. Hydrolysis of cellulose giving rise to monomeric glucose can be achieved *via* catalysis by mineral acids or enzymes. Acidolysis of cellulose by sulfuric acid or hydrochloric acid is a widely used process; the use of concentrated acid requires less harsh conditions. However, this process requires expensive corrosion-resistant reactors and has major waste disposal problems associated with it. In addition, the degradation of products by the acid significantly lowers the glucose yield and interferes with downstream applications.[112] Zhao *et al.* reported a novel method for cellulose hydrolysis catalysed by mineral acids in [C$_4$C$_1$Im]Cl, where the hydrolysis rate of cellulose was dramatically accelerated at 100 °C under atmospheric pressure and without pretreatment. The authors reported a 39% glucose yield within 1 hour, catalysed by 3.2 wt.% dilute sulfuric acid, using Avicel, α-cellulose or Sigmacell as the feedstock.[112]

Enzymatic hydrolysis of cellulose with IL pretreatment has made some interesting progress in recent years.[113] Cellulase-facilitated enzymatic hydrolysis of cellulose to glucose in aqueous media suffers from a slow reaction rate due to the highly crystalline structure of cellulose and inaccessibility of enzyme adsorption sites. Varanasi *et al.* investigated the enhancement of cellulose saccharification kinetics by commercial enzyme using a [C$_4$C$_1$Im]Cl pretreatment step followed by rapid precipitation with water or alcohol anti-solvent. The authors found that the resultant regenerated cellulose was amorphous in structure, allowing a greater number of sites for enzyme adsorption with an approximate 50-fold enhancement of hydrolysis kinetics as compared to untreated cellulose (Avicel PH-101).[54] Cao *et al.* found that the regeneration process from [C$_4$C$_1$Im]Cl could significantly influence the crystallinity and digestibility of cellulose; almost amorphous cellulose was obtained by regenerating using acetone (DRC-a), while partial cellulose II structure could be found in these regenerated samples from water and ethanol. Above 90% of generated cellulose could be converted into glucose by commercial enzyme cocktails after pretreatment with [C$_4$C$_1$Im]Cl (100 °C for 4 h) for DRC-a and regenerated cellulose without drying as compared to 9.7% for MCC (9.7%).[114] Recently, it has been reported that the hydrolysis of cellulose in IL could be operated by using an IL-tolerant enzyme system.[115] As described in the previous section, Shi *et al.* demonstrated a one-pot, wash-free process that combines IL pretreatment and saccharification into a single vessel.[65] This study opened avenues for developing more efficient and cost-effective processes for product recovery and IL recycling.

3.3.3   Dissolution and Depolymerization of Lignin in ILs

In contrast to cellulose, the dissolution of lignin in ILs has received less attention. The lignin macromolecule is primarily connected through carbon–carbon and carbon–oxygen bonds between building blocks of phenylpropane monomers. It can be dissolved in several organic solvents, including DMSO and dioxane, and in alkaline solutions.[116] A systematic study of dissolution of softwood lignin in different ILs was performed in 2007. It was discovered that the solubility of lignin was mainly influenced by the nature of anions. For the [C$_4$C$_1$Im]$^+$-containing ILs the order of lignin solubility for various anions was [MeSO$_4$]$^-$ > Cl$^-$ ~ Br$^-$ ≫ [PF$_6$]$^-$. The results indicate that ILs containing strongly hydrogen-bonding anions such as [MeSO$_4$]$^-$ are efficient solvents for lignin.[117] It was proposed that the available terminal hydroxyl groups in lignin interacted with ILs by hydrogen bonds, which disrupt the internal network in lignin.[118,119] Similar to the dissolution of cellulose in ILs, the role of the cation was also noted. Replacing the $[C_4C_1Im]^+$ in $[C_4C_1Im][Cl]$ with 1-benzyl-3-methylimidazolium ($[BzC_1Im]^+$) yields improved lignin solubilization, indicating the presence of $\pi–\pi$ stacking interactions between the IL and lignin.[120] A density functional theory calculation of the interactions between a lignin model compound (1-(4-methoxyphenyl)-2-methoxyethanol) and imidazolium chloride suggested that hydrogen bonding and $\pi–\pi$ interactions between lignin and IL cation play an important role in the dissolution.[121] However, it is noted that lignin can be dissolved in DMSO and dioxane where there is no $\pi–\pi$ interaction involved, pointing out the non-essential nature of the $\pi–\pi$ interaction.[122]

It has been shown that lignin degradation occurs during IL pretreatment or dissolution.[123,124] The relative impact on reducing lignin molecular weight by various anions was established and it was found that sulfate-based ILs facilitated the most comprehensive breakdown of lignin.[123] In 2013, Varanasi *et al.* conducted a survey of renewable chemicals produced from the breakdown of lignin during ionic liquid pretreatment.[125] Later in 2014, Sathitsuksanoh *et al.* characterized different lignin streams during the pretreatment of wheat straw, *Miscanthus* and pine with $[C_2C_1Im]OAc$ using solution-state two-dimensional (2D) NMR and size exclusion chromatography (SEC). The results suggested that the lignin isolated from the pretreatment of three different feedstocks were depolymerized.[126] Wen *et al.* carried out a detailed study of chemical transformations of lignin in $[C_2C_1Im][OAc]$ under different pretreatment conditions. Results showed an increase in phenolic hydroxyl groups as a result of cleavage of the β-O-4 linkage. The cleavage of the β-O-4 linkage and degradation of the β-β, β-5 linkages caused a decrease in molecular weight.[124] The depolymerization of lignin during IL pretreatment could facilitate lignin upgrading to industry relevant chemicals, thereby potentially improving biorefinery economics. The depolymerization of lignin can be augmented by the use of different catalysts. Many in-depth literature reports and reviews are available in this area – readers can consult these reviews for further interest.[127,128]

3.4　Challenges and Opportunities

3.4.1　Techno-Economic Analysis (TEA)

Although ILs exhibit many technical merits compared with other pretreatment technologies, pretreatment using current commercially available ILs is economically challenging. Klein-Marcuschamer *et al.* have conducted techno-economic analysis of IL-based biomass pretreatment processes, and reported that in order to make it a practical reality, several key factors should be addressed (Figure 3.4):

(1) reduction of IL price and volume
(2) increasing solid loading during IL pretreatment, saccharification and fermentation, as TEA indicates that higher solid loadings lower both the capital expense (CAPEX) and operating expense (OPEX) of the process

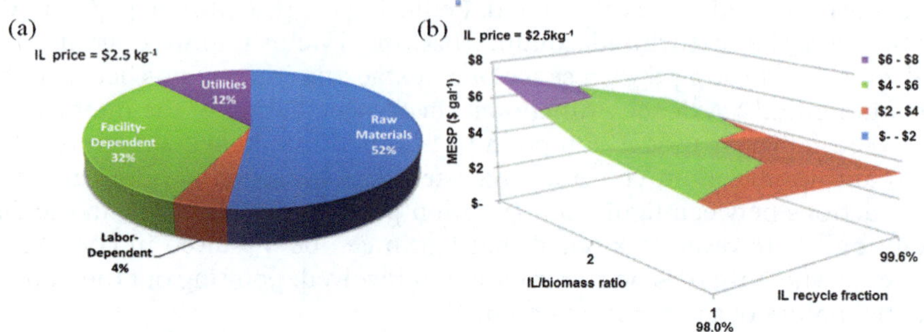

Figure 3.4   (a) OPEX and (b) minimum ethanol selling price (MESP) as a function of solid loading and IL recycle fractions at given IL price of $2.5 per kg.[62] (Reproduced with permission from John Willey & Sons.)

(3) development of biocompatible ILs or novel IL-tolerant enzyme mixtures and lowering enzyme costs
(4) development of efficient IL recycling and product recovery technologies
(5) development of a systems approach for integrating and optimizing feedstock preparation, IL pretreatment, and downstream saccharification and fermentation, and product recovery unit operations.[62]

The cost of the IL itself is one of the most significant factors determining the MESP and thus the techno-economics of a biorefinery using IL pretreatment as a platform technology. Based on previous reports, IL price varies significantly depending on type and vendor ($1.00–$800 per kg). However, if an IL is being considered as a component of an industrial process (for example, as the solvent for a biomass pretreatment process), it is important to investigate and optimize in terms of both cost and environmental impact of the synthetic route (at manufacturing scale) to produce the IL. To reduce the cost of IL, both anion and cation should be available from cheap and renewable resources. For example, as detailed in a patented process, synthesis of $[C_2C_1Im][OAc]$ requires either costly imidazole as a staring material or multi-step synthesis routes. Chen *et al.* reported IL pretreatment using triethylammonium hydrogensulfate ($[HNEt_3][HSO_4]$). Synthesis of this new IL only requires a simple stoichiometric mixing of two cheap starting materials: triethylamine and sulfuric acid (neither costs more than $2.0 per kg in tonne quantities).[129] Ohno *et al.* reported the preparation of ILs derived from amino acids, and the follow-up work to combine amino acid-based anions with bio-derived cations such as cholinium and amino acid or their ester salts.[130] Many of these ILs have been shown to be highly effective at pretreating biomass.[72–74,131] Recently, Socha *et al.* described a process for synthesizing ILs from materials derived from lignin and hemicellulose.[75] With respect to overall sugar yield, experimental evaluation of these compounds showed that they perform comparably to traditional ILs in biomass pretreatment.

In addition to being less expensive, biocompatible and biodegradable ILs are highly desirable in terms of both compatibility with biorefinery downstream processes and environmental concerns, *i.e.* toxicity to the aquatic ecosphere.[132,133] Many of the imidazolium-based ILs are not compatible with commercial cellulases/hemicellulose platforms. In addition, downstream microbial fermentation is inhibited by residual imidazolium-based ILs, thus requiring further separation of sugar from the IL stream. In contrast, ILs composed of bio-derived anions and cations have been shown to be more compatible with enzyme and microbes.[134] These ILs often contain a cholinium cation combined with amino acid-based anions ([Ch][AA] ILs) or carboxylic acid-based anions ([Ch][CA] ILs).[72,130,135–137] Cholinium ILs are intrinsically less expensive (due to the lower cost of the cation starting material), more biocompatible and biodegradable[138] whilst still displaying a high pretreatment efficiency compared with imidazolium ILs based on our recent study.[74]

The use of IL–water mixtures as pretreatment agents reduces the energy inputs and costs associated with IL recycling, facilitating scale-up and downstream processing. Doherty *et al.* reported blending of 5 and 10% (w/w) of water with $[C_2C_1Im][OAc]$, $[C_4C_1Im][OAc]$ and $[C_4C_1Im][MeSO_4]$ to pretreat wood flour.[132] A decrease in glucose and xylose yields was observed with higher percentages of water and the authors attributed the reduction in pretreatment effectiveness to the fact that the addition of water decreased the ability of the ILs to disrupt cellulose crystallinity, and affect both intra- and inter-crystalline swelling, as well as fibre size reduction. However, Fu *et al.* observed a strong correlation between cellulose digestibility, lignin removal and crystallinity index of cellulose for the pretreatment of triticale straw in aqueous $[C_2C_1Im][OAc]$ solution.[139] This work led to a later optimization study of pretreatment with aqueous IL solutions, whereby an RSM (response surface methodology) statistical approach was undertaken with wheat straw and $[C_2C_1Im][OAc]$ as the biomass and IL respectively. Results indicated that efficiencies comparable to pretreatment with pure ILs could be achieved at optimized conditions, 158 °C, 3.6 h with 50.5% water content (w/w).[140] It seems that a high pretreatment temperature and long reaction time are preferable to compensate the loss of pretreatment effectiveness due to water addition. However, the solvent property of IL–water mixtures and correlations between cellulose digestibility, cellulose solvation and lignin depolymerization during IL–water pretreatment of lignocellulosic biomass are not well understood. Shi *et al.* recently examined the pretreatment of switchgrass with aqueous $[C_2C_1Im][OAc]$, using both experimental and computational approaches. Results indicated that the chemical composition and crystallinity of the pretreated biomass, and the corresponding lignin dissolution and depolymerization, were dependent on $[C_2C_1Im][OAc]$ concentration that correlated strongly with cellulose digestibility. In addition, the hydrogen bond basicity of the $[C_2C_1Im][OAc]$–water mixtures was found to be a good indicator for cellulose dissolution, lignin depolymerization and sugar yields. Molecular dynamics simulations provided molecular level explanations on cellulose I_β dissolution at various $[C_2C_1Im][OAc]$–water loadings.[141]

3.4.2   Process Integration and Optimization

System complexity associated with IL recycling, biomass–solute separation and downstream processing are the main challenges in the commercialization of IL pretreatment technology.[40,62] In a conventional approach to IL-based bioprocessing, IL pretreatment is a separate unit operation preceding downstream saccharification and fermentation (Figure 3.5a). This pretreatment configuration typically requires extensive washing of the biomass following the pretreatment to remove residual amounts of IL, which can inhibit subsequent saccharification and fermentation.[142–144] Excessive water use and waste disposal associated with washing poses a challenge for the scale-up of any IL pretreatment technology.

The configuration in Figure 3.5b outlines a one-pot, wash-free scheme that combines IL pretreatment and saccharification, followed by direct extraction of sugars, recovery of lignin and recycling of [C₂C₁Im][OAc] in order to minimize costs and enhance sustainability. The new configuration could lead to approximately 2–15 fold reductions in water use compared with the conventional approach of separate pretreatment and saccharification steps. A wash-free configuration both reduces costs associated with energy-intensive evaporation or reverse osmosis recycling of ILs and downsizes the water footprint of the biorefinery by greatly reducing grey water generation.[145] Furthermore, eliminating water washing with a one-pot pretreatment and saccharification process could simplify the downstream sugar/lignin recovery and IL recycling and greatly improve the economics of IL pretreatment technology.

3.4.3   IL Recycling and Product Recovery

Beyond biomass pretreatment technologies, the key to an economically viable and scalable IL-based biorefinery is the development of efficient and cost-competitive separation processes to recover bioproducts and to recycle

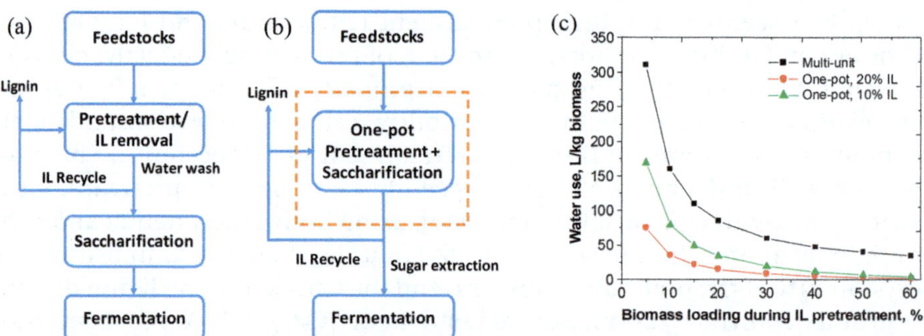

Figure 3.5   One-pot pretreatment and saccharification reduces water use.[65] (Reproduced with permission from Royal Society of Chemistry.) (a) Conventional pretreatment process; (b) one-pot process; (c) comparison of water use in one-pot and multi-unit conventional IL-pretreatment technologies.

the IL. The economic feasibility of an IL-based pretreatment technology demands near complete recovery of the IL and by-products. To address this issue, a range of methods including extraction, evaporation, electrodialysis and pervaporation have been reported in the literature. In 2010, Brennan *et al.* employed the chemical affinity of boronates for sugars to extract 90–97% of the mono- and disaccharides from $[C_2C_1Im][OAc]$–biomass liquor following pretreatment. This technology provides an effective method to extract fermentable sugars from pretreated lignocellulosic biomass.[146] In the same year, the use of aqueous kosmotropic salt solutions to form a three-phase system that precipitates the biomass, forming IL-rich and salt-rich phases was described, *via* which a 95% IL recovery could be achieved.[147] An alternative technique is the use of membranes, which have been applied for the separation of IL from biomass following processing. For example, high separation selectivity can be achieved *via* pervaporation using perfluorinated hollow fibre membranes, which also demonstrate high chemical and thermal resistance, high water permeance and anti-fouling features. In comparison to costly distillation/chromatography techniques, pervaporation could be used as an efficient and simple method for the separation of biofuels and solvents in bioprocesses.[148,149] Additionally, membrane (micro-, ultra- and nano-) filtration, reverse/forward osmosis, biphasic extraction and electro-dialysis separation technologies could be potentially integrated into biorefinery concepts (Table 3.3). Since most of the membrane-based separation technologies employ simple IL–water mixture, future investigations may start with synthetic solute systems, and extend to hydrolysate or fermentation broths from real lignocelluloses. Meanwhile, energy consumption, operating and capital costs have to be carefully analysed and compared with the alternative distillation/chromatography techniques under a biorefinery scheme.

3.5 Conclusions

Lignocellulosic biomass is an abundant renewable feedstock with great geological diversity and availability. There are great opportunities for the production of various commodities such as biofuels, chemicals and biomaterials. The extraordinary potential of ILs in facilitating the fractionation and separation of biomass components in a biorefinery concept has been proven. The progress made in last few years has demonstrated that ILs can be successfully employed as pretreatment or bioprocessing media in a biorefinery to obtain cellulose, hemicellulose and lignin fractions from a variety of biomass feedstocks and mixtures with product purities and efficiencies equal or superior to the currently employed acid-based pretreatment method for the production of second-generation biofuels and biochemicals. The IL pretreatment chemistry largely depends on the selection of ILs, the processing conditions and recovery methods. It determines how the biomass is altered during the process and the pretreatment efficacy. IL pretreatment technology is evolving rapidly and this is augmented by the development of new ILs, process development, and nature of ILs being a designer solvent. Despite the

Table 3.3 Comparison of IL separation methods.[a]

Entry	IL system	Separation description	Recovery rate/%	Ref.
1	Hydrolysate/H_2O [C_2C_1Im][OAc]	(1) Boronate complexes (2) Extraction	90–97% of mono- and disaccharides	146
2	Lignin/[C_2C_1Im][OAc]	(1) Extraction: alkali lignin (200 mg), IL (2 mL), iso-propanol (40 mL), 25 °C; (2) Centrifugation; (3) UV analysis	N/A	177,180
3	Hydrolysate/H_2O/[C_2C_1Im][OAc]-ethanolamine (60/40 vol%)	(1) Anti-solvent precipitation: deionized water: acetone (1:1, v/v); (2) Evaporation, 80 °C	N/A	177
4	Hydrolysate/H_2O/[C_4C_1Im]Cl	(1) Anti-solvent precipitation: deionized water; (2) Electrodialysis: 500 mL with 6.90 g L^{-1} of [C_4C_1Im]Cl, electrical potential (10 V), room temperature	74.1	167
5	H_2O/[C_4C_1Im][OAc]	(1) Anti-solvent precipitation: deionized water:acetone (1:1, v/v); (2) Aqueous biphasic extraction, H_2O/IL/K_3PO_4, pH 12–13	95	147
6	Lignin/H_2O/ethanol/[C_4C_1Im]Cl/$CrCl_36H_2O$	(1) Anti-solvent precipitation: ethanol/water; (2) Vacuum evaporation: 60 °C, 24 h, vacuum oven	N/A	179

[a]N/A: Not available.

enormous potential and growth, there are certain challenges that have to be addressed in order to make IL pretreatment an economically viable technology in the biorefinery context. The next critical steps for biorefinery implementation are to:

(1) demonstrate the technology at larger scales and under industrially relevant conditions in order to understand any scale-up effect
(2) design new ILs that are cheap, biocompatible and renewable
(3) explore and develop cost effective ways for IL recycling and product recovery
(4) conduct techno-economic and life cycle analysis to evaluate and guide the improvement of this technology.

Acknowledgements

The work conducted by the Joint BioEnergy Institute is supported by the Office of Science, Office of Bio-logical and Environmental Research of the US Department of Energy under contract DE-AC02-05CH11231.

References

1. ASTM. *Standard E170 Standard technology relating to biotechnology (2002)*, http://www.astm.org, West Conshohocken, PA, 1995.
2. D. Sandor, R. Wallace and S. Peterson, *NREL TP-150-42120. National Renewable Energy Laboratory*, http://www.nrel.gov/docs/fy08osti/42120.pdf, 2008.
3. L. R. Brown, *Earth Policy Institute*, 2006.
4. M. J. Groom, E. M. Gray and P. A. Townsend, *Conserv. Biol.*, 2008, **22**, 602–609.
5. T. Searchinger, R. Heimlich, R. A. Houghton, F. Dong, A. Elobeid, J. Fabiosa, S. Tokgoz, D. Hayes and T.-H. Yu, *Science*, 2008, **319**, 1238–1240.
6. T. W. Simpson, A. N. Sharpley, R. W. Howarth, H. W. Paerl and K. R. Mankin, *J. Environ. Qual.*, 2008, **37**, 318–324.
7. Y. Sun and J. Y. Cheng, *Bioresour. Technol.*, 2002, **83**, 1–11.
8. D. Fengel and G. Wegener, *Wood: Chemistry, Ultrastructure, Reactions*, de Gruyter, New York, NY, USA, 1983.
9. E. Sjöström, *Wood Chemistry: Fundamentals and Applications*, Academic Press, San Diego, USA, 1993.
10. A. Mittal, R. Katahira, M. E. Himmel and D. K. Johnson, *Biotechnol. Biofuels*, 2011, **4**, 41–55.
11. B. Yang, Z. Dai, S.-Y. Ding and C. E. Wyman, *Biofuels*, 2011, **2**, 421–450.
12. E. Palmqvist and B. Hahn-Hagerdal, *Bioresour. Technol.*, 2000, **74**, 17–24.
13. P. Kumar, D. M. Barrett, M. J. Delwiche and P. Stroeve, *Ind. Eng. Chem. Res.*, 2009, **48**, 3713–3729.

14. M. E. Himmel, S. Y. Ding, D. K. Johnson, W. S. Adney, M. R. Nimlos, J. W. Brady and T. D. Foust, *Science*, 2007, **315**, 804–807.
15. S. A. Rydholm, *Pulping Processes*, Interscience Publishers, John Wiley & Sons, New York, USA, 1965.
16. H. F. Wenzel, *The Chemical Technology of Wood*, Academic Press, New York, USA, 1970.
17. T. A. Hsu, M. R. Ladisch and G. T. Tsao, *Chem. Technol.*, 1980, **10**, 315–319.
18. V. S. Chang and M. T. Holtzapple, *Appl. Biochem. Biotechnol.*, 2000, **84–86**, 5–37.
19. N. Mosier, C. Wyman, B. Dale, R. Elander, Y. Y. Lee, M. Holtzapple and M. Ladisch, *Bioresour. Technol.*, 2005, **96**, 673–686.
20. L. Laureano-Perez, F. Teymouri, H. Alizadeh and B. E. Dale, *Appl. Biochem. Biotechnol.*, 2005, **121**, 1081–1099.
21. V. P. Puri, *Biotechnol. Bioeng.*, 1984, **26**, 1219–1222.
22. P. Alvira, E. Tomas-Pejo, M. Ballesteros and M. J. Negro, *Bioresour. Technol.*, 2010, **101**, 4851–4861.
23. S. E. Jacobsen and C. E. Wyman, *Appl. Biochem. Biotechnol.*, 2000, **84–86**, 81–96.
24. T. A. Hsu, in *Handbook on Bioethanol, Production and Utilization*, ed. C. Wyman, Taylor and Francis, Washington DC (USA), 1996, ch. 10, pp. 179–212.
25. V. B. Agbor, N. Cicek, R. Sparling, A. Berlin and D. B. Levin, *Biotechnol. Adv.*, 2011, **29**, 675–685.
26. F. Carrillo, M. J. Lis, X. Colom, M. López-Mesas and J. Valldeperas, *Process Biochem.*, 2005, **40**, 3360–3364.
27. W. E. Kaar and M. T. Holtzapple, *Biomass Bioenergy*, 2000, **18**, 189–199.
28. T. H. Kim, J. S. Kim, C. Sunwoo and Y. Y. Lee, *Bioresour. Technol.*, 2003, **90**, 39–47.
29. B. C. Saha and M. A. Cotta, *Enzyme Microb. Technol.*, 2007, **41**, 528–532.
30. Y. Chen, M. A. Stevens, Y. Zhu, J. Holmes and H. Xu, *Biotechnol. Biofuels*, 2013, **6**, 1–10.
31. Y. Sun and J. Cheng, *Bioresour. Technol.*, 2002, **83**, 1–11.
32. B. L. Foster, B. E. Dale and J. B. Doran-Peterson, *Appl. Biochem. Biotechnol.*, 2001, **91–93**, 269–282.
33. M. T. Holtzapple, J.-H. Jun, G. Ashok, S. L. Patibandla and B. E. Dale, *Appl. Biochem. Biotechnol.*, 1991, **28–29**, 59–74.
34. E. Y. Vlasenko, H. Ding, J. M. Labavitch and S. P. Shoemaker, *Bioresour. Technol.*, 1997, **59**, 109–119.
35. P. Wasserscheid and T. Welton, *Ionic Liquids in Synthesis*, Wiley-VCH, Weinheim, Germany, 2003.
36. K. R. Seddon, *J. Chem. Technol. Biotechnol.*, 1997, **68**, 351–356.
37. J. G. Huddleston, A. E. Visser, W. M. Reichert, H. D. Willauer, G. A. Broker and R. D. Rogers, *Green Chem.*, 2001, **3**, 156–164.
38. S. Singh, B. A. Simmons and K. P. Vogel, *Biotechnol. Bioeng.*, 2009, **104**, 68–75.

39. C. Li, B. Knierim, C. Manisseri, R. Arora, H. V. Scheller, M. Auer, K. P. Vogel, B. A. Simmons and S. Singh, *Bioresour. Technol.*, 2010, **101**, 4900–4906.

40. S. Singh and B. A. Simmons, in *Aqueous Pretreatment of Plant Biomass for Biological and Chemical Conversion to Fuels and Chemicals*, John Wiley & Sons, Ltd, 2013, pp. 223–238.

41. J. Shi, V. S. Thompson, N. A. Yancey, V. Stavila, B. A. Simmons and S. Singh, *Biofuels*, 2013, **4**, 63–72.

42. P. Walden, *Bull. Acad. Imp. Sci. St.-Petersbourg*, 1914, 405–422.

43. F. H. Hurley and T. P. Wier, *J. Electrochem. Soc.*, 1951, **98**, 203–206.

44. J. S. Wilkes, *Green Chem.*, 2002, **4**, 73–80.

45. S. E. Fry and N. J. Pienta, *J. Am. Chem. Soc.*, 1985, **107**, 6399–6400.

46. J. A. Boon, J. A. Levisky, J. L. Pflug and J. S. Wilkes, *J. Org. Chem.*, 1986, **51**, 480–483.

47. R. Hagiwara, T. Nohira, K. Matsumoto and Y. Tamba, *Electrochem. Solid-State Lett.*, 2005, **8**, A231–A233.

48. N. Papageorgiou, Y. Athanassov, M. Armand, P. Bonhote, H. Pettersson, A. Azam and M. Grätzel, *J. Electrochem. Soc.*, 1996, **143**, 3099–3108.

49. A. B. McEwen, H. L. Ngo, K. LeCompte and J. L. Goldman, *J. Electrochem. Soc.*, 1999, **146**, 1687–1695.

50. H. Sakaebe and H. Matsumoto, *Electrochem. Commun.*, 2003, **5**, 594–598.

51. J. P. Hallett and T. Welton, *Chem. Rev.*, 2011, **111**, 3508–3576.

52. W. L. Hough, M. Smiglak, H. Rodriguez, R. P. Swatloski, S. K. Spear, D. T. Daly, J. Pernak, J. E. Grisel, R. D. Carliss, M. D. Soutullo, J. J. H. Davis and R. D. Rogers, *New J. Chem.*, 2007, **31**, 1429–1436.

53. R. P. Swatloski, S. K. Spear, J. D. Holbrey and R. D. Rogers, *J. Am. Chem. Soc.*, 2002, **124**, 4974–4975.

54. A. P. Dadi, S. Varanasi and C. A. Schall, *Biotechnol. Bioeng.*, 2006, **95**, 904–910.

55. A. P. Dadi, C. A. Schall and S. Varanasi, *Appl. Biochem. Biotechnol.*, 2007, **137–140**, 407–421.

56. H. Zhao, C. I. L. Jones, G. A. Baker, S. Xia, O. Olubajo and V. N. Person, *J. Biotechnol.*, 2009, **139**, 47–54.

57. G. Cheng, P. Varanasi, C. L. Li, H. B. Liu, Y. B. Menichenko, B. A. Simmons, M. S. Kent and S. Singh, *Biomacromolecules*, 2011, **12**, 933–941.

58. D. C. Dibble, C. L. Li, L. Sun, A. George, A. R. L. Cheng, O. P. Cetinkol, P. Benke, B. M. Holmes, S. Singh and B. A. Simmons, *Green Chem.*, 2011, **13**, 3255–3264.

59. N. Sun, M. Rahman, Y. Qin, M. L. Maxim, H. Rodriguez and R. D. Rogers, *Green Chem.*, 2009, **11**, 646–655.

60. R. Arora, C. Manisseri, C. L. Li, M. D. Ong, H. V. Scheller, K. Vogel, B. A. Simmons and S. Singh, *Bioenerg. Res.*, 2010, **3**, 134–145.

61. I. P. Samayam and C. A. Schall, *Bioresour. Technol.*, 2010, **101**, 3561–3566.

62. D. Klein-Marcuschamer, B. A. Simmons and H. W. Blanch, *Biofuels, Bioprod. Biorefin.*, 2011, **5**, 562–569.

63. H. Wu, M. Mora-Pale, J. Miao, T. V. Doherty, R. J. Linhardt and J. S. Dordick, *Biotechnol. Bioeng.*, 2011, **108**, 2865–2875.

64. A. Cruz, C. Scullin, C. Mu, G. Cheng, V. Stavila, P. Varanasi, D. Xu, J. Mentel, Y.-D. Chuang, B. Simmons and S. Singh, *Biotechnol. Biofuels*, 2013, **6**, 52.
65. J. Shi, J. M. Gladden, N. Sathitsuksanoh, P. Kambam, L. Sandoval, D. Mitra, S. Zhang, A. George, S. W. Singer, B. A. Simmons and S. Singh, *Green Chem.*, 2013, **15**, 2579–2589.
66. C. Li, L. Sun, B. Simmons and S. Singh, *Bioenerg. Res.*, 2013, **6**, 14–23.
67. X. Gao, R. Kumar, S. Singh, B. Simmons, V. Balan, B. Dale and C. Wyman, *Biotechnol. Biofuels*, 2014, **7**, 71.
68. C. Li, Q. Wang and Z. K. Zhao, *Green Chem.*, 2008, **10**, 177–182.
69. K. Ogura, K. Ninomiya, K. Takahashi, C. Ogino and A. Kondo, *Biotechnol. Biofuels*, 2014, 7(1–10), 120.
70. T. You, L. Zhang, S. Zhou and F. Xu, *Bioresour. Technol.*, 2014, **167**, 574–577.
71. N. Nasirpour, S. M. Mousavi and S. A. Shojaosadati, *Bioresour. Technol.*, 2014, **169**, 33–37.
72. X. D. Hou, T. J. Smith, N. Li and M. H. Zong, *Biotechnol. Bioeng.*, 2012, **109**, 2484–2493.
73. Q.-P. Liu, X.-D. Hou, N. Li and M.-H. Zong, *Green Chem.*, 2012, **14**, 304–307.
74. N. Sun, R. Parthasarathi, A. M. Socha, J. Shi, S. Zhang, V. Stavila, K. L. Sale, B. A. Simmons and S. Singh, *Green Chem.*, 2014, **16**, 2546–2557.
75. A. M. Socha, R. Parthasarathi, J. Shi, S. Pattathil, D. Whyte, M. Bergeron, A. George, K. Tran, V. Stavila, S. Venkatachalam, M. G. Hahn, B. A. Simmons and S. Singh, *Proc. Natl. Acad. Sci. U. S. A.*, 2014, **111**, E3587–E3595.
76. A. Pinkert, K. N. Marsh, S. Pang and M. P. Staiger, *Chem. Rev.*, 2009, **109**, 6712–6728.
77. M. E. Zakrzewska, E. Bogel-Lukasik and R. Bogel-Lukasik, *Energy Fuels*, 2010, **24**, 737–745.
78. R. C. Remsing, R. P. Swatloski, R. D. Rogers and G. Moyna, *Chem. Commun.*, 2006, 1271–1273.
79. T. G. A. Youngs, J. D. Holbrey, C. L. Mullan, S. E. Norman, M. C. Lagunas, C. D'Agostino, M. D. Mantle, L. F. Gladden, D. T. Bowron and C. Hardacre, *Chem. Sci.*, 2011, **2**, 1594–1605.
80. J. Zhang, H. Zhang, J. Wu, J. Zhang, J. He and J. Xiang, *Phys. Chem. Chem. Phys.*, 2010, **12**, 1941–1947.
81. B. Lu, A. Xu and J. Wang, *Green Chem.*, 2014, **16**, 1326–1335.
82. T. Erdmenger, C. Haensch, R. Hoogenboom and U. S. Schubert, *Macromol. Biosci.*, 2007, **7**, 440–445.
83. J. Vitz, T. Erdmenger, C. Haensch and U. S. Schubert, *Green Chem.*, 2009, **11**, 417–424.
84. H. Zhang, J. Wu, J. Zhang and J. S. He, *Macromolecules*, 2005, **38**, 8272–8277.
85. J.-P. Mikkola, A. Kirilin, J.-C. Tuuf, A. Pranovich, B. Holmbom, L. M. Kustov, D. Y. Murzin and T. Salmi, *Green Chem.*, 2007, **9**, 1229–1237.
86. T. G. A. Youngs, J. D. Holbrey, M. Deetlefs, M. Nieuwenhuyzen, M. F. Costa Gomes and C. Hardacre, *ChemPhysChem*, 2006, **7**, 2279–2281.

87. T. G. A. Youngs, C. Hardacre and J. D. Holbrey, *J. Phys. Chem. B*, 2007, **111**, 13765–13774.

88. H. Liu, K. L. Sale, B. M. Holmes, B. A. Simmons and S. Singh, *J. Phys. Chem. B*, 2010, **114**, 4293–4301.

89. H. Liu, K. L. Sale, B. A. Simmons and S. Singh, *J. Phys. Chem. B*, 2011, **115**, 10251–10258.

90. A. S. Gross, A. T. Bell and J.-W. Chu, *J. Phys. Chem. B*, 2011, **115**, 13433–13440.

91. A. S. Gross, A. T. Bell and J.-W. Chu, *Phys. Chem. Chem. Phys.*, 2012, **14**, 8425–8430.

92. H. Liu, G. Cheng, M. Kent, V. Stavila, B. A. Simmons, K. L. Sale and S. Singh, *J. Phys. Chem. B*, 2012, **116**, 8131–8138.

93. Y. Zhao, X. Liu, J. Wang and S. Zhang, *ChemPhysChem*, 2012, **13**, 3126–3133.

94. Y. Zhao, X. Liu, J. Wang and S. Zhang, *Carbohydr. Polym.*, 2013, **94**, 723–730.

95. B. D. Rabideau, A. Agarwal and A. E. Ismail, *J. Phys. Chem. B*, 2014, **118**, 1621–1629.

96. B. Mostofian, J. Smith and X. Cheng, *Cellulose*, 2014, **21**, 983–997.

97. C. Kennedy, G. Cameron, A. Šturcová, D. Apperley, C. Altaner, T. Wess and M. Jarvis, *Cellulose*, 2007, **14**, 235–246.

98. K. Leppänen, S. Andersson, M. Torkkeli, M. Knaapila, N. Kotelnikova and R. Serimaa, *Cellulose*, 2009, **16**, 999–1015.

99. S. P. S. Chundawat, G. Bellesia, N. Uppugundla, L. da Costa Sousa, D. Gao, A. M. Cheh, U. P. Agarwal, C. M. Bianchetti, G. N. Phillips, P. Langan, V. Balan, S. Gnanakaran and B. E. Dale, *J. Am. Chem. Soc.*, 2011, **133**, 11163–11174.

100. G. T. Beckham, J. F. Matthews, B. Peters, Y. J. Bomble, M. E. Himmel and M. F. Crowley, *J. Phys. Chem. B*, 2011, **115**, 4118–4127.

101. K. Igarashi, T. Uchihashi, A. Koivula, M. Wada, S. Kimura, T. Okamoto, M. Penttilä, T. Ando and M. Samejima, *Science*, 2011, **333**, 1279–1282.

102. G. Cheng, P. Varanasi, R. Arora, V. Stavila, B. A. Simmons, M. S. Kent and S. Singh, *J. Phys. Chem. B*, 2012, **116**, 10049–10054.

103. D. Glas, R. Paesen, D. Depuydt, K. Binnemans, M. Ameloot, D. E. De Vos and R. Ameloot, *ChemSusChem*, 2015, **8**, 82–86.

104. Q. Sun, M. Foston, X. Meng, D. Sawada, S. Pingali, H. O'Neill, H. Li, C. Wyman, P. Langan, A. Ragauskas and R. Kumar, *Biotechnol. Biofuels*, 2014, **7**, 150.

105. Y. Pu, F. Hu, F. Huang, B. Davison and A. Ragauskas, *Biotechnol. Biofuels*, 2013, **6**, 15.

106. Z. Liu, H. Wang, Z. Li, X. Lu, X. Zhang, S. Zhang and K. Zhou, *Mater. Chem. Phys.*, 2011, **128**, 220–227.

107. K. A. Le, R. Sescousse and T. Budtova, *Cellulose*, 2012, **19**, 45–54.

108. K. M. Gupta, Z. Hu and J. Jiang, *RSC Adv.*, 2013, **3**, 4425–4433.

109. K. M. Gupta, Z. Hu and J. Jiang, *RSC Adv.*, 2013, **3**, 12794–12801.

110. X. Geng and W. A. Henderson, *RSC Adv.*, 2014, **4**, 31226–31229.
111. X. Sun, Y. Chi and T. Mu, *Green Chem.*, 2014, **16**, 2736–2744.
112. C. Li and Z. K. Zhao, *Adv. Synth. Catal.*, 2007, **349**, 1847–1850.
113. K. Shilla, K. Millerb, D. S. Clarka and H. W. Blanch, *Bioresour. Technol.*, 2012, **126**, 290–297.
114. X. Cao, X. Peng, S. Sun, L. Zhong, S. Wang, F. Lu and R. Sun, *Carbohydr. Polym.*, 2014, **111**, 400–403.
115. J. I. Park, E. J. Steen, H. Burd, S. S. Evans, A. M. Redding-Johnson, T. Batth, P. I. Benke, P. D'haeseleer, N. Sun, K. L. Sale, J. D. Keasling, T. S. Lee, C. J. Petzold, A. Mukhopadhyay, S. W. Singer, B. A. Simmons and J. M. Gladden, *PLoS One*, 2012, 7(1–10), e37010.
116. G. Cheng, M. S. Kent, L. He, P. Varanasi, D. Dibble, R. Arora, K. Deng, K. Hong, Y. B. Melnichenko, B. A. Simmons and S. Singh, *Langmuir*, 2012, **28**, 11850–11857.
117. Y. Pu, N. Jiang and A. J. Ragauskas, *J. Wood Chem. Technol.*, 2007, **27**, 23–33.
118. W. Ji, Z. Ding, J. Liu, Q. Song, X. Xia, H. Gao, H. Wang and W. Gu, *Energy Fuels*, 2012, **26**, 6393–6403.
119. H. Lateef, S. Grimes, P. Kewcharoenwong and B. Feinberg, *J. Chem. Technol. Biotechnol.*, 2009, **84**, 1818–1827.
120. I. A. Kilpeläinen, H. Xie, A. King, M. Granstrom, S. Heikkinen and D. S. Argyropoulos, *J. Agric. Food Chem.*, 2007, **55**, 9142–9148.
121. B. G. Janesko, *Phys. Chem. Chem. Phys.*, 2011, **13**, 11393–11401.
122. M. M. Hossain and L. Aldous, *Aust. J. Chem.*, 2012, **65**, 1465–1477.
123. A. George, K. Tran, T. J. Morgan, P. I. Benke, C. Berrueco, E. Lorente, B. C. Wu, J. D. Keasling, B. A. Simmons and B. M. Holmes, *Green Chem.*, 2011, **13**, 3375–3385.
124. J.-L. Wen, T.-Q. Yuan, S.-L. Sun, F. Xu and R.-C. Sun, *Green Chem.*, 2014, **16**, 181–190.
125. P. Varanasi, P. Singh, M. Auer, P. Adams, B. Simmons and S. Singh, *Biotechnol. Biofuels*, 2013, **6**, 14.
126. N. Sathitsuksanoh, K. M. Holtman, D. J. Yelle, T. Morgan, V. Stavila, J. Pelton, H. Blanch, B. A. Simmons and A. George, *Green Chem.*, 2014, **16**, 1236–1247.
127. H. Wang, M. Tucker and Y. Ji, *J. Appl. Chem.*, 2013, **2013**, 1–9.
128. G. Chatel and R. D. Rogers, *ACS Sustainable Chem. Eng.*, 2013, **2**, 322–339.
129. L. Chen, M. Sharifzadeh, N. Mac Dowell, T. Welton, N. Shah and J. P. Hallett, *Green Chem.*, 2014, **16**, 3098–3106.
130. K. Fukumoto, M. Yoshizawa and H. Ohno, *J. Am. Chem. Soc.*, 2005, **127**, 2398–2399.
131. K. Fukumoto and H. Ohno, *Angew. Chem., Int. Ed. Engl.*, 2007, **46**, 1852–1855.
132. K. M. Docherty and C. F. Kulpa, *Green Chem.*, 2005, 7, 185–189.
133. T. P. Thuy Pham, C.-W. Cho and Y.-S. Yun, *Water Res.*, 2010, **44**, 352–372.
134. M. Petkovic, D. O. Hartmann, G. Adamova, K. R. Seddon, L. P. N. Rebelo and C. S. Pereira, *New J. Chem.*, 2012, **36**, 56–63.

135. G.-h. Tao, L. He, W.-s. Liu, L. Xu, W. Xiong, T. Wang and Y. Kou, *Green Chem.*, 2006, **8**, 639–646.
136. T. Mourão, L. C. Tomé, C. Florindo, L. P. N. Rebelo and I. M. Marrucho, *ACS Sustainable Chem. Eng.*, 2014, **2**, 2426–2434.
137. S. Shahriari, L. C. Tome, J. M. M. Araujo, L. P. N. Rebelo, J. A. P. Coutinho, I. M. Marrucho and M. G. Freire, *RSC Adv.*, 2013, **3**, 1835–1843.
138. X. D. Hou, Q. P. Liu, T. J. Smith, N. Li and M. H. Zong, *PLoS One*, 2013, **8**(1–7), e59145.
139. D. Fu and G. Mazza, *Bioresour. Technol.*, 2011, **102**, 7008–7011.
141. D. Fu and G. Mazza, *Bioresour. Technol.*, 2011, **102**, 8003–8010.
141. J. Shi, K. Balamurugan, R. Parthasarathi, N. Sathitsuksanoh, S. Zhang, V. Stavila, V. Subramanian, B. A. Simmons and S. Singh, *Green Chem.*, 2014, **16**, 3830–3840.
142. M. Ouellet, S. Datta, D. C. Dibble, P. R. Tamrakar, P. I. Benke, C. L. Li, S. Singh, K. L. Sale, P. D. Adams, J. D. Keasling, B. A. Simmons, B. M. Holmes and A. Mukhopadhyay, *Green Chem.*, 2011, **13**, 2743–2749.
143. F. Ganske and U. Bornscheuer, *Biotechnol. Lett.*, 2006, **28**, 465–469.
144. M. B. Turner, S. K. Spear, J. G. Huddleston, J. D. Holbrey and R. D. Rogers, *Green Chem.*, 2003, **5**, 443–447.
145. Y.-W. Chiu and M. Wu, *Environ. Sci. Technol.*, 2012, **46**, 9155–9162.
146. T. C. R. Brennan, S. Datta, H. W. Blanch, B. A. Simmons and B. M. Holmes, *Bioenerg. Res.*, 2010, **3**, 123–133.
147. K. Shill, S. Padmanabhan, Q. Xin, J. M. Prausnitz, D. S. Clark and H. W. Blanch, *Biotechnol. Bioeng.*, 2011, **108**, 511–520.
148. K. Bélafi-Bakó, N. Dörmő, O. Ulbert and L. Gubicza, *Desalination*, 2002, **149**, 267–268.
149. L. Gubicza, K. Belafi-Bako, E. Feher and T. Frater, *Green Chem.*, 2008, **10**, 1284–1287.
150. R. Torget, C. Hatzis, T. K. Hayward, T. A. Hsu and G. P. Philippidis, *Appl. Biochem. Biotechnol.*, 1996, **57–8**, 85–101.
151. M. Laser, D. Schulman, S. G. Allen, J. Lichwa, M. J. Antal Jr and L. R. Lynd, *Bioresour. Technol.*, 2002, **81**, 33–44.
152. J. A. Pérez, I. Ballesteros, M. Ballesteros, F. Sáez, M. J. Negro and P. Manzanares, *Fuel*, 2008, **87**, 3640–3647.
153. C. G. Liu and C. E. Wyman, *Ind. Eng. Chem. Res.*, 2003, **42**, 5409–5416.
154. T. H. Kim and Y. Y. Lee, *Bioresour. Technol.*, 2005, **96**, 2007–2013.
155. B. C. Saha, L. B. Iten, M. A. Cotta and Y. V. Wu, *Process Biochem.*, 2005, **40**, 3693–3700.
156. Y. Sun and J. J. Cheng, *Bioresour. Technol.*, 2005, **96**, 1599–1606.
157. S. Kim and M. T. Holtzapple, *Bioresour. Technol.*, 2005, **96**, 1994–2006.
158. Y.-G. Liang, B. Cheng, Y.-B. Si, D.-J. Cao, E. R. Nie, J. Tang, X.-H. Liu, Z. Zheng and X.-Z. Luo, *Biomass Bioenergy*, 2014, **71**, 106–112.
159. S. P. S. Chundawat, B. Venkatesh and B. E. Dale, *Biotechnol. Bioeng.*, 2007, **96**, 219–231.
160. J. Li, G. Henriksson and G. Gellerstedt, *Bioresour. Technol.*, 2007, **98**, 3061–3068.

161. S. Banerjee, R. Sen, R. A. Pandey, T. Chakrabarti, D. Satpute, B. S. Giri and S. Mudliar, *Biomass Bioenerg.*, 2009, **33**, 1680–1686.

162. X. Zhao, K. Cheng and D. Liu, *Appl. Microbiol. Biotechnol.*, 2009, **82**, 815–827.

163. R. D. O. de Barros, R. D. Paredes, T. Endo, E. P. D. Bon and S. H. Lee, *Bioresour. Technol.*, 2013, **136**, 288–294.

164. Q. Li, Y. C. He, M. Xian, G. Jun, X. Xu, J. M. Yang and L. Z. Li, *Bioresour. Technol.*, 2009, **100**, 3570–3575.

165. S. Datta, B. Holmes, J. I. Park, Z. W. Chen, D. C. Dibble, M. Hadi, H. W. Blanch, B. A. Simmons and R. Sapra, *Green Chem.*, 2010, **12**, 338–345.

166. T. A. D. Nguyen, K. R. Kim, S. J. Han, H. Y. Cho, J. W. Kim, S. M. Park, J. C. Park and S. J. Sim, *Bioresour. Technol.*, 2010, **101**, 7432–7438.

167. L. T. P. Trinh, Y. J. Lee, J. W. Lee, H. J. Bae and H. J. Lee, *Sep. Purif. Technol.*, 2013, **120**, 86–91.

168. A. Brandt, M. J. Ray, T. Q. To, D. J. Leak, R. J. Murphy and T. Welton, *Green Chem.*, 2011, **13**, 2489–2499.

169. C. L. Li, D. Tanjore, W. He, J. Wong, J. L. Gardner, K. L. Sale, B. A. Simmons and S. Singh, *Biotechnol. Biofuels*, 2013, **6**(1–13), 154.

170. L. Sun, C. L. Li, Z. J. Xue, B. A. Simmons and S. Singh, *RSC Adv.*, 2013, **3**, 2017–2027.

171. D. Groff, A. George, N. Sun, N. Sathitsuksanoh, G. Bokinsky, B. A. Simmons, B. M. Holmes and J. D. Keasling, *Green Chem.*, 2013, **15**, 1264–1267.

172. J. G. Lynam and C. J. Coronella, *Bioresour. Technol.*, 2014, **166**, 471–478.

173. R. Yanez, B. Gomez, M. Martinez, B. Gullon and J. L. Alonso, *J. Chem. Technol. Biotechnol.*, 2014, **89**, 1337–1343.

174. C. Y. Yang and T. J. Fang, *Bioresour. Technol.*, 2014, **164**, 198–202.

175. F. Cheng, H. Wang, G. Chatel, G. Gurau and R. D. Rogers, *Bioresour. Technol.*, 2014, **164**, 394–401.

176. J. Bian, F. Peng, X. P. Peng, X. Xiao, P. Peng, F. Xu and R. C. Sun, *Carbohydr. Polym.*, 2014, **100**, 211–217.

177. P. Weerachanchai and J. M. Lee, *Bioresour. Technol.*, 2014, **169**, 336–343.

178. M. J. Dougherty, H. M. Tran, V. Stavila, B. Knierim, A. George, M. Auer, P. D. Adams and M. Z. Hadi, *PLoS One*, 2014, **9**(6), e100836.

179. H. M. Yu, J. Hu, J. Fan and J. Chang, *Ind. Eng. Chem. Res.*, 2012, **51**, 3452–3457.

180. P. Weerachanchai, K. H. Lim and J.-M. Lee, *Bioresour. Technol.*, 2014, **156**, 404–407.

Biomass Hydrolysis in Ionic Liquids

OMAR MERINO PÉREZ[a], JORGE ABURTO ANELL[a], AND RAFAEL MARTÍNEZ-PALOU*[a]

[a]Dirección de Investigación y Posgrado, Instituto Mexicano del Petróleo, Eje Central Lázaro Cárdenas 152, 07730 México D.F., México
*E-mail: rpalou@imp.mx

4.1 Introduction

Currently, one of the major challenges for the scientific community is the development of renewable energy sources as an alternative to the poor, non-renewable and polluting fossil fuels. The current energy crisis and global warming caused by CO_2 emissions from the combustion of fossil fuels, require the urgent development of sustainable processes based on renewable substrates.[1]

The only sustainable source of organic carbon is biomass from plants, which contain a high content of cellulose. Cellulose, as a renewable polymer, is the most abundant natural and renewable polymer in our environment and is called to be the substrate of choice for the development of biorefineries. In the agriculture and forestry sectors, the widespread availability and biodiversity of residual feedstocks make possible the production of biofuels, biomaterials and chemicals without affecting food production.[2]

Biofuels derived from plant biomass include bio-alcohols (ethanol, butanol, *etc.*), biodiesel, bio-oils and biogas,[3] and other value-added chemicals

RSC Green Chemistry No. 36
Ionic Liquids in the Biorefinery Concept: Challenges and Perspectives
Edited by Rafal Bogel-Lukasik

Published by the Royal Society of Chemistry, www.rsc.org

such as 5-(hydroxymethyl)-2-furaldehyde (HMF), 2,5-furan dicarboxylic acid, 3-hydroxypropanoic acid, glutamic acid, aspartic acid, fumaric acid, succinic acid, levulinic acid, 3-hydroxybutirolactone, 3-hydroxybutirolactone, sorbitol, xylitol, *etc.*[4]

Conversion of lignocellulosic biomass (especially agricultural wastes) into fermentable sugars is a viable approach for the production of renewable fuels. Products from plant biomass, including biofuels such as bio-alcohols (ethanol and butanol), biodiesel, bio-oils and biogas, biomaterials and chemicals, are a 'green' alternative to petroleum-based products.[5]

A biorefinery is a facility that integrates biomass conversion processes and equipment to produce power, fuels and other high value-added chemicals. The main objective of a biorefinery is the efficient production of high volumes of liquid fuels, biomaterials and chemicals in a technically and economically viable way.[6]

In this context, ionic liquids (ILs) have begun playing a very important role in several steps as biomass pre-treatment for the purification of final products.[7,8] The primary conversion of lignocelluloses is generally conducted in water, but ILs are being considered seriously as an alternative reaction media. This chapter focuses on the applications and challenges of ILs as a means to carry out biomass hydrolysis.

4.2 Biomass

Biomass is any organic matter of vegetable or animal origin, including materials from natural and/or artificial (this being a renewable resource) transformation, which should be properly exploited to produce either fuels or materials or chemicals. Biomass includes forest and mill residues, agricultural crops and wastes, wood and wood wastes, animal wastes, livestock operation residues, aquatic plants, fast growing trees, and municipal and industrial wastes. The exploitation of biomass generates a lot of waste and residues that can be used for producing bioenergy and value-added products without affecting the conventional use of the resource.[9]

Lignocellulosic biomass is the key for supplying the sustainable production of chemicals and fuels without impacting the human food supply. Lignocellulosic biomass containing polymers of fermentable hexoses and pentoses is a renewable and sustainable raw material for the production of biofuels and chemicals, as depicted in Figure 4.1.

The biochemical conversion of cellulosic biomass into ethanol fuel involves three basic steps:

  (i) pre-treatment, which increases the access to cellulose enzymes and solubilizes hemicellulose sugar
 (ii) hydrolysis, which features a special enzyme preparation for breaking down cellulose and carrying out its transformation into sugars
(iii) fermentation for diverse products formation, *e.g.* ethanol

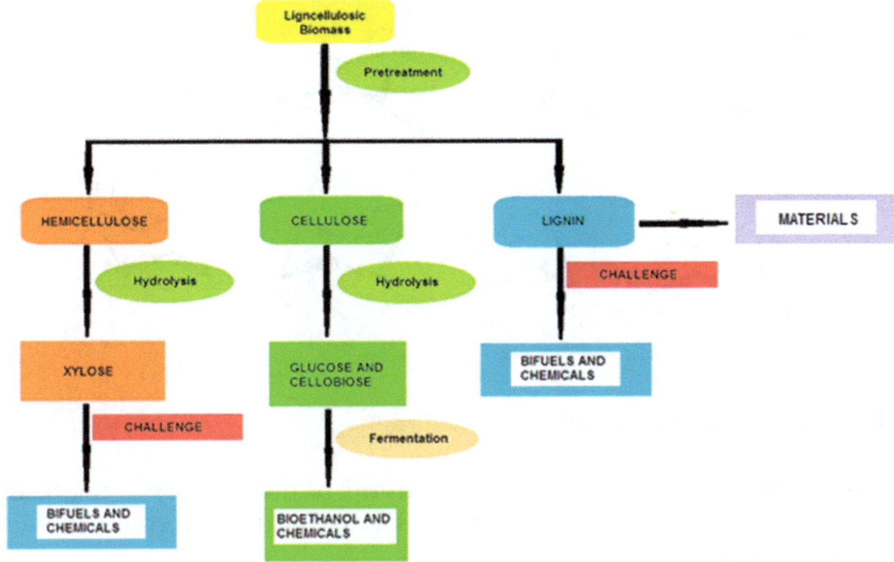

Figure 4.1   Biomass processing into fuels, materials, and chemicals.

In order to make the complete conversion of cellulosic biomass into ethanol more economical and practical, molecular science bases are required. By using numerous enzymes, new biochemical pathways, cellular systems and catalysts can be developed.[10]

Various research institutes around the world have focused on the research and development of the production of biomass-based biofuels, where the main challenges regarding the production of lignocellulosic biomass are the pre-treatment and hydrolysis of the raw material. Higher value-added chemicals such as aromatic compounds as well as fuels are potentially obtainable from lignin, but selective conversion of this recalcitrant polymer remains a challenge.[11] The efficient conversion of xylose obtained from hemicellulose, and lignin transformation into biofuels and chemicals are possible tasks, although major challenges are still unresolved. In the same sense, a number of technical and scientific issues within pre-treatment and hydrolysis remain to be solved. The degradation compounds formed during the valorization of lignocellulosic biomass are processed to yield ethanol or other biorefinery products such as furans, phenolics, organic acids as well as mono- and oligomeric pentoses and hexoses. Depending on the reaction conditions, glucose can be converted into HMF and/or levulinic acid and formic acid.

4.2.1   Biomass Composition

Lignocellulosic materials are composed predominantly of two carbohydrate polymers (cellulose and hemicellulose), of lignin, and to a lesser extent of other compounds (proteins, acids, salts and minerals). Cellulose and

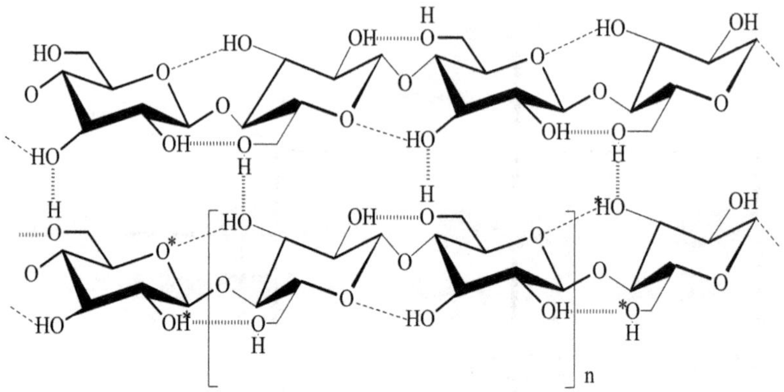

Figure 4.2   Structure of cellulose.

hemicellulose typically account for two thirds of the dry mass, and, being polysaccharides, they can be hydrolysed into sugars and eventually fermented to produce ethanol.

Cellulose makes up 40–60% of the dry biomass, although its content may vary significantly from one plant species to another.[12] Its linear cellobiose polymer (artificially is a double molecule of glucose), the orientation of the chains and the additional hydrogen bonding make this a rigid and difficult to break polymer. During hydrolysis, the polysaccharide is split to release the sugar molecules with the addition of water. This stage is also called saccharification. The product (glucose, a sugar) consists of six carbon atoms (Figure 4.2).

Cellulose in plant biomass is the most important synthesis compound of biofuels and is a fascinating building block for the design of new biomaterials.[13] Hemicellulose (20–40%) consists of polymers of five-carbon sugars (principally xylose) and additionally contains arabinose (five carbon atoms) and galactose, glucose and mannose (all containing six carbon atoms). Due to its amorphous nature, it is relatively easy to hydrolyse. Therefore, hemicellulose is a polysaccharide comprising more than one monomer type, formed by a heterogeneous group of polysaccharides, which in turn are formed by a single type of monosaccharides joined by β-(1–4) bonds, which form a branched linear chain. Among these monosaccharides, glucose, galactose or fructose are found, as shown in Figure 4.3.

Lignin is a complex polymer of phenol derivatives linked by ether-type and C–C bonds. Lignin is a molecule with high molecular weight, which results from the union of various acids, alcohols and phenols. The coupling of these radicals originates a tri-dimensional and amorphous polymer with high heterogeneity, illustrated in Figure 4.4.[14] These complex polymers constitute about 10–25% of the lignocellulosic material, therefore any ethanol production process generates lignin as a residue. Only few organisms are able to degrade this structure to obtain commercially valuable products such as phenols and organic acids.

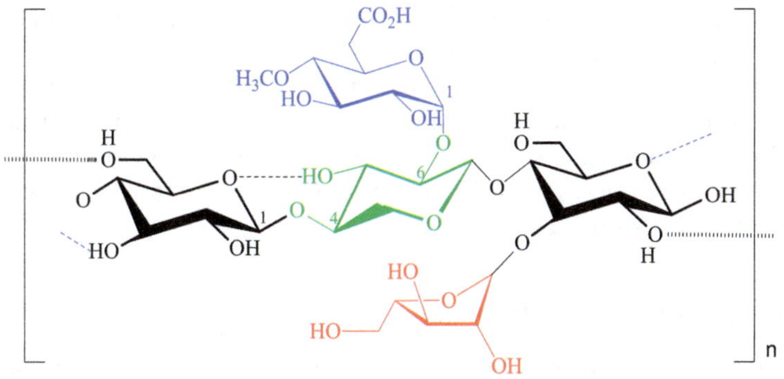

Figure 4.3 Structure of hemicellulose showing L-arabinose (red), D-xylose (green), D-glucose (black) and D-galactose (blue).

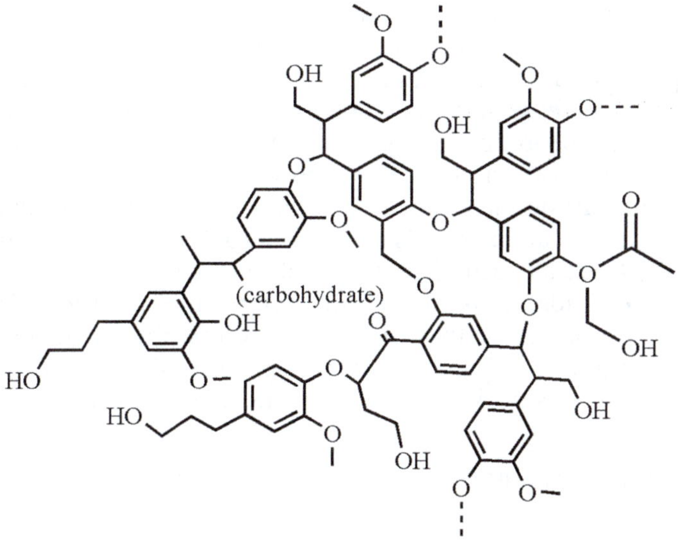

Figure 4.4 Example of lignin structure.

The combination of hemicellulose and lignin creates a protective barrier around the cellulose. This barrier should be removed to make the hydrolysis process efficient since the crystalline structure of cellulose makes it very insoluble, and the polymer in these conditions is very difficult to degrade.

4.2.2 Biomass Processing for Producing Biofuels

As shown in Figure 4.1, biomass processing for producing biofuels involves three main steps: biomass pre-treatment, hydrolysis and fermentation. These steps will be discussed below, with special emphasis on biomass hydrolysis, which is the object of this chapter.[15]

Bioethanol can be produced from three types of raw materials: lignocellulose (from wood, agricultural residues, waste sulfite liquor from pulp and paper mills), sugars (from sugar cane, fruit wastes and sugar beet), and starch (from corn, potatoes and root crops).[16] The production of bioethanol from lignocellulosic materials is one of the most common used methods due to the high diversity and availability of the raw materials and wastes, although more production steps are required.

4.2.2.1   Biomass Pre-Treatment

An efficient lignocellulosic biomass pre-treatment is a crucial step for the valorization of these kinds of raw materials.[17] Pre-treatment aims at the reduction of the cellulose crystallinity, dissociation of the cellulose–lignin complex, increment of the material surface area and reduction of the presence of substances that hinder hydrolysis. An effective pre-treatment must meet other features, such as: low energy consumption, low investment costs using inexpensive and readily recoverable reagents, and the possibility of being applicable to various substrates. Pre-treatments before the hydrolysis stage allow an increase in the rate of ethanol production as they help access to a greater proportion of sugars from cellulose, reaching 30% of ethanol yield. By performing a pre-treatment process, hydrolysis inhibitors present in the hemicellulosic material can also be removed.[18] Unfortunately, most pre-treatments feature some disadvantages such as the formation of compounds which act as natural biomass inhibitors and the possible generation of toxic degradation products, which in turn inhibit the hydrolysis and fermentation processes.[19] In more general terms, the technologies for the pre-treatment of lignocellulosic materials are classified as biological, physical (including mechanical), thermochemical and chemical.[20]

4.2.2.2   Biomass Hydrolysis

The process to convert biomass polymers into fermentable sugar is called hydrolysis. Hydrolysis aims at breaking down cellulose into sugar components using different enzyme preparations or acids.[21]

4.2.2.2.1   Enzymatic Hydrolysis of Biomass. Enzymatic hydrolysis is a process catalysed by a group of enzymes called cellulases, which feature a mixture of different enzymatic activities whose combined actions lead to the degradation of cellulose. The released saccharide compounds are the feedstock for fermentation.[22] Enzymes are naturally occurring proteins that act as catalysts with high specificity. Enzymatic hydrolysis of lignocellulose has been studied as a method to depolymerize biomass into fermentable sugars for further conversion to biofuels and biochemical products.[23] The use of enzymes in the hydrolysis of cellulose is more advantageous than the use of chemical catalysts because enzymes are highly specific and can work under mild conditions; however, the enzymatic process is hindered by the high cost of the commercially available enzymes meant for this purpose. When hydrolysis is performed at high-solid loadings (≥15% of solids, w/w), enzymatic hydrolysis potentially offers lower processing costs because sugar and ethanol concentrations are increased.[24] Commercialization of the enzymatic process is hindered because the costs of the currently available enzyme isolation and purification are very high.[25]

4.2.2.2.2 Acid Hydrolysis of Biomass. Acid hydrolysis is a chemical process which, by using acid catalysts, transforms the polysaccharide chains, which form biomass (hemicellulose and cellulose), into their basic monomers. Such hydrolysis can be performed by using various kinds of acids such as sulfurous, hydrochloric, sulfuric, phosphoric, nitric and formic acids. Industrial processes for acid hydrolysis can be grouped into two types: those that employ concentrated acids and those that use dilute acids. With hydrolysis processes that involve the use of strong acids, and which operate at low temperatures, high yields can be obtained (above 90% of the potential glucose). Despite this, the large amount of acid used in the impregnation of the material to be treated and the high cost of its recovery, in addition to the concomitant corrosive effects of strong acids, which requires high investment in equipment, make the process unprofitable.[26]

4.2.2.3 Ionic Liquids in Biomass Processing

Ionic liquids (ILs) are gaining wide recognition as potential environmental solvents due to their unique properties and applications in organic synthesis,[27–29] catalysis,[30] biocatalysis,[31,32] synthesis of nanomaterials,[33] synthesis of polymers,[34] and in industry.[35–38] ILs are a class of salts which have a structure featuring an organic cation and an anion that can be inorganic (Cl^-, Br^-, BF_4^-, PF_6^- and metal salts) or organic (AcO^-, BzO^-, CF_3COO^-, Tf^-, NTf_2^-) (Figure 4.5).

The application of ILs has opened new opportunities for the efficient utilization of lignocellulosic materials in areas such as fractionation, preparation of cellulose composites and derivatives, analysis, distillation, and purification and removal of pollutants. As seen through this book, ILs are now also widely employed in biomass dissolution,[39–43] pre-treatment[44–47] and obviously in biomass hydrolysis,[48] as is discussed in the following sections.

4.3 Biomass Hydrolysis in Ionic Liquids

Hydrolysis of cellulose to fermentable sugars is an essential step in any practical cellulosic-ethanol process before the microbial action to produce ethanol. Biocatalytic transformations in ILs have been performed using a range of different enzymes and some whole cell preparations, primarily in biphasic aqueous systems using hydrophobic dialkylimidazolium ILs. The results are encouraging, with activity levels generally equalling or surpassing the best molecular organic solvent/water alternative for a number of commercially

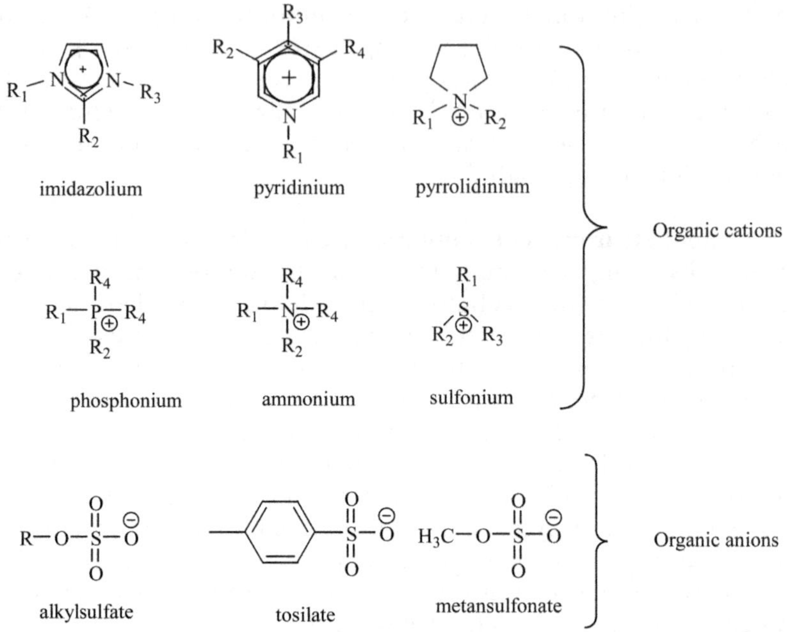

Figure 4.5 Typical organic cations and anions of ILs.

useful enzymes, including lipases and lyases. Increasing interest is also being developed by the potential utility of ILs as sole solvents for enzyme-catalysed reactions, making use of their exceptional solvation capacities and frequent high degree of biocompatibility.[49]

4.3.1 Acid Hydrolysis of Biomass in Ionic Liquids

An interesting approach to sugar production by using ILs is the application of acid catalysts to produce sugars and other compounds *in situ* through the hydrolysis of polysaccharides as a more economical alternative to enzymatic hydrolysis.[50] Li and Zhao reported for the first time in 2007, the application of an IL/acid catalyst for direct hydrolysis of lignocellulosic biomass without pre-treatment. In this work, hydrolysis was carried out by adding catalytic amounts of H_2SO_4 to cellulose dissolved in [C_4MIM][Cl]. A H_2SO_4/cellulose mass ratio of 0.92 produces total reducing sugars (TRS) and glucose in 59 and 36% yields, respectively, within 3 min. Further reduction of the acid/cellulose mass ratio to 0.46 produced higher yields after 42 min, and when the mass ratio dropped to 0.11, the yields of TRS and glucose reached 77 and 43%, respectively, in 9 h.[51] From this discovery, an avalanche of papers related to this strategy have been published, yielding a rapid progress in the development of this methodology applied to both cellulose and lignocellulosic materials, as summarized in Table 4.1.

Interestingly, in 2010 the hydrolysis of cellulose dissolved in [C_2MIM][Cl] was reported in high total reducing sugar yield (up to 97%) under relatively

Table 4.1 The acid-assisted hydrolysis of biomass in ILs.

IL	Catalyst	Biomass	Results	Ref.
$[C_4MIM][Cl]$	Macroreticulated styrene-divynyl resins functionalized with sulfonic groups (Amberlyst 15, 35 and 70), Nafion, sulfated-ZrO_3, zeolite Y, SiO_2–Al_2O_3 and ZSM-S	Cellulose, microcrystalline cellulose wood (spruce)	The acid resins with relatively large pore Amberlyst 15 and 35 were the solid catalyst with the best performance for depolymerization of α-cellulose. The reactions was very selective in Amberlyst 15, in the first step, cellooligomers are formed, which were subsequently broken down into sugars	52
$[C_4MIM][Cl]$, $[C_6MIM][Cl]$, $[C_4MIM][Br]$, $[AMIM][Cl]$, $[C_4MIM][HSO_4]$, $[SbMIM][HSO_4]$	HCl, H_2SO_4, H_3PO_4, Maleic acid	Corn stalk, rice straw, pine wood and bagasse	$[C_4MIM]Cl/HCl$ was the best combination. TRS yields were up to 66%, 74%, 81% and 68% for hydrolysis in the presence of 7 wt.% HCl at 100 °C under atmospheric pressure within 60 min. Different sets of ILs and acids afforded similar results albeit longer reaction time were needed	53
$[C_4MIM][Cl]$	H-form zeolites (HFZ) with a lower Si/Al molar ratio and a larger surface and sulfated ion-exchanging resin NKC-9.	Cellulose	HZK was better catalyst than NKC-9. MW-assisted IL hydrolysis accelerated the process. A typical hydrolysis reaction with Avicel cellulose produced glucose with yield close to 37% within 8 min	54
$[C_2MIM][Br]$	Maleic acid, ethanodioic acid, sulfamic acid, 1,1,1,-trifluoroethanoic acid, methanesulfonic acid and H_2SO_4	Cellobiose, cellulose, hemicellulose and lignocellulosic biomass (*Miscanthus* grass)	The results showed that the rate of the two competing reactions, polysaccharide hydrolysis and sugar decomposition, varies with acid strength, and that for acids with an aqueous pK_a thus allowing hydrolysis to be performed with a high selectivity in glucose. The reaction requires very mild conditions in comparison to the same reaction with aqueous acids and when the acid strength is lower than $pK_a = 0.5$, hydrolysis yielding monosaccharides is favoured with respect to the competing decomposition of glucose	55

(*continued*)

Table 4.1 (*continued*)

IL	Catalyst	Biomass	Results	Ref.
[C$_4$MIM][Br]	CF$_3$COOH	Loblolly pine wood and cellulose	Almost the entire carbohydrate fraction of the starting material can be converted into water-soluble products. Unlike in aqueous-phase reactions, the presence of the lignin matrix did not hinder the hydrolysis process. At the same time, most of the lignin fraction remained as a solid residue, so almost complete sugar–lignin fractionation was achieved. Nearly complete conversion of the carbohydrate fraction into water-soluble products was readily observed at 120 °C	56
[C$_2$MIM][Cl]	HCl	Corn stover	A high-yielding chemical process for the hydrolysis of biomass into monosaccharides is reported. Adding water gradually to a chloride ionic liquid-containing catalytic acid led to a nearly 90% yield of glucose from cellulose and 70–80% yield of sugars from untreated corn stover. Ion-exclusion chromatography allowed recovery of the ionic liquid and delivered sugar feedstocks that support the vigorous growth of ethanologenic microbes.	57
[AMIM][Cl] – 1-allyl-3-methyl-imidazolium chloride	HCl	*Eucalyptus grandis*, southern pine and Norway spruce thermo-mechanical pulp (N. spruce TMP)	The acidic pre-treatment of these wood species in IL resulted nearly complete hydrolysis of cellulose and hemicelluloses and in a significant amount of lignin degradation. Aqueous reactions (under identical acid concentrations) showed a remarkably lower efficiency, demonstrating that ILs offer a unique environment for the acid-catalysed dehydration chemistry	58
[C$_2$MIM][Cl]	HCl	*Miscanthus*	It was determined that while there is a small co-inhibition effect associated with the simultaneous hydrolysis of the cellulosic and hemicellulosic portions of *Miscanthus*, the largest rate decreases were observed for the hydrolysis of the hemicellulosic portion	59

[C$_4$MIM][Cl]	A sulfonated carbon material was prepared as catalyst by incomplete hydrothermal carbonization of glucose followed by sulfonation containing SO$_3$H, COOH, and phenolic OH groups	Cellulose	The glucose-derivative exhibited high catalytic performance for the hydrolysis of cellulose and was used during 5 cycles without loss on activity. TRS yield of 72.7% was obtained in the IL at 110 °C in 240 min reaction time	60
[C$_2$MIM][Cl], [C$_2$MIM][Br], [EPy][Br]-*N*-ethylpyridinium bromide, [BPy][Br]-*N*-butylpyridinium bromide, [N$_{2,2,2,2}$][Cl]-tetraethylammonium chloride	HCl	*Chlorella* biomass (algal biomass)	This strategy showed the great potential to produce fermentable sugars from algal biomass. After 3 h of dissolution in [EMIM]Cl and then 3 h of hydrolysis in 7 wt.% HCl at 105 °C, 75% of *Chlorella* biomass was dissolved, with nearly 90% of total sugar releasing from *Chlorella* biomass. [EMIM]Cl and sugars in the hydrolysate could be recovered by using ion-exclusion chromatography, with the recovery of 94% of glucose and 87% of xylose and arabinose	61
[C$_4$MIM][Cl]	HCl	Cellulose	Selection of the ideal ionic liquid ([EMIM]Cl), optimisation the dissolution time and temperature (15 min, 408 K) and the reaction procedure (addition of water during the reaction time) led to a very high selectivity to glucose + cellobiose (99.6%)	62
[C$_4$MIM][Br]	Sulfonated poly(styrene-co-divinylbenzene) (SPS–DVB)	Cellulose	BMIMBr and sulfate polymer catalyst can be recovered and reused without further treatment. The results revealed that there may be an ion-exchange process between the acidic sites of sulfonated catalyst and IL during the hydrolysis of cellulose	63
[C$_4$MIM][Br]	pH adjustment with aqueous HCl (10%, v/v)	Corn stalk	A selective two-stage hydrolysis was developed through pH adjustment in the IL. In the first stage, the matrix of corn stalk was disrupted and hydrolysed in the IL at pH 4.5 and 90 °C to obtain xylose with 23.1% yield. In the second stage, cellulose-rich materials in solid residues were further hydrolysed in the IL at pH 2–3 and 90 °C to produce glucose with 26.9% yield, and pure lignin was also obtained	64

mild conditions (≤140 °C, 1 atm) after 3 hours without using either mineral acid or acid IL. At 140 °C longer reaction times resulted in sharply reduced TRS yields due to degradation of the resulting reducing sugars, while at 120 °C high content of reducing sugars are obtained after 5 h that do not suffer decomposition during 24 hours. In the same work, HMF was reported to be obtained with high conversion (up to 89%) when $CrCl_2$ was added to the IL. Other ILs were surveyed for the hydrolysis of cellulose under the same set of conditions (120 °C, 24 h and 1 equiv. of H_2O). Thus, switching the IL from [EMIM][Cl] to [C₄MIM][Cl] with 1 or 4 equiv. of H_2O resulted in a considerable drop in the TRS yield (52% and 58%, respectively). Apparently, the more acidic the [RMIM][Cl]–H_2O mixture (pH = 4.37 for R = *n*-Bu *vs.* 5.12 for R = Et, 1:1 wt. ratio, RT), the faster the sugar degradation, as depicted in Figure 4.6.[65]

In 2013, Sun *et al.* described a novel process that uses the phase separation behaviour of imidazolium ILs/alkali/water solutions in tandem with acid catalysed hydrolysis to extract the sugars liberated from switchgrass from the aqueous IL solutions. The process is based in an aqueous biphasic system (ABS) in the presence of concentrated kosmotropic salts (K_3PO_4, K_2HPO_4, K_2CO_3, KOH, NaOH or Na_2HPO_4) first reported by Rogers *et al.*[39] At a certain concentration of kosmotropic salts, an aqueous phase containing chaotropic IL can phase separate with the salt phase. The amount of sugar produced from this biphasic system was proportional to the extent of biomass dissolved. Pre-treatment at high temperatures (*e.g.* 160 °C, 1.5 h) was more effective in producing glucose. Sugar extraction into the alkali phase

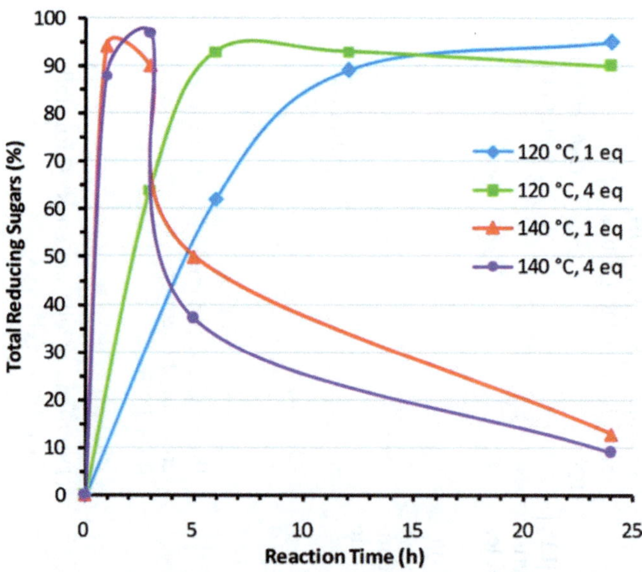

Figure 4.6 Plot of TRS at different times at 120 °C and 140 °C and 1 and 4 equiv. of water. (Reprinted with permission from ref. 65. Copyright 2014 American Chemical Society.)

was dependent on both the amount of sugar produced by acidolysis and the alkali concentration in the aqueous extracting phase. Maximum yields of 53% glucose and 88% xylose were recovered in the alkali phase, based on the amounts present in the initial biomass.[66] Recently, Qian *et al.* have proposed a novel polymeric catalyst based on poly(styrene sulfonic acid). Polymer chains were synthesized *via* surface initiated atom-transfer radical polymerization, growing from the substrate surface and used to catalyse biomass hydrolysis. Neighbouring poly(vinyl imidazolium chloride) were synthesized *via* UV-initiated free radical polymerization also grown from the substrate surface, and help dissolve lignocellulosic biomass and enhance the catalytic activity of the PSSA chains. These catalysts were used for the hydrolysis of cellulose in [C_2MIM][Cl] and aqueous solutions with yields of 97% and 32% of TRS.[67]

4.3.2 Hydrolysis of Biomass in Acid Ionic Liquids

The use of ILs in conjunction with mineral acids has been a good alternative for the hydrolysis of biomass without pre-treatment and preventing enzymatic hydrolysis which is so far the biggest economic challenge for profitability of biofuel production process; however, the use of mineral acids still presents difficulties associated with their high corrosivity, separation from fermentable sugars and difficulties that may represent the recovery of ILs.[68]

The application of Brønsted acid ILs, which act as both the solvent and catalyst to dissolve and to hydrolyse cellulose, is a smart and cheap strategy that avoids both the use of mineral acids and the enzymatic saccharification. The acid strength can be adjusted by changing their structure, such as the functional groups, anions or cations, and no neutralization and separation of the acid catalyst are required. Additionally, a higher concentration of acidic active sites (*i.e.* $-SO_3H$ and $-HSO_4^-$) can accelerate the reaction at more mild conditions with the corresponding energy saving. Functionalized ILs can also be easily immobilized on solid carriers.[69] Amarasekara and Owereh were the first to envision that incorporation of the acidic function into the IL would yield a more efficient process. Thus, the application of Brønsted acidic ILs in the hydrolysis of cellulose by 1-(1-propylsulfonic)-3-methylimidazolium chloride was reported by these researchers in 2009. With this procedure high yield of TRS (62%) was obtained in 1 hour of preheating at 70 °C and 30 min of heating at 70 °C after adding water. Three Brønsted acid ILs based on methyl imidazolium (**1a,b**), pyridinium (**2**) and triethanolammonium (**3**) (Figure 4.7), were evaluated for this purpose under mild reaction temperatures, with compound **1a** being the most efficient to hydrolyse Sigmacell cellulose.[70]

From these results, other acid ILs have been evaluated for their ability to dissolve and hydrolyse different cellulosic materials, examples are [C_4SO_3HMIM][HSO_4], [C_1COOHMIM][Cl], [BMIM][HSO_4], low cost 1-*H*-3-methylimidazolium chloride ([HMIM][Cl]) with TRS yields of 53.27 mg from 0.2 g of soybean straw and 50.03 mg from 0.2 g of corn straw,[71]

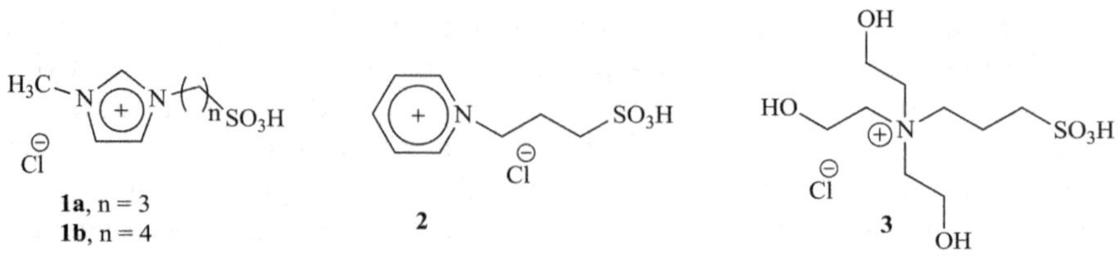

Figure 4.7   Brønsted acidic ILs employed in the hydrolysis of cellulose.

1-(4-sulfobutyl)-3-methylimidazolium chloride [SBMIM][Cl] for dissolution and hydrolysis of fibre sludge,[72] SO$_3$H-functional ILs with HSO$_4^-$ as anion for the hydrolysis of soybean isoflavone glycosides with the conversion of glycitin more than 90%,[73] [C$_3$SO$_3$HMIM][Cl] for cellulose degradation under moderate temperature and pressure,[74] and for the hydrolysis of cellobiose,[75] as well as an immobilized functionalized acidic IL modified silica catalyst was reported for the hydrolysis of cellulose.[76] Very recently, interesting work was carried out using starch-based industrial waste (potato starch) as the lignocellulosic residue and [C$_4$SO$_3$HMIM][Cl] as the catalyst under conventional heating, MW and low-frequency ultrasound. The depolymerization under microwave irradiation offered the highest TRS content within 60 min regardless of the starting material to reduce the reaction time by reaching the required temperature in a short time period. In the case of ultrasound at 80 °C, a parasite phenomenon called 'vaporous cavitation' appears and dramatically decreases the efficiency of acoustic cavitation. [AMIM][Cl] was more suitable for the dissolution of potato starch; however, [SBMIM][Cl] dissolved potato starch and depolymerized the starting materials into reducing sugars in one step in an aqueous system, playing the role of dual solvent/catalyst. Temperature was a relevant factor for the depolymerization of starch in conventional heating. The yield of reducing sugars under the optimum conditions (conventional heating in aqueous [SBMIM][Cl] – 33% (w/w) of H$_2$O, a solution of 20 wt.%, 120 min of stirring at 80 °C) reached 43% for a complex wet matrix–wet potato sludge.[77]

4.3.3   Biomass Hydrolysis Toward Furfural Derivatives

HMF is one of the main products of the degradation of cellulose. This compound is considered as a potential platform chemical in the future.[78] It can be converted to 2,5-dimethylfuran, which is a biofuel, and to other important molecules such as levulinic acid, 2,5-furandicarboxylic acid, dihydroxymethylfuran, 2,5-diformylfuran and 5-hydroxy-4-keto-2-pentenoic acid.[79,80] The hydrolysis of biomass to produce HMF in ILs was first demonstrated in 2007 by Zhao *et al.* In this work, the catalytic conversion of sugars (glucose and fructose) giving high yield (near 70%) to HMF employing 1-alkyl-3-methylimidazolium chloride/CrCl$_3$ as catalysts without using acids was described. Using a wide range of metal halides, HMF yields ranging from 63 to 83% were achieved in 3 hours when using 6 mol% loading (based on sugar) of

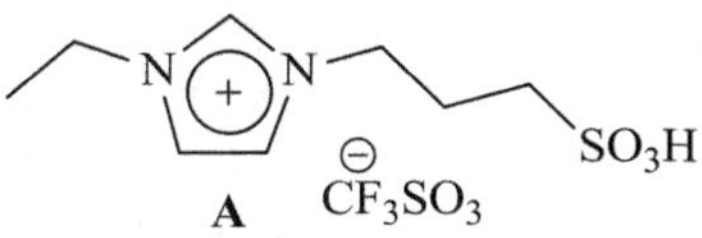

Figure 4.8 Structure of the best performance acid IL for the hydrolysis of microcrystalline cellulose.

$CrCl_2$, $CrCl_3$, $FeCl_2$, $FeCl_3$, $CuCl$, $CuCl_2$, VCl_3, $MoCl_3$, $PdCl_2$, $PtCl_2$, $PtCl_4$, $RuCl_3$ or $RhCl_3$ with very low yields of levulinic acid and α-angelica lactone (less than 0.08%). Other metal halides were not effective for this transformation; for example, the alkali chlorides, $LaCl_3$ and $MnCl_2$.[81]

Many other studies have demonstrated that metal salts, *i.e.* $CrCl_2$ and $CrCl_3$,[82–84] and $ZrCl_4$ under microwave irradiation,[85] $SnCl_4$,[86] $GeCl_4$ and $LaCl_3$,[87] iron,[88] and cobalt,[89] $InCl_3$,[90] *N,N*-dimethylacetamide, Ru/C and formic acid,[91] the pairs $CuCl_2$/$CrCl_2$,[92] and $CrCl_3$/LiCl in [EMIM][Cl],[93] and other acid ILs such as [C_4SO_3HMIM][HSO_4], [C_4SO_3HMIM][Cl] and [C_2MIM][HSO_4] act as catalysts in the depolymerization of cellulose to HMF.[94,95] Tao *et al.* studied 16 imidazolium and pyridinium derivatives acid ILs in $MnCl_2$ for the production of HMF and furfural from cellulose. In this study, the authors showed that the IL acidity plays a very important role in the IL performance; but the structure also influences in the reaction activity. Thus, the almost most acidic IL **A** shown in Figure 4.8, was the best catalyst with and without $MnCl_2$. The efficiency in all cases increases when the IL is combined with $MnCl_2$. The authors suggested that the metal ion in the IL could form the complex $[MCl_m(SO_4)_n]^{2n-}$ which promoted rapid conversion of the α-anomers of glucose into the β-anomers through hydrogen bonding between the oxygen atom in SO_4^{2-} or Cl^- in the cases of metal chlorides and the OH groups.[96]

In 2013, others reported the conversion of cellulose into HMF in [BMIM][Cl] without any co-solvent in one step by conventional heating by oil at atmospheric pressure by one step. 94% TRS (include HMF and other reducing chemicals produced from degradation of cellulose) and 53% HMF yield could be obtained with metallic ILs catalysts such as $CuCr([PSMIM][SO_4])_5$ and $Cr([PSMIM][HSO_4])_3$ obtained by addition of $CrCl_3$ or $CrCl_3$–$CuCl_2$ into 1-(3-sulfonic acid) propane-3-methylimidazole hydrosulfate ([PSMIM][HSO_4]) in molar ratio of 3:1 and 2:2:5, respectively. The last catalyst showed the best performance to effectively degrade cellulose. The maximum yield of HMF and TRS is 53% and 94%, respectively, with the conversion of cellulose of 95%. The yield was achieved using 0.1 g MCC, 0.05 g $Cr([PSMIM][HSO_4])_3$/2.0 g [BMIM][Cl], 120 °C and 5 h. The catalytic system can be reused and the HMF yield can be improved with the cyclic utilization.[97]

An ecologically viable catalytic pathway was proposed recently for furfural production without the use of inorganic acids. In this work, solid acids such as $H_3PW_{12}O_{40}$, Amberlyst-5 and NKC-9 (macroporous styrene-based sulfonic acid resin) were used as catalysts for the production of furfural from xylose, xylan and lignocellulosic biomass in [BMIM][Cl] under microwave irradiation at atmospheric pressure. A surprisingly high furfural yield of 93.7% from xylan was obtained by $H_3PW_{12}O_{40}$ at 160 °C in 10 min.[98]

Recently, Zhang, Du and Quan also showed that the furfural yield is improved when $MnCl_2$ is added to the reaction mixture when the polymer-bound sulfonic acid ($PEG\text{-}OSO_3H$) is used in an IL as the catalyst for the dehydration of biomass to furfural.[99]

Very recently, the direct conversion of fructose into HMF and alkyl levulinate was achieved by making use of IL-based polyoxometalate salts (IL-POMs) as an efficient, environmently friendly and recyclable solid acid catalyst. Phosphotungstic acid-derived IL-POM shows the highest catalytic performance in both the HMF and ethyl levulinate (EL) formation after optimizing the reaction conditions. High HMF and EL yields of up to 99% and 82%, respectively, are obtained from fructose under the investigated conditions. Moreover, the generality of the catalyst is further demonstrated by processing representative di- and poly-saccharides such as sucrose and inulin with good yields to HMF (76% from inulin and 48% from sucrose) and to EL (67% from inulin and 45% from sucrose), again under mild conditions, thereby eliminating the separate hydrolysis step before the dehydration reaction. The catalyst recycling experiment indicates that the adsorption and accumulation of oligomeric products on the catalyst surface results in a partial deactivation of the catalyst. The mechanism research reveals that a major pathway for EL formation involves a fructose-to-HMF transformation followed by HMF etherification and rehydration of HMF-ether to give EL.[100]

4.3.4   Biomass Hydrolysis to Levulinic Acid

4-Oxopentanoic acid, better known as levulinic acid (LA) is a ketoacid that can be obtained by hydrolysis of HMF according the reaction shown in Figure 4.9.

LA is a versatile building block for fuel additives, polymer precursors, pharmaceuticals, herbicides and chemical intermediates. Production of LA from cellulose has become one of the key steps for biomass refining. The effectiveness of ILs for the selective conversion of cellulose to LA has been reported in various recent works.[88,101–106] In 2013, Ya'aini and Amin studied the catalytic conversion of lignocellulosic biomass to LA in ILs: [EMIM] [Cl] conducted with a hybrid catalyst containing equal $CrCl_3$ and HY zeolite weight ratios using a wet impregnation method. Initially, optimization of cellulose as a model compound was carried out using two-level full factorial design with two centre points. Under optimum process conditions, 46.0%

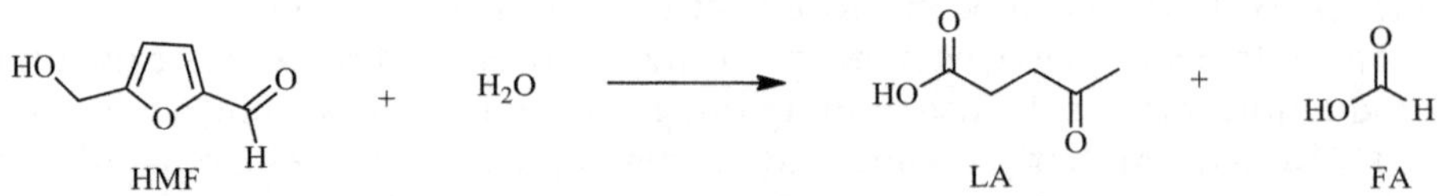

Figure 4.9   Hydrolysis of HMF to produce LA.

of LA yield was obtained from cellulose. Subsequently, utilization of ligno-cellulosic biomass gave a yield of 15.5% and 15.0% of LA from empty fruit bunch (EFB) and kenaf, respectively, at the optimum conditions. In the presence of IL under the same process conditions, 20.0% and 17.0% of LA were obtained for EFB and kenaf, respectively.[107] In a short communication, Ren, Zhou and Liu developed a procedure for a highly selective conversion of cellulose to LA with high yield (below 55%) *via* microwave-assisted synthesis in SO_3H-functionalized ILs (SFILs) with different anions. In order to establish the relationship between SFILs structures and their catalytic activities, the Brønsted acidities of SFILs were determined by the Hammett method using UV-vis spectroscopy with 4-nitroaniline as indicator. They observed that the catalytic activities of SFILs depend on the anions and decrease in the order: $HSO_4^- > CH_3SO_3^- > H_2PO_4^-$, which is in good agreement with their acidity order as depicted in Table 4.2.

The SFILs are efficient catalysts for cellulose conversion into LA and the subsequent esterification, which facilitates the separation of ethyl levulinate and reuse of ILs.

The ratio of water and IL to cellulose plays an important role in the selective formation of LA with respect to glucose. This approach offers significant improvements for the production of LA from cellulose and provides an environmentally friendly route to biomass utilization.[107] In other interesting work, response surface analysis with a four-factor-five-level central composite design was applied to optimize the hydrolysis conditions for the conversion of bamboo (*Phyllostachys Praecox* f. *preveynalis*) shoot shell (BSS) to LA catalysed by $[C_4MIM][HSO_4]$ IL. The effects of the four main reaction parameters, time, temperature, initial concentration of $[C_4MIM][HSO_4]$ and X_{BSS} (initial BSS intake), on the hydrolysis reaction for the yield of LA were analysed. The analysis of variance of the results indicated that the yield of LA in the range studied was significantly ($P < 0.05$) affected by the four factors.

Table 4.2 Hammett acidity (H_0) and catalytic activity of SFILs.[a]

SFILs	A_{max}	[I] (%)	[IH⁺] (%)	H_0[b]	Yield (%)	
—	0.38	100	0	—	0	0
$[C_3SO_3HMIM]$ $[H_2PO_4]$	0.31	82	18	1.65	3.5	10.5
$[C_3SO_3HMIM]$ $[CH_3SO_3]$	0.26	68	32	1.32	36.3	0
$[C_3SO_3HMIM]$ $[HSO_4]$	0.24	63	37	1.22	44.5	0
$[C_4SO_3HMIM]$ $[HSO_4]$	0.23	61	39	1.18	41.4	0
$[C_3SO_3HPy]$ $[HSO_4]$	0.23	61	39	1.18	40.5	0
$[C_3SO_3HN_{111}]$ $[HSO_4]$	0.23	61	39	1.18	43.2	0

[a]Reaction conditions: 250 mg cellulose, 3.3 mmol SFIL, 2.000 g H_2O, MW, 160 °C, 30 min.
[b]H_0 = pK(I)eq. + log ([I]/[IH⁺]). Indicator: 4-nitroaniline (pK(I)eq. = 0.99).

The optimized reaction conditions were as follows: temperature of 145 °C, time of 103.8 min, of concentration of 0.9 mol·L^{-1} and X_{BSS} of 2.04% (by mass). A high yield 71 ± 0.41 mol% was obtained at the optimum conditions which are in good agreement with the model prediction 73.8 mol% based on available C_6 sugars in BSS or 17.9 wt.% based on the mass of BSS.[108]

4.3.5   Conversion of Lignin

Lignin is the third most abundant biomass component after cellulose and hemicellulose, accounting for 18–40 wt.% of dry wood. Lignin is a complex phenylpropanoid polymer with wide potential of applicability. It is very difficult to dissolve and process due to its resistance to degradation, which is why the processes for transforming lignin into biofuel is still a challenge. Studies for the preparation, isolation and transformation of this natural polymer are one of the topics currently receiving more attention within the biorefinery concept. The isolated lignin is required not only for the production of heat or fuel but also for manufacturing of several commodities, such as emulsifiers, binders, dispersants, sequestrants and polymers.[109] It is known that lignin hinders the enzymatic hydrolysis and an extensive delignification should be attained to improve hydrolysis[110] and also lignin provokes a chemical inhibition in ILs-assisted catalytic hydrolysis of cellulose.[111] The production of monomeric phenols by thermochemical conversion of biomass is well known.[112] Lignin is commonly obtained through steam explosion process, but also by lignocellulosic biomass fractionation using ILs. After biomass regeneration, lignin generally is partially extracted in the IL/anti-solvent mixture.[113] The pre-treatment efficiency is dependent on the IL, lignocellulosic biomass (type, moisture, size and load), temperature, length of pre-treatment and anti-solvent used.[44,114]

Recently it has been demonstrated that inexpensive protic ILs (PIL) can be employed for the simple extraction of lignin from lignocellulosic biomass. After the lignin-extraction step, the PIL is easily recovered using distillation, leaving the separated lignin and cellulose-rich residues available for further processing. Biopolymer solubility tests indicate that increasing the xylan (*i.e.* hemicellulose) solubility in the PIL results in greater fibre disruption/penetration, which significantly enhances the effectiveness of the lignin extraction.[115]

ILs have not only played an important role in the separation and removal of lignin from lignocellulosic materials, but also their transformation into high-value products, such as the oxidation of lignin using ILs as strategy to produce renewable chemicals. In this sense, researchers from Texas University at Austin studied the catalytic degradation of lignin model compounds in acidic imidazolium-based ILs, looking at the hydrolytic cleavage of β-O-4 ether bonds in the model compounds, guaiacyl-glycerol-β-guaiacyl ether (GG) and veratryl-glycerol-β-guaiacyl ether (VG) in [BMIM][Cl] with metal chlorides and water. $FeCl_3$, $CuCl_2$ and $AlCl_3$ were found to be effective and functioned catalytically in cleaving the β-O-4 bond of GG, although a number of other metal chlorides are considerably less active. $AlCl_3$ functioned more effectively in cleaving the β-O-4 bond of VG than did $FeCl_3$ and $CuCl_2$. After 120 min at 150 °C, GG conversion reached 100%, and about 70% of the β-O-4 bonds of GG were hydrolysed, liberating guaiacol, in the presence of $FeCl_3$ and $CuCl_2$, while about 80% of the β-O-4 bonds of GG were hydrolysed in the presence of $AlCl_3$ with 100% GG conversion. About 75% of the β-O-4 bonds of VG were hydrolysed in the presence of $AlCl_3$ after 240 min at 150 °C. The acidity of each IL was determinate using 3-nitroaniline as an indicator to measure the Hammett acidity (H_0). The most acidic IL is [HMIM][Cl], with an H_0 value of 1.48. The H_0 values of other ILs, [HMIM][BF$_4$], [HMIM][HSO$_4$], [HMIM][Br], and [BMIM][HSO$_4$], are 1.70, 1.99, 2.04, and 2.08, respectively, thus all ILs evaluated were strongly acidic but the relative acidity did not correlate with the ability of the IL to catalyse β-O-4 ether bond hydrolysis. The reactivity of the model compounds in the ILs depend not only on the acidity, but also on the nature of the ions and their interaction with the model compounds.

While [HMIM][BF$_4$] and [HMIM][HSO$_4$] always produced lower yields of guaiacol, the activities of [HMIM][Cl], [HMIM][Br], and [BMIM][HSO$_4$] with respect to cleaving the β-O-4 ether bond varied significantly with temperature. [HMIM][Cl] at 150 °C produced the highest yield of guaiacol (82.5%). The authors suggested that the ability of the anion to hydrogen bond with the model compound is a major contributor to the ability of an acidic IL to effectively catalyse hydrolysis of the β-O-4 ether linkage with stronger coordination leading to a chemical environment more conducive to ether bond hydrolysis.[116]

The same research group studied an optional strategy for the degradation of lignin using [BMIM][Cl] with metal chlorides and water. $FeCl_3$, $CuCl_2$ and $AlCl_3$ were found to be effective and functioned catalytically in cleaving the β-O-4 bond of the model compound guaiacyl-glycerol-β-guaiacyl ether. $AlCl_3$ was the most effective and functioned catalytically in cleaving the β-O-4 bond.[117] They subsequently demonstrated that the acid IL 1-*H*-3-methylimidazolium chloride can act as both solvent and catalyst for the depolymerization of oak wood under mild conditions.[118]

Recent research about the use of ILs for lignin dissolution have also focused on the various types of reactions utilized for the analysis and conversion of lignin to useful chemicals as simpler monosaccharides, which can then be converted to fuels and other chemicals. In this sense, Sievers *et al.* studied the depolymerization of cellulose and hemicellulose from loblolly pine wood in [BMIM][Cl], which is capable of dissolving carbohydrates and lignin. In the presence of an acid catalyst, the carbohydrate fraction was converted into water-soluble products under milder conditions than reported for similar reactions in the aqueous phase. The water-soluble products included monosaccharides, oligosaccharides, furfural and 5-hydroxymethylfurfural (HMF). The lignin fraction is recovered as a solid residue. It is found by ^{13}C CP MAS NMR spectroscopy that chemical modifications of lignin occurred only to a very moderate extent.[119]

4.3.6 Enzymatic Hydrolysis in Ionic Liquids

As described in the sections above, the pre-treatment of cellulose and lig-nocellulose with ILs greatly enhanced enzymatic (cellulase) hydrolysis rates compared to untreated substrates. However, one of the disadvantages of ILs is their strong tendency to inactivate enzymes. For this reason, in most cases it is imperative to do a thorough flushing of the medium after pre-treatment to remove the small amounts of ILs co-precipitated with recovered cellulose prior to enzymatic hydrolysis.[120]

In the last years, several reports have described enzyme compatible ILs. Using these ILs, the enzymatic saccharification can be carried out *in situ* without removing the ILs. For example, Kamiya *et al.* investigated in 2008 the *in situ* enzyme saccharification of cellulose in an IL ([C$_2$MIM][diethyl-phosphate]) that is compatible with the enzyme employed, with the aim of eliminating the need to recover regenerated cellulose. Cellulase was directly added to the aqueous/IL mixture containing cellulose at 40 °C. The authors observed little cellulase activity when the volume of IL to water was greater than 3:2; however, decreasing the volume ratio to 1:4 (IL:water) enhanced cellulase activity and resulted over 70% of the starting cellulose being con-verted to glucose and cellobiose.[121]

Engel *et al.* carried out a point by point analysis of how IL affects the enzy-matic hydrolysis with extremophilic cellulases of native and modified cel-lulose. They investigated the stability of hyperthermophilic enzymes in the presence of the IL 1-ethyl-3-methylimidazolium acetate ([C$_2$MIM][OAc]) and compared it to the industrial benchmark *Trichoderma viride* cellulase. The endoglucanase from a hyperthermophilic bacterium, *Thermatoga maritima*, and a hyperthermophilic archaeon, *Pyrococcus horikoshii*, were over expressed in *E. coli* and purified to homogeneity. Under their optimum conditions, both hyperthermophilic enzymes showed significantly higher [C$_2$MIM][OAc] tol-erance than *T. viride* cellulase. Using differential scanning calorimetry they determined the effect of the IL on protein stability and the results indicate that higher concentrations of IL correlated with lowered protein stability. Both hyperthermophilic enzymes were active on [C$_2$MIM][OAc] pre-treated Avicel and corn stover. Furthermore, these enzymes can be recovered with little loss in activity after exposure to 15% [C$_2$MIM][OAc] for 15 h.[122]

Researchers from DOE Joint Genome Institute and Joint BioEnergy Institute in collaboration with other universities have found that certain hyperthermophilic[123] and halophilic enzymes[124] are good candidates to be IL-tolerant cellulolytic enzymes.

Recently, an efficient, easy and economical method for the recovery of IL and lignin from biomass pre-treatment was published. It may be extremely difficult to develop an IL biomass pre-treatment process with a positive over-all energy balance that includes IL recovery from wash liquids by distillation. This underscores the need for an efficient process of recovering IL from pre-treated solids other than by washing and distillation, or for minimizing solid products, such as the *in situ* hydrolysis of polysaccharides into fermentable sugars. This procedure consists of the use of a mixture of acetone:ethanol:IL (4–6:1:1) that precipitates cellulose and lignocellulosic biomass from solutions of the IL [C$_2$MIM][AcO] without the formation of intermediate gel phases and removes lignin and most residual IL content in pre-treated corn stover to less than 0.2 wt.%.[125]

4.4   Future Perspectives

The employment of ILs in biomass processing and especially in biomass hydrolysis is still a relatively new area of research that could open the door to sustainable and economically viable development of the overall concept of a biorefinery; however, there are still many challenges to be met to achieve these goals and to put these potential applications into industrial reality.

In order to be environmentally and economically viable for industrial applications, the IL should be inexpensive, non-toxic and preferably derived from renewable materials, biodegradable, recoverable and recyclable without a loss of activity, and preferably derived from renewable raw materials. The IL's toxicity toward microorganisms and enzymes must also be studied before considering an IL for a scalable pre-treatment. The choice of IL should be a compromise between cost, toxicity, solubilizing power and enzyme compatibility.

In the case of hydrolysis, great progress has been made in this area, but more basic research is needed, especially in process development and optimization. Examples of where more research is needed is in the recovery of ILs, hemicellulose and lignin from the ILs after pre-treatment and in the separation of the sugars from the aqueous IL and recovery of the IL after acid hydrolysis.

For an effective integrated bioprocess which would use enzymes *in situ* with ILs, additional research on enzyme compatible ILs is required because the cheapest ILs and also the best prototypes to dissolve cellulose are those containing the anion Cl$^-$ but lamentably these ILs are poison for the enzymes.

Brønsted acid ILs could potentially provide an effective one step (hydrolysis and fermentation) method for liberating fermentable sugars from biomass without the use of expensive enzyme cocktails; however, in this case also, the separation of the sugars from the aqueous IL and the develop of an efficient process to recover and recycle the non-toxic IL is still an unsolved problem.

Despite all the challenges yet to be solved, it is our point of view that ILs have a promising future for the development of the concept of biorefineries.

References

1. R. E. Sims, W. Mabee, J. N. Saddler and M. Taylor, *Bioresour. Technol.*, 2010, **101**, 1570–1580.
2. T. Damartzis and A. Zabaniotou, *Renewable Sustainable Energy Rev.*, 2011, **15**, 366–378.

3. G. W. Huber, S. Iborra and A. Corma, *Chem. Rev.*, 2006, **106**, 4044–4098.

4. R. A. Sheldon, *Green Chem.*, 2014, **16**, 950–963 and references cited therein.

5. P. Claassen, J. Van Lier, A. L. Contreras, E. Van Niel, L. Sijtsma, A. Stams, S. De Vries and R. Weusthuis, *Appl. Microbiol. Biotechnol.*, 1999, **52**, 741–755.

6. A. Stark, *Energy Environ. Sci.*, 2011, **4**, 19–32.

7. N. Sun, H. Rodríguez, M. Rahman and R. D. Rogers, *Chem. Commun.*, 2011, **47**, 1405–1421.

8. H. Tadesse and R. Luque, *Energy Environ. Sci.*, 2011, **4**, 3913–3929.

9. P. McKendry, *Bioresour. Technol.*, 2002, **83**, 47–54.

10. D. J. Hayes, *Catal. Today*, 2009, **145**, 138–151.

11. J. Zakzeski, P. C. Bruijnincx, A. L. Jongerius and B. M. Weckhuysen, *Chem. Rev.*, 2010, **110**, 3552–3599.

12. C. H. Haigler, *Biosynthesis and Biodegradation of Cellulose*, CRC Press, 1990.

13. Y. Habibi, *Chem. Soc. Rev.*, 2014, **43**, 1519–1542.

14. F. S. Chakar and A. J. Ragauskas, *Ind. Crops Prod.*, 2004, **20**, 131–141.

15. M. S. Singhvi, S. Chaudhari and D. V. Gokhale, *RSC Adv.*, 2014, **4**, 8271–8277.

16. H. Chen and L. Wang, in *Biomass Now – Sustainable Growth and Use*, ed. M. D. Matovic, INTECH, Rijeka, Croatia, 2013, ch. 14.

17. D. Klein-Marcuschamer, B. A. Simmons and H. W. Blanch, *Biofuels, Bioprod. Biorefin.*, 2011, **5**, 562–569.

18. P. Vasudevan, S. Sharma and A. Kumar, *J. Sci. Ind. Res.*, 2005, **64**, 822.

19. R. Luque, L. Herrero-Davila, J. M. Campelo, J. H. Clark, J. M. Hidalgo, D. Luna, J. M. Marinas and A. A. Romero, *Energy Environ. Sci.*, 2008, **1**, 542–564.

20. P. Kumar, D. M. Barrett, M. J. Delwiche and P. Stroeve, *Ind. Eng. Chem. Res.*, 2009, **48**, 3713–3729.

21. R. A. Nieves, R. J. Todd, R. P. Ellis and M. E. Himmel, in *Enzymatic Conversion of Biomass for Fuels Production*, ed. M. E. Himmel, J. O. Baker and R. P. Overend, American Chemical Society, 1994, ch. 11, vol. 566, pp. 236–243.

22. A. Verardi, I. De Bari, E. Ricca and V. Calabrò, in *Bioethanol*, ed. M. A. Pinheiro Lima, InTech, Croatia, 2011, pp. 95–112.

23. Y. Sun and J. Y. Cheng, *Bioresour. Technol.*, 2002, **83**, 1–11.

24. A. A. Modenbach and S. E. Nokes, *Biomass Bioenergy*, 2013, **56**, 526–544.

25. A. W. Bhutto, K. Qureshi, K. Harijan, G. Zahedi and A. Bahadori, *RSC Adv.*, 2014, **4**, 3392–3412.

26. M. Galbe and G. Zacchi, *Appl. Microbiol. Biotechnol.*, 2002, **59**, 618–628.

27. R. Martínez-Palou, *Mol. Diversity*, 2010, **14**, 3–25.

28. J. P. Hallett and T. Welton, *Chem. Rev.*, 2011, **111**, 3508–3576.

29. R. Martínez-Palou, *J. Mex. Chem. Soc.*, 2007, **51**, 252–264.

30. Y. Gu and G. Li, *Adv. Synth. Catal.*, 2009, **351**, 817–847.

31. S. Muginova, A. Galimova, A. Polyakov and T. Shekhovtsova, *J. Anal. Chem.*, 2010, **65**, 331–351.

32. R. Bogel-Lukasik, N. M. T. Lourenco, P. Vidinha, M. D. R. G. da Silva, C. A. M. Afonso, M. N. da Ponte and S. Barreiros, *Green Chem.*, 2008, **10**, 243–248.

33. Z. Li, Z. Jia, Y. Luan and T. Mu, *Curr. Opin. Solid State Mater. Sci.*, 2008, **12**, 1–8.

34. J. Lu, F. Yan and J. Texter, *Prog. Polym. Sci.*, 2009, **34**, 431–448.

35. J. Aburto, D. M. Márquez, J. C. Navarro and R. Martínez-Palou, *Tenside, Surfactants, Deterg.*, 2014, **51**, 313–317.

36. R. Martínez-Palou, N. V. Likhanova and O. Olivares-Xomelt, in *Developments in Corrosion Protection*, INTECH, Rijeka, Croatia, 2014, ch. 19, pp. 431–446.

37. R. Martínez-Palou and R. Luque, *Energy Environ. Sci.*, 2014, **7**, 2414–2447.

38. N. V. Plechkova and K. R. Seddon, *Chem. Soc. Rev.*, 2008, **37**, 123–150.

39. R. P. Swatloski, S. K. Spear, J. D. Holbrey and R. D. Rogers, *J. Am. Chem. Soc.*, 2002, **124**, 4974–4975.

40. M. E. Zakrzewska, E. Bogel-Lukasik and R. Bogel-Lukasik, *Energy Fuels*, 2010, **24**, 737–745.

41. A. A. Rosatella, R. F. M. Frade and C. A. M. Afonso, *Curr. Org. Synth.*, 2011, **8**, 840–860.

42. A. Pinkert, K. N. Marsh, S. Pang and M. P. Staiger, *Chem. Rev.*, 2009, **109**, 6712–6728.

43. V. Myllymaki and R. Aksela, WO2005/017001, 2005.

44. A. M. da Costa Lopes, K. G. João, A. R. C. Morais, E. Bogel-Lukasik and R. Bogel-Lukasik, *Sustainable Chem. Processes*, 2013, **1**, 3.

45. A. Brandt, J. P. Hallett, D. J. Leak, R. J. Murphy and T. Welton, *Green Chem.*, 2010, **12**, 672–679.

46. O. Merino, R. Martínez-Palou, J. Labidi and R. Luque, in *Biofuels and Chemicals with Microwave*, ed. Z. Fang, R. L. Smith Jr and X. Qi, 2015, pp. 197–224.

47. R. Pezoa, V. Cortinez, S. Hyvarinen, M. Reunanen, J. Hemming, M. E. Lienqueo, O. Salazar, R. Carmona, A. Garcia, D. Y. Murzin and J. P. Mikkola, *Cellul. Chem. Technol.*, 2010, **44**, 165–172.

48. A. M. da Costa Lopes and R. Bogel-Lukasik, *ChemSusChem*, 2015, **8**, 947–965.

49. M. Erbeldinger, A. J. Mesiano and A. J. Russell, *Biotechnol. Prog.*, 2000, **16**, 1129–1131.

50. T. Vancov, A.-S. Alston, T. Brown and S. McIntosh, *Renewable Energy*, 2012, **45**, 1–6.

51. C. Li and Z. K. Zhao, *Adv. Synth. Catal.*, 2007, **349**, 1847–1850.

52. R. Rinaldi and F. Schuth, *ChemSusChem*, 2009, **2**, 1096–1107.

53. C. Z. Li, Q. Wang and Z. K. Zhao, *Green Chem.*, 2008, **10**, 177–182.

54. Z. Zhang and Z. K. Zhao, *Carbohydr. Res.*, 2009, **344**, 2069–2072.

55. L. Vanoye, M. Fanselow, J. D. Holbrey, M. P. Atkins and K. R. Seddon, *Green Chem.*, 2009, **11**, 390–396.

56. C. Sievers, M. B. Valenzuela-Olarte, T. Marzialetti, I. Musin, P. K. Agrawal and C. W. Jones, *Ind. Eng. Chem. Res.*, 2009, **48**, 1277–1286.

57. J. B. Binder and R. T. Raines, *Proc. Natl. Acad. Sci. U. S. A.*, 2010, **107**, 4516–4521.

58. B. Li, I. Filpponen and D. S. Argyropoulos, *Ind. Eng. Chem. Res.*, 2010, **49**, 3126–3136.

59. S. Dee and A. T. Bell, *Green Chem.*, 2011, **13**, 1467–1475.

60. H. X. Guo, X. H. Qi, L. Y. Li and R. L. Smith, *Bioresour. Technol.*, 2012, **116**, 355–359.

61. N. Zhou, Y. Zhang, X. Gong, Q. Wang and Y. Ma, *Bioresour. Technol.*, 2012, **118**, 512–517.

62. S. Morales-delaRosa, J. M. Campos-Martin and J. L. Fierro, *Chem. Eng. J.*, 2012, **181**, 538–541.

63. G. Fan, C. Liao, T. Fang, M. Wang and G. Song, *Fuel Process. Technol.*, 2013, **116**, 142–148.

64. W. p. Yin, X. Li, Y. l. Ren, S. Zhao and J. j. Wang, *J. Appl. Polym. Sci.*, 2013, **129**, 472–479.

65. Y. Zhang, H. Du, X. Qian and E. Y.-X. Chen, *Energy Fuels*, 2010, **24**, 2410–2417.

66. N. Sun, H. Liu, N. Sathitsuksanoh, V. Stavila, M. Sawant, A. Bonito, K. Tran, A. George, K. L. Sale and S. Singh, *Biotechnol. Biofuels*, 2013, **6**, 1–15.

67. X. Qian, J. Lei and S. R. Wickramasinghe, *RSC Adv.*, 2013, **3**, 24280–24287.

68. H. Ohno and Y. Fukaya, *Chem. Lett.*, 2009, **38**, 2–7.

69. A. Bordoloi, S. Sahoo, F. Lefebvre and S. Halligudi, *J. Catal.*, 2008, **259**, 232–239.

70. A. S. Amarasekara and O. S. Owereh, *Ind. Eng. Chem. Res.*, 2009, **48**, 10152–10155.

71. X. M. Hu, Y. B. Xiao, K. Niu, Y. Zhao, B. X. Zhang and B. Z. Hu, *Carbohydr. Polym.*, 2013, **97**, 172–176.

72. J. Holm, U. Lassi and A. Hernoux-Villiere, *Biomass Bioenergy*, 2013, **56**, 432–436.

73. Q. Yang, Z. Wei, H. Xing and Q. Ren, *Catal. Commun.*, 2008, **9**, 1307–1311.

74. A. S. Amarasekara and B. Wiredu, *Ind. Eng. Chem. Res.*, 2011, **50**, 12276–12280.

75. A. S. Amarasekara and B. Wiredu, *Int. J. Carbohydr. Chem.*, 2012, **2012**, 1–6.

76. A. S. Amarasekara and O. S. Owereh, *Catal. Commun.*, 2010, **11**, 1072–1075.

77. A. Hernoux-Villière, J.-M. Lévêque, J. Kärkkäinen, N. Papaiconomou, M. Lajunen and U. Lassi, *Catal. Today*, 2014, **223**, 11–17.

78. J. J. Bozell and G. R. Petersen, *Green Chem.*, 2010, **12**, 539–554.

79. A. A. Rosatella, S. P. Simeonov, R. F. Frade and C. A. Afonso, *Green Chem.*, 2011, **13**, 754–793.

80. M. E. Zakrzewska, E. Bogel-Lukasik and R. Bogel-Lukasik, *Chem. Rev.*, 2011, **111**, 397–417.

81. H. Zhao, J. E. Holladay, H. Brown and Z. C. Zhang, *Science*, 2007, **316**, 1597–1600.
82. H. Li, Q. Zhang, X. Liu, F. Chang, Y. Zhang, W. Xue and S. Yang, *Bioresour. Technol.*, 2013, **144**, 21–27.
83. C. Li, Z. Zhang and Z. K. Zhao, *Tetrahedron Lett.*, 2009, **50**, 5403–5405.
84. Z. Zhang and Z. K. Zhao, *Bioresour. Technol.*, 2010, **101**, 1111–1114.
85. B. Liu, Z. Zhang and Z. K. Zhao, *Chem. Eng. J.*, 2013, **215**, 517–521.
86. S. Q. Hu, Z. F. Zhang, Y. X. Zhou, B. X. Han, H. L. Fan, W. J. Li, J. L. Song and Y. Xie, *Green Chem.*, 2008, **10**, 1280–1283.
87. T. Ståhlberg, M. G. Sørensen and A. Riisager, *Green Chem.*, 2010, **12**, 321–325.
88. F. Tao, H. Song and L. Chou, *ChemSusChem*, 2010, **3**, 1298–1303.
89. F. Tao, H. Song and L. Chou, *Carbohydr. Res.*, 2011, **346**, 58–63.
90. H. Li, Q. Zhang, X. Liu, F. Chang, D. Hu, Y. Zhang, W. Xue and S. Yang, *RSC Adv.*, 2013, **3**, 3648–3654.
91. S. De, S. Dutta and B. Saha, *ChemSusChem*, 2012, **5**, 1826–1833.
92. Y. Su, H. M. Brown, G. Li, X.-d. Zhou, J. E. Amonette, J. L. Fulton, D. M. Camaioni and Z. C. Zhang, *Appl. Catal., A*, 2011, **391**, 436–442.
93. P. Wang, H. Yu, S. Zhan and S. Wang, *Bioresour. Technol.*, 2011, **102**, 4179–4183.
94. S. Lima, P. Neves, M. M. Antunes, M. Pillinger, N. Ignatyev and A. A. Valente, *Appl. Catal., A*, 2009, **363**, 93–99.
95. F. Jiang, Q. J. Zhu, D. Ma, X. M. Liu and X. W. Han, *J. Mol. Catal. A: Chem.*, 2011, **334**, 8–12.
96. F. Tao, H. Song and L. Chou, *Bioresour. Technol.*, 2011, **102**, 9000–9006.
97. L. Zhou, R. Liang, Z. Ma, T. Wu and Y. Wu, *Bioresour. Technol.*, 2013, **129**, 450–455.
98. L. Zhang, H. Yu and P. Wang, *Bioresour. Technol.*, 2013, **136**, 515–521.
99. Z. Zhang, B. Du, Z.-J. Quan, Y.-X. Da and X.-C. Wang, *Catal. Sci. Technol.*, 2014, **4**, 633–638.
100. J. Chen, G. Zhao and L. Chen, *RSC Adv.*, 2014, **4**, 4194–4202.
101. M. Selva, M. Gottardo and A. Perosa, *ACS Sustainable Chem. Eng.*, 2012, **1**, 180–189.
102. D. W. Rackemann and W. O. Doherty, *Biofuels, Bioprod. Biorefin.*, 2011, **5**, 198–214.
103. N. A. S. Ramli, N. Ya'aini and N. A. S. Amin, *Int. J. Nano Biomater.*, 2014, **5**, 59–74.
104. S. Dutta and S. Pal, *Biomass Bioenergy*, 2014, **62**, 182–197.
105. S. Yin, Y. Pan and Z. Tan, *Int. J. Green Energy*, 2011, **8**, 234–247.
106. N. Ya'aini and N. A. S. Amin, *BioResources*, 2013, **8**, 5761–5772.
107. H. Ren, Y. Zhou and L. Liu, *Bioresour. Technol.*, 2013, **129**, 616–619.
108. C. S. Zhou, X. J. Yu, H. L. Ma, R. H. He and S. Vittayapadung, *Chin. J. Chem. Eng.*, 2013, **21**, 544–550.
109. C. Bonini, M. D'Auria, L. Ernanuele, R. Ferri, R. Pucciariello and A. R. Sabia, *J. Appl. Polym. Sci.*, 2005, **98**, 1451–1456.

110. S. D. Zhu, Y. X. Wu, Q. M. Chen, Z. N. Yu, C. W. Wang, S. W. Jin, Y. G. Ding and G. Wu, *Green Chem.*, 2006, **8**, 325–327.
111. H. J. Lee, B. Sanyoto, J. W. Choi, J. M. Ha, D. J. Suh and K. Y. Lee, *Cellulose*, 2013, **20**, 2349–2358.
112. C. Amen-Chen, H. Pakdel and C. Roy, *Bioresour. Technol.*, 2001, **79**, 277–299.
113. N. Sun, M. Rahman, Y. Qin, M. L. Maxim, H. Rodriguez and R. D. Rogers, *Green Chem.*, 2009, **11**, 646–655.
114. A. M. da Costa Lopes, K. G. Joao, D. F. Rubik, E. Bogel-Lukasik, L. C. Duarte, J. Andreaus and R. Bogel-Lukasik, *Bioresour. Technol.*, 2013, **142**, 198–208.
115. E. C. Achinivu, R. M. Howard, G. Li, H. Gracz and W. A. Henderson, *Green Chem.*, 2014, **16**, 1114–1119.
116. B. J. Cox, S. Y. Jia, Z. C. Zhang and J. G. Ekerdt, *Polym. Degrad. Stab.*, 2011, **96**, 426–431.
117. S. Jia, B. J. Cox, X. Guo, Z. C. Zhang and J. G. Ekerdt, *ChemSusChem*, 2010, **3**, 1078–1084.
118. B. J. Cox and J. G. Ekerdt, *Bioresour. Technol.*, 2012, **118**, 584–588.
119. C. Sievers, M. B. Valenzuela-Olarte, T. Marzialetti, I. Musin, P. K. Agrawal and C. W. Jones, *Ind. Eng. Chem. Res.*, 2009, **48**, 1277–1286.
120. M. B. Turner, S. K. Spear, J. G. Huddleston, J. D. Holbrey and R. D. Rogers, *Green Chem.*, 2003, **5**, 443–447.
121. N. Kamiya, Y. Matsushita, M. Hanaki, K. Nakashima, M. Narita, M. Goto and H. Takahashi, *Biotechnol. Lett.*, 2008, **30**, 1037–1040.
122. P. Engel, R. Mladenov, H. Wulfhorst, G. Jager and A. C. Spiess, *Green Chem.*, 2010, **12**, 1959–1966.
123. S. Datta, B. Holmes, J. I. Park, Z. W. Chen, D. C. Dibble, M. Hadi, H. W. Blanch, B. A. Simmons and R. Sapra, *Green Chem.*, 2010, **12**, 338–345.
124. T. Zhang, S. Datta, J. Eichler, N. Ivanova, S. D. Axen, C. A. Kerfeld, F. Chen, N. Kyrpides, P. Hugenholtz and J.-F. Cheng, *Green Chem.*, 2011, **13**, 2083–2090.
125. D. C. Dibble, C. L. Li, L. Sun, A. George, A. R. L. Cheng, O. P. Cetinkol, P. Benke, B. M. Holmes, S. Singh and B. A. Simmons, *Green Chem.*, 2011, **13**, 3255–3264.

Relevance of Ionic Liquids and Biomass Feedstocks for Biomolecule Extraction

ANDRÉ M. DA COSTA LOPES[a,b], LUÍSA BIVAR ROSEIRO[a], AND RAFAL BOGEL-LUKASIK*[a]

[a]Laboratório Nacional de Energia e Geologia, Unidade de Bioenergia, 1649-038 Lisboa, Portugal; [b]LAQV/REQUIMTE, Departamento de Química, Faculdade de Ciências e Tecnologia, Universidade Nova de Lisboa, 2829-516 Caparica, Portugal
*E-mail: rafal.lukasik@lneg.pt

5.1 Introduction

Biomass feedstocks, such as wood, food residues and perennial plant material, usually contain significant biomolecules, such as phenolics, terpenes, alkaloids, sterols and polysaccharides, among other phytochemicals, depending on the type of biomass. These are valuable phytochemicals that can be used in the pharmaceutical, food, nutraceutical and chemical industries.

Environmentally friendly solvent and separation systems together with biomass feedstocks provide a rich area for interdisciplinary research within the biorefinery concept. The use of biomass for fuels, energy (heat and electricity) and chemical products such as biomolecules can be seen as a sustainable alternative to conventional feedstocks, when used in combination with environmentally sound production and processing techniques. In order

RSC Green Chemistry No. 36
Ionic Liquids in the Biorefinery Concept: Challenges and Perspectives
Edited by Rafal Bogel-Lukasik

Published by the Royal Society of Chemistry, www.rsc.org

to successfully compete as a sustainable energy source, the value of biomass must be maximized through the production of valuable co-products in the biorefinery. Specialty chemicals and other bio-based products can be extracted from biomass prior to or after the conversion process, thus increasing the overall profitability and sustainability of the biorefinery.

Efficient separation of these biomolecules is a great challenge due to the low concentrations of these compounds in the biomass matrix and sometimes their heat sensitivity. These compounds have been commonly obtained using conventional extraction methods with organic solvents such as methanol, ethyl acetate, chloroform and others, which require their complete removal by a subsequent evaporation process before being used as nutraceuticals or ingredients for the food, cosmetic or pharmaceutical industries.[1] Also, with increasing safety considerations for operating personnel and consumers, extraction by organic solvents is a challenge due to the volatility, flammable and toxicity of the solvents.[2] For these reasons, alternative solvents within the green chemistry concept, such as subcritical water, supercritical CO_2 and ionic liquids (ILs), have recently been developed. In particular, the exploitation of ionic liquids for extraction has gained great interest. ILs can improve the selectivity and the extraction yields of bioactive compounds in samples as well as alleviating the environmental pollution compared to the conventional organic solvents. ILs great interest as separation media is due to their unique physical and chemical properties of low vapour pressure, high thermal and chemical stability, non-flammability, high polarities and good solvating properties.[3]

Within the different types of biomass feedstock, lignocellulosic biomass, particularly agricultural and forestry residues and energy crops, are potential sources of biomolecules due to their main components. These are cellulose, hemicellulose and lignin, which have a complex composition and a variety of functionalized groups that are the basis of several product lines (fibre composite materials, starch- and protein-derived products) already in the market.[4] Lignocellulosic materials thus represent the most abundant organic residues in the world. Within biorefinery processes, the aim is to depolymerize the polysaccharides (cellulose and hemicellulose) in lignocellulosic biomass to monosaccharides that can be further converted to different products.

As said before, phytochemicals from lignocellulosic biomass comprise a number of compounds such as polyphenols, polysaccharides, essential oils and others. Moreover, lignocellulosic biomass contains valuable bioactive compounds, such as phenolics, phytosterols, alcohols and fatty acids, that can find applications in food, cosmetic and pharmaceutical areas.[1]

According to Behera *et al.*,[5] a variety of ILs has been identified for their potential to enhance the valorization of lignocellulosic biomass, such as bagasse,[6] wheat straw,[7] *etc.* Recently, it was discovered that wood can be completely dissolved in some ILs.[8] The use of ILs to deal with the lignocellulose recalcitrance is a promising new pre-treatment for these biomasses. High efficiency for the solvation of cellulose, lignin and even wood has been proved in an increasing range of ILs based on many different concepts, including those of dialkyl imidazolium derivatives and organic super-base derived approaches.

Utilization of various IL treatments in combination with heat have proven to dissociate wood polysaccharides to monosaccharides quite well.[9] Also, lignin fractionation from biomass has been demonstrated with conventional ILs. This typically requires high temperature processing of extracts ($\geq$100 °C) and may result in the build-up of soluble residuals (extractives, sugars, soluble lignin derivatives, *etc.*) in the ILs during their recovery and reuse. This might, however, constitute a problem, as the high cost of the ILs demands their total recovery, which is difficult to achieve due to small IL losses during processing and minor amounts of thermal degradation of the ions which occurs during the handling of ILs at elevated temperatures for long periods of time.

Recently, Achinivu *et al.*[10] demonstrated that protic ionic liquid (PILs) salts formed in a one-step reaction from low cost acid (acetic acid) and base (amine) reagents, which typically melt below ambient temperature, can be used at a relatively low temperature (<100 °C) as an alternative to conventional (chemically intensive) lignin removal methods for biomass.[10] Nevertheless, Varanasi *et al.*[11] developed an IL-based process that depolymerizes lignin and converts the low molecular weight lignin fractions into a variety of renewable chemicals from biomass. They have also observed that the lignin breakdown products generated are dependent on the type of starting biomass, and significant levels were produced on dissolution at 160 °C for 6 h. Their study also highlights that the generated chemicals (phenols, guaiacols, syringols, eugenol, catechols), their oxidized products (vanillin, vanillic acid, syringaldehyde) and their easily derivatized hydrocarbons (benzene, toluene, xylene, styrene, biphenols and cyclohexane) have already relatively high market value as commodity and specialty chemicals, green building materials, nylons and resins.

Other types of biomass residues have been successfully used by different ILs for their assisted extraction of relevant bioactive compounds. IL-assisted sample pre-treatment techniques, such as liquid–liquid extraction, liquid-phase micro-extraction, solid-phase micro-extraction and aqueous two-phase system extraction have also been published. Co-solvent mixtures of ILs and polar covalent molecules were also recently applied to different biomasses, including algal biomass, for the extraction of bio-oils[12,13] in order to evaluate the improvement of the lipid extraction efficiency. This recent interest in IL-assisted extraction from several different types of biomass is related to the fact that conventional techniques of extraction for their bioactive compounds are time-consuming, laborious, and may involve bulk amounts of volatile/hazardous organic solvents. Also, the selectivity, yield of extraction and purity grade using ILs might improve considerably, especially if associated with microwave- or ultrasonic-assisted extractions, when compared to conventional extraction procedures. Such is the case of shrimp waste, from which the extraction of astaxanthin, a powerful antioxidant, increased 98%

with an optimized method using ILs, when compared to the conventional method.[14] Additionally, using ILs, it was also possible to obtain high molecular weight purified chitin (a very important polysaccharide due to its bioactivity, biocompatibility and low toxicity), from the same waste. Recently, in a similar way, poultry feathers were used for IL-assisted extraction of keratin (with important applications in biomaterials, biomedical, flocculants and adhesive) with 75.1% yield and a dissolution rate of 96.7%.[15]

5.2 Plant Biomass

The diversity of organic compounds present in plants is extremely valuable as it can support a majority of human needs. The bioactive properties inherent to those compounds raise the potentiality of this type of biomass for further valorization. Therefore, the selective extraction of phenolic compounds, terpenes and alkaloids, among others, is imperative. The utilization of ILs for that purpose has been successfully achieved and it arises as an alternative technology instead of conventional processes using hazardous and toxic extracting solvents. The main results regarding the extraction of biomolecules with ILs from a diverse range of plants are given in Table 5.1.

In this chapter the following abbreviations of ILs are used:

Cations: [xyim] = x-alkyl-y-alkylimidazolium, where x and y are: a – allyl, b – butyl, bz – benzyl, C_2CN – propionitril, COOHm – carboxymethyl, d – decyl, do – dodecyl, e – ethyl, h – hexyl, m – methyl, NH_2p – propylamine, OHe – hydroxyethyl, OHp – hydroxypropyl, o – octyl, p – propyl, SO_3Hb – 1-(butyl-4-sulfonic), t – tetradecyl, v – vinyl; [bmpyrr] – 1-butyl-3-methylpyrrolidinium; [bpy] – *N*-butylpyridinium; [Ch] – cholinium; [DEA] – *N,N*-diethylammonium; [DIPE] – *N,N*-diisopropylethylammonium; [DMA] – *N,N*-dimethylammonium; [DMAPA] – 3-(dimethylamino)-1-propylammonium; [DMCEA] – *N,N*-dimethyl(cyanoethyl)ammonium; [DMEA] – *N,N*-dimethylethanolammonium; [DMED] – *N,N*-dimethylethylenediammonium; [DMHEEA] – *N,N*-dimethyl-*N*-(2-hydroxyethoxyethyl)ammonium; [HEA] – 2-hydroxyethylammonium; [MHEA] – *N*-methyl-2-hydroxyethylammonium; [pyr] – pyridinium; [TBA] – tributylammonium; [TEA] – triethylammonium; [TMG] – tetramethylguanidinium; [TPA] – tripropylammonium.

Anions: [ABS] – alkylbenzenesulfonate; [Ace] – acesulfamate; $[AlCl_4]$ – tetrachloroaluminate; $[BF_4]$ – tetrafluoroborate; [Br] – bromide; [BS] – benzenesulfonate; [CARB] – *N',N'*-dimethylcarbamate; $[CF_3SO_3]$ – trifluoromethanesulfonate; [Cl] – chloride; $[ClO_4]$ – perchlorate; [Dec] – decanoate; [DEP] – diethylphosphate; [DS] – dodecylsulfonate; $[EtSO_4]$ – ethylsulfate; [FA] – formate; [GA] – glycolate; [Gly] – glycinate; $[H_2PO_4]$ – dihydrogenphosphate; [Hex] – hexanoate; $[HSO_4]$ – hydrogensulfate; [I] – iodide; [Lac] – lactate; [Lys] – lysine; $[MeSO_3]$ – methylsulfonate; $[MeSO_4]$ – methylsulfate; $[NO_3]$ – nitrate; [OAc] – acetate; [Oct] – octanoate; [PA] – propionate; $[PF_6]$ – hexafluorophosphate; [Phe] – phenylalanine; $[N(CN)_2]$ – dicyanamide; $[NTf_2]$ – bis(trifluoromethylsulfonyl)imide; [SA] – succinate; [Sac] – saccharinate; [SCN] – thiocyanate; $[SO_4]$ – sulfate; [TFA] – trifluoroacetate; [TsO] – tosylate.

Table 5.1 Plant biomass used for biomolecule extraction with ionic liquids.

Biomass		Extraction technology	Target biomolecules		Ionic liquids	Extraction yield	Ref.
Name	Type		Class	Name			
Acacia catechu	Whole plant	SLE	Phenolic compounds	Catechin; ellagic acid; gallic acid; pyrocatechol	[DMA][CARB]	85[a]	28
Arctium lappa	Leaves	ILUMAE	Phenolic compounds	Caffeic acid Chlorogenic acid Quercetin	[bmim][BF$_4$]; [bmim][Br]; [bmim][Cl]; [bmim][H$_2$PO$_4$]; [emim][Br]; [hmim][Br]; [omim][Br]; [SO$_3$Hbmim][Br]	0.281[b]/[bmim][Br] 1.293[b]/[bmim][Br] 0.276[b]/[bmim][Br]	23
Arnebia euchroma	Whole plant	ILUAE	Naphthoqui-nones	Shikonin β,β′-Dimethy-lacrylshikonin	[bmim][BF$_4$]; [emim][BF$_4$]; [hmim][BF$_4$]; [hmim][PF$_6$]; [omim][BF$_4$]; [omim][PF$_6$]	82.35[a][hmim][BF$_4$] 76.92[a]/[hmim][BF$_4$]	50
Bauhinia championii	Whole plant	ILMAE	Phenolic compounds	Kaempferol Myricetin Quercetin	[bmim][BF$_4$]; [bmim][Br]; [bmim][Cl]; [bmim][H$_2$PO$_4$]; [bmim][PF$_6$]; [bmim]$_2$[SO$_4$]; [COOHmmim][Cl]; [emim][Br]; [hmim][Br]	0.0718[b]/[bmim][PF$_6$] 0.0497[b]/[bmim][Br] 0.2375[b]/[bmim][Br]	32
Cajanus cajan	Leaves	ILMAE	Phenolic compounds	Stilbene acid derivative Apigenin Biochanin Cajaninstilbene acid Cajanol Cajanuslactone Formononetin Longistyline C Pinostrobin	[bmim][BF$_4$]; [bmim][Br]; [bmim][Cl]; [bmim][H$_2$PO$_4$]; [bmim][HSO$_4$]; [emim][Br]; [hmim][Br]; [omim][Br]	0.267[b]/[bmim][Br] 0.159[b]/[bmim][Br] 0.405[b]/[bmim][Br] 10.793[b]/[bmim][Br] 0.369[b]/[bmim][Br] 0.328[b]/[bmim][Br] 0.318[b]/[bmim][Br] 7.952[b]/[bmim][Br] 5.548[b]/[bmim][Br]	31
Cajanus cajan	Roots	ILNPCAE	Phenolic compounds	Apigenin Genistein Genistin	[omim][Br]	0.291[b] 0.496[b] 0.482[b]	19

(continued)

Table 5.1 (*continued*)

Biomass		Extraction technology	Target biomolecules		Ionic liquids	Extraction yield	Ref.
Name	Type		Class	Name			
Camellia sinensis	Leaves/aerial parts	SLE	Phenolic compounds	Polyphenols	[amim][Cl]; [bmim][Cl]; [bzmim][Cl]; [bmim][H$_2$PO$_4$]; [Ch][Cl]; [Ch][Hex]; [Ch][NTf$_2$]; [Ch][OAc]; [emim][N(CN)$_2$]; [emim][Cl]; [emim][EtSO$_4$]; [emim][Lac]; [emim][OAc]; [emim][OTf]; [hmim][Cl]; [omim][Cl]; [OHemim][Cl]	≈35[c]/[Ch][OAc]	52
			Steroids	Saponins		≈65[c]/[bzmim][Cl]	
Catharanthus roseus	Leaves	ILUAE	Alkaloids	Vindoline Catharanthine Vinblastine	[amim][Br]; [bmim][BF$_4$]; [bmim][Br]; [bmim][Cl]; [bmim][ClO$_4$]; [bmim][HSO$_4$]; [bmim][I]; [bmim][NO$_3$]; [bmim][TsO]; [emim][Br]; [hmim][Br]; [omim][Br]	n.d./[omim][Br] n.d./[omim][Br] n.d./[omim][Br]	38
Chamaecyparis obtuse	Leaves	SLE	Phenolic compounds	Amentoflavone Dihydrokaempferol Myricetin Quercitrin	[bmim][Br]; [dmim][BF$_4$]; [dmim][Br]; [dmim][Cl]; [dmim][NTf$_2$]; [dmim][PF$_6$]; [domim][Br]; [emim][Br]; [hmim][Br]; [omim][Br]; [SO$_3$Hbmim][Br]; [tmim][Br]	0.76[b]/[dmim][Br] 2.41[b]/[dmim][Br] 3.15[b]/[dmim][Br] 3.47[b]/[dmim][Br]	24
Chamaecyparis obtuse	Leaves	SLE	Phenolic compounds	Amentoflavone Myricetin Quercitrin	poly[vbmim][BF$_4$]; poly[vbmim][BS]; poly[vbmim][DS]; poly[vbmim][Lac]; poly[vbmim][NTf$_2$]; [vbmim][PF$_6$]	≈1.20[b]/poly[vbmim][Lac] ≈0.60[b]/poly[vbmim][Lac] ≈0.1 [b]/poly[vbmim][Lac]	45
Coleus forskohlii	Roots	ILUAE	Terpenes	Forskolin	[bmim][BF$_4$]; [bmim][Br]; [bmim][Cl]; [bpy][BF$_4$]; [HEA][FA]; [TMG][Lac]	0.508[d]/[TMGL][Lac]	44
Cortex fraxini	Dry tegument	ILUAE		Aesculetin Aesculin	[bmim][BF$_4$]; [bmim][Br]; [bmim][Cl]; [bmim][ClO$_4$]; [bmim][HSO$_4$]; [bmim][I]; [bmim][TsO]; [dmim][Br]; [domim][Br]; [emim][Br]; [hmim][Br]; [omim][Br]	n.d./[bmim][Br] n.d./[bmim][Br]	36

Cynanchum bungei	Roots	ILUAE	Acetophenones	2,4-Dihydroxyace-tophenone	[bmim][BF$_4$]; [emim][BF$_4$]; [hmim][BF$_4$]; [pmim][BF$_4$]; [pmim][Br]; [pmim][I]	0.2054^b/[bmim][BF$_4$]	53
				2,5-Dihydroxyace-tophenone		0.0217^b/[bmim][BF$_4$]	
				4-Hydroxyace-tophenone		0.2862^b/[bmim][BF$_4$]	
				Baishouwubenzo-phenone		0.6326^b/[bmim][BF$_4$]	
Cynanchum paniculatum	Roots	ILMAE SLE	Phenolic compounds	Paeonol	[bmim][Cl]	2.51^a 1.95^a	30
Dioscorea zin-giberensis	Rhyzomes	ILUMAE	Steroids	Diosgenin	[bmim][BF$_4$]; [bmim][Br]; [bmim][Cl]; [emim][Br]; [emim][BF$_4$]; [pmim][BF$_4$]	10.24^b/[emim][BF$_4$]	51
Dryopteris fragrans	Whole plant	ILMAE	essential oils	Terpenes	[amim][Cl]; [bmim][Br]; [bmim][Cl]; [emim][OAc]	0.91^d/[emim][OAc]	47
Foeniculum vulgare	Whole plant	ILUASEME		Estragole *p*-Anisaldehyde *trans*-Anethole	[bmim][PF$_6$]; [hmim][NTf$_2$]; [hmim][PF$_6$]	>95^a/[hmim][PF$_6$] >94^a/[hmim][PF$_6$] >95^a/[hmim][PF$_6$]	96
Flos sophorae	Whole plant	ILMAE	Phenolic compounds	Rutin	[bmim][BF$_4$]; [bmim][Br]; [bmim][Cl]; [bmim][TsO]	171.82^b/[bmim][Br]	22
		ILUAE				171.08^b/[bmim][Br]	
		ILHE				170.21^b/[bmim][Br]	
		ILME				163.28^b/[bmim][Br]	
Galla chinensis	Galls	ILUMAE	Phenolic compounds	Tannins	[bmim][BF$_4$]; [bmim][Br]; [bmim][Cl]	630.2^b/[bmim][Br]	29
Glaucium flavum	Whole plant	SLE	Alkaloids	*S*-(+)-glaucine	[bmim][Ace]; [bmim][Br]; [bmim][Cl]; [bmim][Sac]; [dmim][Ace]; [hmim][Ace]; [omim][Ace]	≈85^c/[bmim][Ace]	40
Glaucium flavum	Whole plant	SLE	Alkaloids	*S*-(+)-glaucine	[bmim][Ace]	≈100^e	39

(continued)

Table 5.1 (*continued*)

Biomass		Extraction technology	Target biomolecules		Ionic liquids	Extraction yield	Ref.
Name	Type		Class	Name			
Ginkgo biloba	Leaves	SLE	Organic acids	Shikimic acid	[bmim][Cl]	2.3^d	54
Glycyrrhiza glabra (extract)	Roots	ILUAE	Phenolic compounds	Glabridin	[bmim][BF$_4$]; [bmim][NTf$_2$]; [bmim][OAc]; [bmim][PF$_6$]; [hmim][NTf$_2$]; [hmim][PF$_6$]; [omim][PF$_6$]	95.72^a/[bmim][NTf$_2$]	35
Ilex paraguar-iensis	Leaves/aerial parts	SLE	Phenolic compounds	Polyphenols	[amim][Cl]; [bmim][Cl]; [bzmim][Cl]; [bmim][H$_2$PO$_4$]; [Ch][Cl]; [Ch][Hex]; [Ch][NTf$_2$]; [Ch][OAc]; [emim][N(CN)$_2$]; [emim][Cl]; [emim][EtSO$_4$]; [emim][Lac]; [emim][OAc]; [emim][CF$_3$SO$_3$]; [hmim][Cl]; [omim][Cl]; [OHemim][Cl]	$\approx 30^c$/[emim][Cl]/[emim][OTf]	52
			Steroids	Saponins		$\approx 55^c$/([bzmim][Cl])	
Iris tectorum	Roots	ILUAE	Phenolic compounds	Iristectorin A	[bmim][BF$_4$]; [bmim][Br]; [omim][Br]	5.28^b/[omim][Br]	27
				Iristectorin B		2.88^b/[omim][Br]	
				Tectoridin		37.45^b/[omim][Br]	
Ligusticum chuanxiong	Whole plant	ILMAE	Lactones	Senkyunolide I	[DMCEA][PA]; [DMHEEA][PA]	$\approx 15^f$/([DMCEA][PA])	46
				Senkyunolide H		$\approx 3^f$/([DMCEA][PA])	
				Z-ligustilide		$\approx 175^f$/([DMCEA][PA])	
Magnoliae officinalis	Cortex	ILUAE	Phenolic compounds	Honokiol and magnolol	[bmim][BF$_4$]; [bmim][PF$_6$]	– [bmim][PF$_6$]	20
Nelumbo nucifera	Leaves	ILMAE	Alkaloids	Nuciferine	[emim][Br]; [bmim][BF$_4$]; [bmim][Br]; [bmim][Cl]; [bmim][PF$_6$]; [hmim][Br]; [omim][Br]	– [hmim][Br]	37
				N-nuciferine		– [hmim][Br]	
				O-nuciferine		– [hmim][Br]	
Ocimum basilicum	Whole plant	ILUASEME	Phenolic compounds	Estragole	[bmim][PF$_6$]; [hmim][PF$_6$]; [hmim][NTf$_2$]	$>95^a$/[hmim][PF$_6$]	96
				p-Anisaldehyde		$>95^a$/[hmim][PF$_6$]	
				trans-Anethole		$>95^a$/[hmim][PF$_6$]	

Panax ginseng	Roots	ILUAE	Steroids	Ginsenosides	[bmim][Br]; [emim][Br]; [hmim][Br]; [pmim][BF$_4$]; [pmim][Br]; [pmim][Cl]	17.81^b/[pmim][Br]	97
Psidium guajava	Leaves	ILMAE	Phenolic compounds	Ellagic acid Gallic acid Quercetin	[bmim]$_2$[SO$_4$]; [bmim][BF$_4$]; [bmim][Br]; [bmim][Cl]; [bmim][H$_2$PO$_4$]; [bmim][N(CN)$_2$]; [bpy][Cl]; [emim][BF$_4$]; [emim][Br]; [hmim][Br]	74.8^a/[bmim][BF$_4$] 91.0^a/[bmim][Cl] 74.5^a/[bmim][H$_2$PO$_4$]	17
Pueraria lobata	Roots	ILUAE	Phenolic compounds	Puerarin	[bmim][BF$_4$]; [bmim][Br]; [COOHmim][BF$_4$]; [OHemim][BF$_4$]; [OHemim][Cl]	82.91^a/[bmim][Br]	34
Pueraria lobata	Roots	ILMAE	Phenolic compounds	Isoflavonoids	[bmim][Br]	10.09^d	33
Rhizma polygoni	Rhizome	ILMAE	Phenolic compounds	*trans*-Resveratrol	[bmim][Br]	92.8^a	16
Rosmarinus officinalis	Leaves	ILMASED	Terpenes Phenolic compounds	Carnosic acid Rosmarinic acid	[bmim][BF$_4$]; [bmim][Br]; [bmim][Cl]; [bmim][NO$_3$]; [dmim][Br]; [emim][Br]; [hmim][Br]; [omim][Br]	33.29^b/[omim][Br] 3.97^b/[omim][Br]	49
			Essential oils	α-Pinene; 1,8-cineole; camphene; camphor		21.5^b/[omim][Br]	
Salvia miltiorrhiza	Whole plant	ILUAE	Terpenes	Cryptotanshinone Tanshinone I Tanshinone II A	[bmim][Cl]; [emim][Cl]; [hmim][Cl]; [omim][Cl]	84.3^a[omim][Cl] 96.2^a[omim][Cl] 94.3^a[omim][Cl]	42
Salvia miltiorrhiza	Roots	ILUPE	Terpenes	Cryptotanshinone Dihydrotanshinone Miltirone Tanshinone I Tanshinone II A	[omim][PF$_6$]	9.3^b 4.06^b 0.593^b 20.3^b 37.4^b	41
Saururus chinensis	Whole plant	ILMAE ILUAE ILHE ILME	Phenolic compounds	Rutin	[bmim][BF$_4$]; [bmim][Br]; [bmim][Cl]; [bmim][TsO]	4.879^b/[bmim][Br] 4.861^b/[bmim][Br] 4.853^b/[bmim][Br] 4.561^b/[bmim][Br]	22

(*continued*)

Table 5.1   (*continued*)

| Biomass | | Extraction | Target biomolecules | | | | |
Name	Type	technology	Class	Name	Ionic liquids	Extraction yield	Ref.
Smilax china	Tuber	ILMAE	Phenolic compounds	Quercetin *trans*-Resveratrol	$[bmim]_2[SO_4]$; $[bmim][BF_4]$; $[bmim][Br]$; $[bmim][Cl]$; $[bmim][H_2PO_4]$; $[bmim][N(CN)_2]$; $[bpy][Cl]$; $[emim][BF_4]$; $[emim][Br]$; $[hmim][Br]$	$59.6^a/[bmim][Br]$ $58.3^a/[bmim][H_2PO_4]$	17
Stephaniae tetrandrae	Roots	ILUAE	Alkaloids	Fangchinoline Tetrandrine	$[bmim][BF_4]$		95
Terminalia chebula	Whole plant	SLE	Phenolic compounds	Catechin; ellagic acid; gallic acid; pyrocatechol	[DMA][CARB]	75^a	28

[a]% (mass of biomolecule/initial mass content of biomolecule in the original biomass × 100).
[b]Yield (mg of biomolecule/g of biomass).
[c]Yield (mass of biomolecule/mass of biomolecule extracted by Soxhlet method).
[d]% (g of biomolecule/100 g of biomass).
[e]Concentration (mg mL^{-1}).
[f]Concentration (mg L^{-1}); SLE – solid–liquid extraction; ILUMAE – ionic liquid-based ultrasound/microwave-assisted extraction; ILUAE – ionic liquid-based ultrasound-assisted extraction; ILMAE – ionic liquid-based microwave-assisted extraction; ILNPCAE – ionic liquid-based negative pressure cavitation-assisted extraction; ILHE – ionic liquid-based heating extraction; ILME – ionic liquid-based marinated extraction; ILUASEME – ionic liquid-based ultrasound-assisted surfactant-emulsified micro-extraction; ILMASED – ionic liquid-based microwave-assisted simultaneous extraction and distillation; ILUPE – ionic liquid-based ultrahigh pressure extraction.

5.2.1 Phenolic Compounds

The first application of ILs in the extraction of biomolecules from plants was reported by Li and co-workers in 2007.[16] *Trans*-resveratrol from *Rhizoma Polygoni Cuspidati* was obtained by ionic liquid-based microwave-assisted extraction (ILMAE) process and three ILs, namely [bmim][Cl], [bmim][Br] and [bmim][BF$_4$], were evaluated as the extracting agents. The results showed that the extraction of *trans*-resveratrol was more efficient using [bmim][Br] (47.9%) in comparison to [bmim][Cl] (35.8%), [bmim][BF$_4$] (17.9%) or even water (1.4%). Actually, some phenolic compounds, including *trans*-resveratrol, demonstrate low solubility or insolubility in water.[17] The higher solubility of *trans*-resveratrol in ILs in contrast to water was referred to the IL strong dissolving power, which establishes multiple interactions, such as π–π, ionic/charge–charge and hydrogen bonding, mainly between the imidazolium cation of the IL and *trans*-resveratrol.[16,18]

This multiple interaction ability of ILs was also reported for the extraction of other phenolic compounds, such as rutin,[19] magnolol and honokiol.[20] Furthermore, the higher acidic property of [bmim][Br], provided by the Br⁻ anion, was reported to have a major impact on the extraction of *trans*-resveratrol giving the highest extraction yield among the examined ILs.[16] It has also been stated that IL efficiently absorbs and dissipates microwave energy through sample, thus the applied microwave irradiation displays an important role by intensifying those interactions.[17] The extraction of several phenolic compounds from *Smilax china* tubers and *Psidium guajava* leaves was also evaluated by Li and co-workers[17] using ILMAE technology, with similar and other types of ILs. ILs with different cations and anions were tested, namely 1-*n*-butyl-3-methylimidazolium based ILs with [Br], [Cl], [BF$_4$], [SO$_4$], [N(CN)$_2$] and [H$_2$PO$_4$] anions. ILs composed of [Br] and [H$_2$PO$_4$] demonstrated higher efficiency on the extraction of phenolics from *P. guajava* leaves, while for *S. china* tubers higher extraction yields of *trans*-resveratrol and quercetin were achieved with ILs composed of Br⁻ and BF$_4$⁻ anions. The authors verified that alkyl chain length of the imidazolium cation influences the extraction ability of ILs for phenolic compounds. Among [emim][Br], [bmim][Br] and [hmim][Br] ILs, [bmim][Br] was the most efficient, where the inherent hydrogen bonding and hydrophobic properties of this IL favoured the extraction.[17]

However, for a different class of cation, such as [bpy], higher extraction yields were obtained. The higher extraction ability of [bpy][Cl] in comparison to [bmim][Cl] was related to the electron-rich aromatic system of the [bpy] cation, which is capable of producing strong polarity, π–π and n–π interactions with the phenolic compounds.[17,18,21] Rutin was also successfully extracted from the plant *Saururus chinensis* and from *Flos Sophorae* flowers using ILMAE process.[22] Among the examined ILs, [bmim][Br] and [bmim][TsO] demonstrated highest efficiency in the extraction. Similar to previous work,[16] the higher acidity of [bmim][Br] or [bmim][TsO] in comparison to the other ILs ([bmim][Cl] and [bmim][BF$_4$]) was stated to be the key factor for enhanced rutin extraction.[22]

Considering the high solubility of rutin in [bmim][Br], several extraction technologies, such as ionic liquid-based heating extraction (ILHE), ionic liquid-based marinated extraction (ILME) and ionic liquid-based ultrasound-assisted extraction (ILUAE), were compared with ILMAE process. The data showed higher extraction efficiency and shorter time needed to reach maximum extraction of rutin after using ILMAE in detriment to the other studied technologies.[22]

Magnoliae officinalis cortex was used for the extraction of hydrophobic polyphenols, such as the lignans magnolol and honokiol by a different approach.[20] Herein, an ILUAE process was developed using [bmim][PF$_6$] IL and ethanol as co-solvent instead of water (regularly used). The ultrasound provided plant tissue disruption and an accelerating solvent penetration through plant cells, decreasing the extraction time.[22] A clear choice for an hydrophobic IL, such as [bmim][PF$_6$], would be expected to extract hydrophobic magnolol and honokiol. The results showed that ILUAE using [bmim][PF$_6$]/ethanol mixture was a more efficient process for the extraction of both magnolol and honokiol when compared to that obtained with ILUAE using [bmim][BF$_4$]/water mixture, or even to traditional heating reflux extraction (HRE) with ethanol.[20]

Burdock leaves (*Arctium lappa* L.) were used as a source of chlorogenic acid, caffeic acid and quercetin. These were obtained by a process involving simultaneous ultrasound and microwave-assisted extraction (ILUMAE) with ILs.[23] In this work, an anion-dependent extraction was also observed, but a different explanation was given. It has been mentioned that anion acidity favours the extraction of targeted compounds,[16,20] but the anion also displays an important role in the miscibility of IL with water. For instance, [H$_2$PO$_4$] anion was shown to decrease the efficiency of the IL in the extraction of polyphenols, due to difficult miscibility of [bmim][H$_2$PO$_4$] with water. On the other hand, highly water-miscible ILs [bmim][Br], [bmim][Cl] and [bmim][BF$_4$] definitely were the most efficient. Regarding the cation effect on the extraction, the same conclusion can be drawn. The results showed that increasing the alkyl chain length of imidazolium cation from ethyl to butyl dramatically increased the extraction efficiency. However, lower extraction was observed when the alkyl chain length increased from hexyl to octyl. Therefore, it is clear that chain length influences the extraction ability of IL and the alteration of IL miscibility with water could be a reason. Furthermore, incorporating functional groups in alkyl chain, such as sulfonic groups, decreases the extraction efficiency by 30–40%. It was observed that [SO$_3$Hbmim][Br] was slightly miscible with water, being the main cause of this weak extraction. Even though, differences in the extraction of each phenolic compound were observed for the most efficient ILs and the reason lies in the different solubilities in such media. Still, [bmim][Br] and [bmim][BF$_4$] presented the highest efficiencies on the overall extraction of chlorogenic acid, caffeic acid and quercetin.[23] Furthermore, using ILUMAE has the advantage to provide both effects of ultrasound and microwave energies enhancing the extraction. ILUMAE with [bmim][Br] presented 17% higher extraction efficiency than HRE and a significant reduction of extraction time from 5 h to 30 s. A higher performance of ILUMAE was also observed in comparison to regular UMAE.[23]

The utilization of water as co-solvent is frequently used in ILUAE and ILMAE processes. Row and co-workers studied the extraction effect of different co-solvents with ILs in *Chamaecyparis obtuse* leaves to obtain several flavonoids, namely dihydrokaempferol, quercitrin, amentoflavone and myricetin. Methanol and ethanol presented the best results due to the enhanced capacity to establish hydrogen bonds with hydroxyl groups of flavonoids. On the other hand, the low polar solvents acetone, acetonitrile, ethyl acetate, dichloromethane and hexane were demonstrated to be less efficient.[24]

Cajanus cajan L.[19] was used for the extraction of three flavonoids (genistin, genistein and apigenin) by another original methodology named as ionic liquid-based negative pressure cavitation-assisted extraction (ILNP-CAE). The apparatus used for this extraction is shown and described in Figure 5.1.

Basically, the process of cavitation is generated by negative pressure, and nitrogen is in a continuous flow through the solid/liquid system increasing the turbulence, collision and mass transfer between the IL and sample

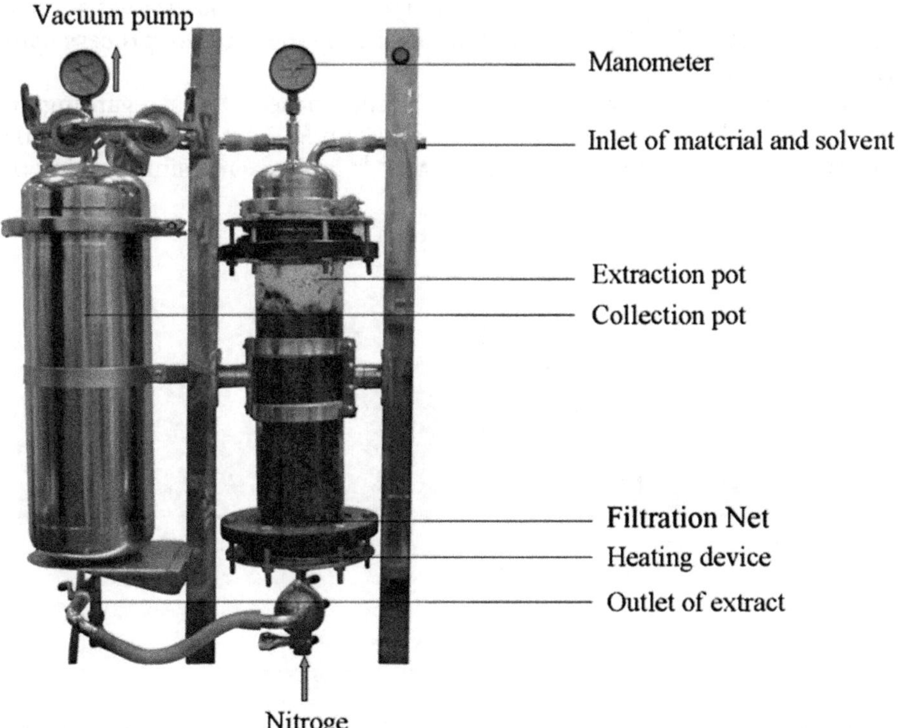

Figure 5.1   Equipment for ILNPCAE process. Reprinted from ref. 19 with permission from Elsevier.

matrix.[19,25,26] Figure 5.2 shows the extraction yields of genistin, genistein and apigenin for the examined ILs and it can be seen that [omim][Br] presented the best results.[19] This IL was also the optimum for the extraction of other flavonoids, such as tectoridin, iristectorin B and iristectorin A, from *Iris tectorum* Maxim after ILUAE process.[27]

Tannins were also attempted to be extracted from two different plants, namely catechu (*Acacia catechu*) and myrobolan (*Terminalia chebula*) using the distillable protic [DMA][CARB] IL.[28] At room temperature, the IL was mixed with the plant sources and subsequently an IL distillation was performed. Water was then added to the resulting solid, extracting other phenolic compounds such as ellagic acid, pyrocatechol and catechin hydrate. In fact, the contact of IL with plants allowed the alteration of the physicochemical properties of the desired tannins by an acid/base reaction. The researchers observed that extracted products with [DMA][CARB] were more basic compared to those extracted only with water. Therefore, the compounds were extracted in their conjugate base form and have been recovered as the dimethylammonium salt. This means that [DMA][CARB] has the ability to deprotonate tannins. This abnormal characteristic was found to be beneficial, once the presence of tannins as salts could be responsible for antifungal and antibacterial actions, which was confirmed by microbial assays of the extract products. In another study, *Galla chinensis* was successfully approached for the extraction of tannins by the ILUMAE process using [bmim][Br] IL.[29]

In general, several studies verified similar observations regarding the extraction of a variety of phenolic compounds from plants using ILs, either by applying ILMAE,[30–33] and ILUAE[34–36] processes or simply using the

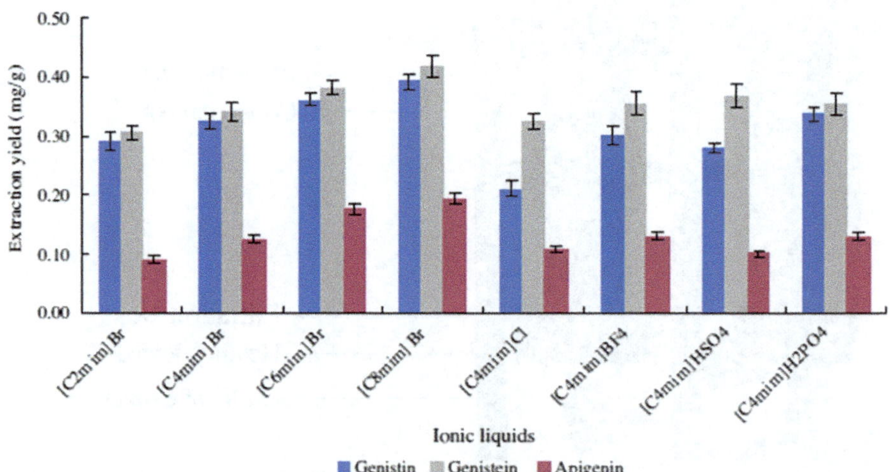

Figure 5.2   Extraction yields of genistin, genistein and apigenin obtained for several ILs after ILNPCAE process. [C2mim]=[emim]; [C4mim]=[bmim]; [C6mim]=[hmim]; and [C8mim]=[omim]. Reprinted from ref. 19 with permission from Elsevier.

conventional solid–liquid extraction (SLE) methodology.[24] The main achievements of these works are compiled in Table 5.1.

5.2.2 Alkaloids

Plants are also rich in alkaloids and their extraction for multiple pharmaceutical and medical applications has been constantly required. For example, lotus leaf *Nelumbo nucifera* Gaertn. was used successfully with ILMAE[37] for the extraction of the three alkaloids, *N*-nornuciferine, *O*-nornuciferine and nuciferine, that have a profile of action associated with dopamine receptor blockade. Different imidazolium ILs were used, changing the cation and/or the anion, and it seemed that the extraction of those alkaloids was anion dependent. Similarly to the conclusion drawn in the extraction of polyphenols,[23] Ma *et al.*[37] stated that anions greatly influence the miscibility of IL with water, influencing the extraction of alkaloids. Among four anions studied ([Cl], [Br], [BF$_4$], [PF$_6$]), only the former three demonstrated good extraction efficiency, because of their hydrophilic property. On the other hand, the hydrophobic [PF$_6$] is not able to extract alkaloids, as it barely mixes with water. The authors[37] also verified that the change in the alkyl chain of the imidazolium cation from ethyl to hexyl enhances the extraction of alkaloids, but for the change from hexyl to octyl a pronounced decrease in the extraction yield was observed. Herein, it is important to notice that [hmim] cation increases the hydrophobicity degree of IL, but the miscibility with water is not compromised, contrary to [omim] cation. Structurally, alkaloids also present hydrophobic regions, in which hexyl side chains attached to imidazolium cation ring could favourably interact with it.[37]

Regarding the cation ability to extract alkaloids, different results were demonstrated by Yang *et al.*[38] in *Catharanthus roseus* leaves for the extraction of vindoline, catharanthine and vinblastine, after applying ILUAE methodology. The data showed that increasing the alkyl chain from ethyl to octyl allows better extraction yields, confirming that the hydrophobicity favours the extraction of the three alkaloids. However, using [amim] cation, which only possesses a three-carbon atom side-chain in the cation, presented the highest extraction efficiency among all cations tested. In this specific case, authors found an explanation on the unsaturated vindoline, catharanthine and vinblastine, which are better dissolved by an unsaturated IL, such as [amim][Br]. The choice for [Br] anion was made after preliminary study of anion effect on the extraction of alkaloids demonstrating to have higher efficiency than other inspected anions ([Cl], [I], [HSO$_4$], [TsO], [ClO$_4$], [NO$_3$] and [BF$_4$]). Furthermore, in order to retain the real effect of [amim][Br], water and NaBr salt were used as blanks for the extraction process. Much lower extraction yields were obtained in comparison to ILUAE process, confirming that both the cation and the anion of [amim][Br] display important roles in the extraction of biomolecules. The researchers[38] also observed an accumulative alkaloid concentration in process after process, and determined three cycles as worthy for an efficient extraction of the three alkaloids from *C. roseus* leaves, as observed in Figure 5.3. This discloses an important issue about IL recycling and reuse, which can be successfully accomplished on the extraction of biomolecules.[38]

In order to explain the mechanism behind the extraction of alkaloids from plants with ILs, Bogdanov *et al.*[39,40] performed a kinetic study from plant *Glaucium flavum* Crantz involving the extraction of *S*-(+)-glaucine using [bmim][Ace] IL or methanol.[39,40] After several and simple solid/liquid extraction experiment trials, it was observed that the glaucine yield was always higher with [bmim][Ace] than methanol, independent of temperature. Actually, maximum glaucine extraction could be achieved at room temperature. This is justified by the high solubility of glaucine in IL due to multiple interactions, as previously described in other works.[16,18,19,22] The results of the kinetic study allowed the demonstration of a fast diffusion of IL in the first moments of extraction. A better wetting property of [bmim][Ace] at the initial moment occurs in contrast to methanol, due to the higher hydrogen bonding ability of IL. It was assumed that IL affected the cell tissues in a greater extent than methanol and reduced the cell wall resistance, leading to higher extraction yields of glaucine. Furthermore, the researchers predicted a possible extraction mechanism based on two consecutive steps: (i) penetration of the [bmim][Ace] into the cellular structure of *Glaucium flavum* Crantz, followed by solvation of glaucine by [bmim][Ace] anions, and (ii) convection of the glaucine through the porous structure of the residual solids of the plant and its transfer into the IL aqueous solution.[39] Although an unusual IL ([bmim][Ace]) was used in this work,[39] it is important to notice that this extraction mechanism could explain the extraction of several other biomolecules from plant materials using ILs with similar properties.

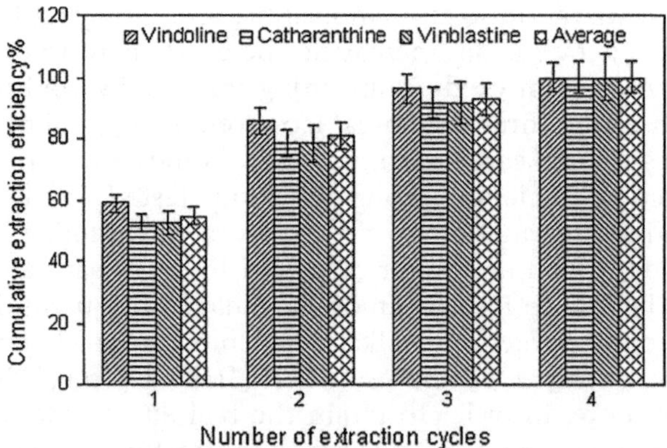

Figure 5.3 Influence of extraction cycles on the extraction of vindoline, catharanthine and vinblastine with [amim][Br]. Reprinted from ref. 38 with permission from Elsevier.

5.2.3   Terpenes

Salvia miltiorrhiza Bunge was used for the extraction of several diterpene tanshinones by advanced and exclusive technology IL-based ultrahigh-pressure extraction (ILUPE).[41] Among several studied anions, the hydrophobic [PF$_6$] and [NTf$_2$] showed high extraction efficiencies of tanshinones as expected, while hydrophilic anions were delayed. Nevertheless, the high costs in the synthesis of [NTf$_2$]-based ILs allowed researchers to choose those containing the [PF$_6$] anion. Regarding the hydrophobicity effect, longer alkyl chain lengths of imidazolium cation, such as [omim] also favoured the extraction of tanshinones, for the same reasons mentioned in previous work.[42] However, a lower extraction yield was achieved for alkyl chain lengths higher than octyl, as high viscous solutions were produced hindering the penetration of IL into the sample. Finally, when pressure was applied up to 300 MP during extraction, an increase of tanshinone extraction yield was observed. Basically, the differential pressure between the inner and the exterior of the plant cells is very large under high pressure conditions, allowing the disruption of cell walls, as demonstrated in Figure 5.4. This clearly enhances IL to permeate through the broken membranes into cells and subsequently to extract tanshinones. Actually, high pressure allows disruption of the hydrogen, electrostatic, van der Waals and hydrophobic bonds, but the covalent bonds are not broken, which means that the biomolecule structure does not change under the ILUPE process.[41,43] Maximum amounts of tanshinones were extracted from *S. miltiorrhiza* Bunge in only 2 min. Another remarkable achievement is the extraction of the target biomolecules by using this technology at room temperature.[41]

Coleus forskohlii roots were used to provide forskolin, another diterpene with biological activities, which was extracted by ILUAE process using [TMG][Lac], a different class of IL.[44] Among other types of examined cations, such as imidazolium, ammonium and pyridinium, the cation demonstrated a higher ability to extract forskolin. Furthermore, as depicted in previous work,[45]

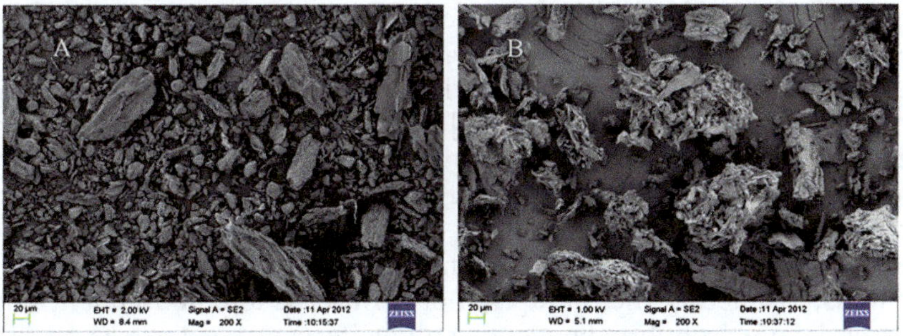

Figure 5.4   Acquired images from scanning electron microscope of *S. miltiorrhiza* Bunge samples before (A) and after (B) ILUPE process using [omim][PF$_6$] IL. Reprinted from ref. 41 with permission from Elsevier.

the [Lac]⁻ anion possesses high hydrogen basicity allowing it to establish a strong hydrogen bonding network. Joining both guanidinium cation and [Lac] anion produces a strong basic nature IL (pH = 10.9), which favours the extraction of forskolin.[44]

5.2.4   Lactones

Lactones from *Ligusticum chuanxiong* Hort (senkyunolide I, senkyunolide H and Z-ligustilide), were extracted using ammonium-based ILs including [DMHEEA][PA] and [DMCEA][PA] by the ILMAE process.[46] The strong interactions between ILs and the plant, along with the assisted microwave irradiation, allowed fast extraction of all lactones within 1 and 5 minutes for [DMHEEA][PA] and [DMCEA][PA] ILs, respectively. The reuse of protic ILs was investigated and the extraction yields dramatically decreased after three cycles. The reason lied on the accumulation of other compounds from *L. chuanxiong* Hort during the recycling cycles, which compete with lactones for solvation by both cation and anion of ILs. Furthermore, an increase of IL viscosity was observed after each cycle, thus decreasing the mass transfer even further. Therefore, a back-extraction of IL was tried with hexane to remove some of the accumulated compounds. However, along with contaminants, hexane also removed Z-ligustilide from IL, thus, more selective organic solvents should be examined for this purpose.[46]

5.2.5   Essential Oils

Essential oils from *Dryopteris fragrans* were obtained by another ground-breaking technology based on the integration of ILMAE process followed by microwave-assisted hydro-distillation (MHD).[47] At the first stage, the powdered plant was subjected to ILMAE process, where the microwave irradiation along with the penetration of IL within the plant matrix allowed the release of essential oils. Four ILs, [bmim][Br], [bmim][Cl], [amim][Cl] and [emim][OAc], which possess the ability to pre-treat biomass,[48] were used as pre-treatment agents of *D. fragrans*. After hydro-distillation, the yield of extracted essential oils decreased in the following order: [emim][OAc] > [amim][Cl] > [bmim][Cl] > [bmim][Br] > deionized water. For optimized conditions, maximal extraction of essential oils reached 0.91% with [emim][OAc], contrasting to 0.28% obtained with water. Jiao *et al.*[47] explained the enhanced extraction efficiency of ILs, mostly [emim][OAc], in two consecutive stages: (i) the IL is allowed to significantly dissolve cellulose from plant cell walls, promoting the disruption of plant cells and subsequent release of essential oils from broken glands and oleiferous receptacles; and (ii) the hydro-distillation process further leads to fast separation of essential oils from crude extract, due to water–essential oil emulsions mediated by the surfactant property of IL.[47]

5.2.6 Simultaneous Extraction of Biomolecules

The maximal exploitation of biomass should always be taken into account in the research and development of new processes. Therefore, the simultaneous extraction of different type of biomolecules and further separation is paramount. For instance, an innovative methodology, named ionic liquid-based microwave-assisted simultaneous extraction and distillation (ILMASED), was developed and applied to *Rosmarinus officinalis* (rosemary) for the simultaneous extraction of carnosic acid and rosmarinic acid and the distillation of essential oils.[49] This integrates the conventional two-step methodology in only one step. In this work, 1-alkyl-3-methylimidazolium-type ILs with different anions ([Cl], [Br], [NO$_3$] and [BF$_4$]) and different alkyl chain lengths in the cation were studied. The data showed different extraction efficiencies for both carnosic and rosmarinic acids. Better extraction performance of both compounds was attained with [Br] anion, while [NO$_3$] and [BF$_4$] anions were specifically efficient for the extraction of carnosic acid and rosmarinic acid, respectively. Therefore, the extraction of phenolic acids with ILs was considered as anion-dependent.

In contrast, the four anions studied showed no significant effect on the extraction of essential oils. Therefore, similarly to the extraction of some phenolics previously described in other works,[19,27] [omim][Br] demonstrated higher efficiency for the extraction of both carnosic and rosmarinic acids.

Plant source materials allowed the IL-assisted extraction of other type of compounds, such as naphthoquinones,[50] steroids,[41,51,52] acetophenones[53] and organic acids.[54] Similar conclusions were presented and the main results are also depicted in Table 5.1.

5.3 Seeds

Seeds, an embryonic stage of plants, are also a rich source of multiple biomolecules. Generally, the type of biomolecules being investigated is very similar to those found in plants. For example, alkaloids are one of the major interesting compounds from seeds, which have been obtained by extraction with ILs.[55-57] However, other types of biomolecules, such as lipids and organic acids, have been also object of study for the extraction with IL technologies.[13,58] A summary of the main data concerning the extraction of biomolecules with ILs from seeds is shown in Table 5.2.

5.3.1 Alkaloids

The embryonic seed of *Nelumbo nucifera* Gaertn. was studied using the ILMAE process[55] for the extraction of the alkaloid liensinine and its analogues isoliensinine and neferine. In this specific case, imidazolium-based ILs composed of [BF$_4$] anion demonstrated higher extraction efficiency than those with [Cl], [Br] and [PF$_6$] anions. In fact, the hydrophobicity of [bmim][PF$_6$]

Table 5.2 Seeds used for biomolecule extraction with ionic liquids.

| Biomass | Extraction technology | Target biomolecules | | Ionic liquids | Extraction yield | Ref. |
		Class	Name			
Brassica napus	SLE	Lipids	Lipids	[emim][MeSO$_4$]	44^a	13
Camptotheca acuminata	ILMAE	Alkaloids	Camptothecin 10-Hydroxycamptothecin	[amim][Br]; [bmim][BF$_4$]; [bmim][Br]; [bmim][Cl]; [bmim][ClO$_4$]; [bmim][HSO$_4$]; [bmim][NO$_3$]; [bzmim][Br];	≈0.5^b/[omim][Br] ≈0.1^b/[omim][Br]/[bmim][Br]	61
	ILUAE	Alkaloids	Camptothecin 10-Hydroxycamptothecin	[emim][Br]; [hmim][Br]; [omim][Br]; [pmim][Br]	– [omim][Br] – [omim][Br]	60
Fructus forsythiae	ILMAE	Essential oils	α-Pinene; α-terpineol; β-pinene; camphor; limonene; linanol	[amim][Cl]; [bmim][Br]; [bmim][Cl]; [emim][OAc]	9.58^c/[emim][OAc]	62
Illicium verum	SLE + reaction	Organic acids	Shikimic acid Shikimic acid ethyl ester	[eim][HSO$_4$]; [emim][HSO$_4$]; [SO$_3$Hbmim][Br]; [SO$_3$Hbmim][Cl]; [SO$_3$Hbmim][H$_2$PO$_4$]; [SO$_3$Hbmim][HSO$_4$]; [SO$_3$Hbmim][NTf$_2$]	≈10^a/[SO$_3$Hbmim][H$_2$PO$_4$] 12.7^a/[SO$_3$Hbmim][H$_2$PO$_4$]	58
Jatropha	SLE	Terpenes	Phorbol esters	[emim][MeSO$_4$]; [emim][OAc]	5.8^b/[emim][MeSO$_4$]	63
Jatropha	SLE	Lipids	Lipids Proteins	[emim][MeSO$_4$]	49.9^a 8–10^a	13
Kamani	SLE	Lipids	Lipids	[emim][MeSO$_4$]	38.0^a	13
Nelumbo nucifera	ILMAE	Alkaloids	Isoliensinine, liensinine, neferine	[bmim][BF$_4$]; [bmim][Br]; [bmim][Cl]; [bmim][PF$_6$]; [emim][BF$_4$]; [hmim][BF$_4$]; [omim][BF$_4$]	– [bmim][BF$_4$]/[hmim][BF$_4$]	55
Paullinia cupana	SLE	Alkaloids	Caffeine	[bmim][Cl]; [bmim][TsO]; [bmpyrr][Cl]; [emim][Cl]; [emim][OAc]; [OHemim][Cl]	9.43^a/[bmim][Cl]	57
Piper nigrum	ILUAE	Alkaloids	Piperine	[bmim][BF$_4$]; [bmim][Br]; [bmim][H$_2$PO$_4$]; [bmim][PF$_6$]; [SO$_3$Hbmim][Br]; [hmim][BF$_4$]	3.6^a/[bmim][BF$_4$]	56
Pongamia	SLE	Lipids	Lipids	[emim][MeSO$_4$]	11.0^a	13

^{a}% (g of biomolecule/100 g of biomass).
^{b}Yield (mg of biomolecule/g of biomass).
^{c}Yield (mL of biomolecule/g of biomass); SLE – solid–liquid extraction; ILMAE – ionic liquid-based microwave-assisted extraction; ILUAE – ionic liquid-based ultrasound-assisted extraction.

was found to hinder the extraction of target alkaloids. Similarly, as observed in the study of the extraction of phenolic compounds from burdock leaves,[23] increasing the alkyl chain length of imidazolium cation reduces the miscibility of IL with water and increases the viscosity of the mixture. Those factors compromised the extraction of alkaloids from *N. nucifera* seeds.[55] Similar conclusions were drawn in the extraction of the alkaloid piperine from white pepper by an ILUAE process.[56] The extraction efficiency increased with the hydrophilicity of the anions as the following order: $[PF_6] < [H_2PO_4] < [Br] < [BF_4]$, where the hydrophilic [bmim][BF_4] was the most efficient IL. Again, increasing alkyl chains reduced the IL ability to extract piperine. Furthermore, the incorporation of sulfonic groups in cation decreases the miscibility with water and the hydrogen bonding ability of IL to extract piperine.[56] This was also observed in a previously revised study.[23]

Caffeine is a very well-known alkaloid present in coffee beans, but it is also present in guaraná seeds (*Paullinia cupana*) with two or three times higher concentration.[57] An extraction methodology of caffeine from guaraná seeds using aqueous ILs without any assisted energy (ultrasound or microwave) was successfully performed. In the first screening trials, [bmim][Cl] was used and contrasted to the utilization of pure water and NaCl aqueous solutions (Figure 5.5). Using NaCl as the extracting agent, the extraction yields obtained were the lowest. In fact, NaCl acted as the salting-out agent, reducing the solubility saturation of caffeine in the extracting solution. On the other hand, although extracting caffeine with water was better than using

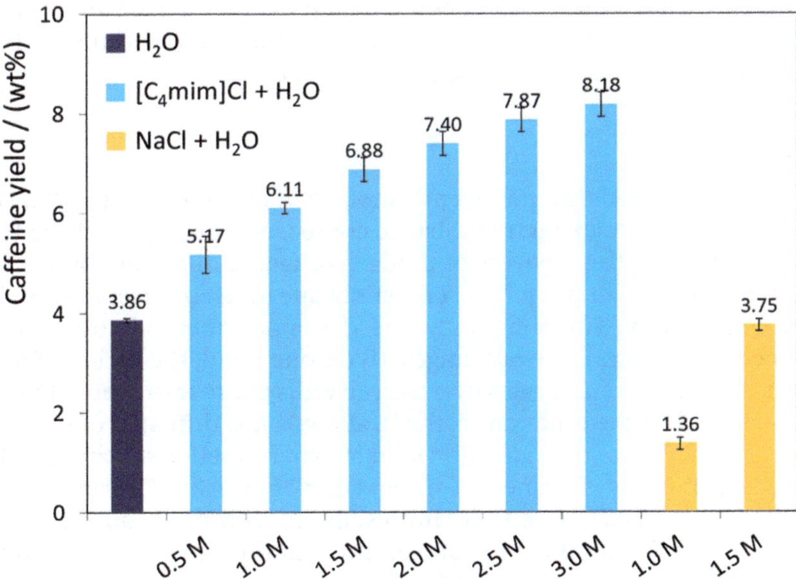

Figure 5.5   Yields of caffeine for the extractions with water, diverse concentrated [bmim][Cl] aqueous solutions and NaCl aqueous solutions C₄mim=bmim. Reproduced from ref. 57 with permission from The Royal Society of Chemistry.

NaCl solution, the highest yields were obtained with IL aqueous solution.[57,59] Thereafter, several ILs, including [bmim][Cl], [bmim][TsO], [bmpyrr][Cl], [emim][Cl], [OHemim][Cl] and [emim][OAc], were screened in aqueous solutions for the extraction of caffeine. Claudio *et al.*[57] observed that the anion does not play a significant effect in the extraction of caffeine. Still, the best result was attained with [bmim][TsO], due to the aromatic nature of its anion, which could perform π–π non-covalent interactions with the target compound. However, the same argument is limited when comparing [bmim][Cl] and [bmpyrr][Cl], as no significant difference was observed in the extraction yields. Nevertheless, the authors chose [bmim][Cl] for further experiments rather than [bmim][TsO], since the differences in the extraction efficiencies were not significant, the former IL is more stable, and the synthesis process is cleaner and cheaper.[57] At optimized conditions, a maximum 9% extraction of caffeine yield was obtained with [bmim][Cl]. Further recycling and reuse of IL was performed up to three times and remarkably around 9% wt. of caffeine was always extracted with the overall solution. Thus, a three times higher caffeine concentration was reached in the overall solution without reaching saturation.[57] Several organic solvents were examined to extract caffeine from IL solutions and butanol was found to be a good candidate for extracting caffeine, thus replacing the more toxic and hazardous ethyl acetate currently used for this purpose.[57]

Camptotheca acuminata seeds were used in two different works to extract two alkaloids with remarkable anticancer activity (camptothecin and 10-hydroxycamptothecin) by ILUAE[60] and ILUMAE[61] processes. In both cases, [omim][Br] demonstrated the highest efficiency among several examined ILs for the extraction of those alkaloids, which means that the ability of ILs for extraction is not dependent of the type of irradiation energy.

5.3.2   Lipids

Different type of seeds, namely from Canola, *Jatropha*, Kamani and *Pongamia*, were subjected to an extracting mixture composed of an hydrophilic IL and a polar solvent for the recovery of lipids.[13] In fact, a different approach was undertaken, where an IL was chosen mainly due to its immiscibility with the target compounds. Fundamentally, the developed methodology could be explained in two simultaneous stages: (i) on one hand, the action of a polar solvent, such as methanol, acts over the cell walls in order to expose lipids out of the cells; (ii) on the other hand, the IL allows a fast diffusion of lipid molecules to the surface of the extraction system, where a self-associating and separate phase of lipids is formed, as observed in Figure 5.6. Therefore, [emim][MeSO$_4$] IL was chosen due to its immiscibility with lipids and the relative low viscosity and solubility in polar solvents. Among several examined polar solvents, methanol promoted the highest extraction yield of lipids. Furthermore, the good miscibility with [emim][MeSO$_4$] and the advantage of using it as a reactant in the direct transesterification of lipids to fatty acid methyl esters (FAMES), makes methanol a good co-solvent for this technology.

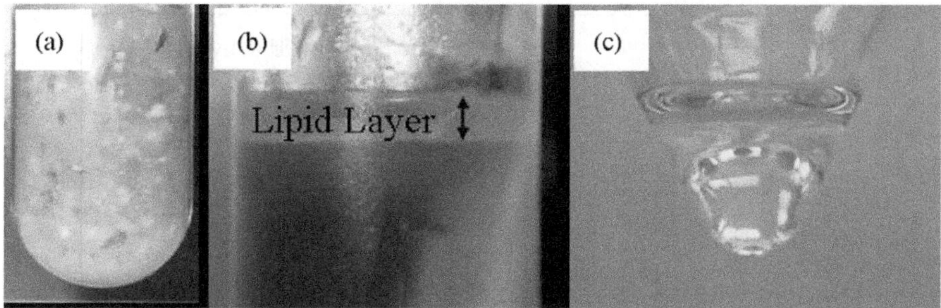

Figure 5.6　Extraction of lipids from seeds with [emim][MeSO$_4$] and polar solvent mixture. (a) Seed suspension in the co-solvent mixture; (b) formation of self-associating and separate phase of lipids in the surface during extraction; and (c) recovery of lipid fraction. Reprinted from ref. 13 with permission from Elsevier.

5.3.3　Organic Acids

Star anise seeds (*Illicium verum*) were used for the extraction of shikimic acid, an important biochemical metabolite in plants, by a different type of ILs with simultaneous transformation into pharmaceutical intermediates, such as shikimic acid ethyl ester.[58] Ressmann *et al.*[58] verified that using Brønsted acidic ILs, a simultaneous dual role of solvent and catalyst could be displayed. For screening trials, star anise seed powder was added to a Brønsted acidic IL/anhydrous ethanol mixture, and the reaction was performed at 80 °C for 24 h. As observed in Figure 5.7, [SO$_3$Hemim][HSO$_4$] and [SO$_3$Hemim][NTf$_2$] were the most efficient at performing both solvent and catalyst actions.

Seeds were also used for the extraction of other type of compounds, such as essential oils[62] and terpenes,[63] using ILs. Similar conclusions were drawn and the principal results are depicted in Table 5.2.

5.4　Lignocellulosic Feedstocks

Lignocellulosic residues are one of the most abundant and renewable sources in the world that can be used as raw materials for further valorization into biomolecules. The composition of this feedstock is based on three main macromolecules, namely cellulose, hemicellulose and lignin.

- Cellulose, which is solely composed of glucose carbohydrate units, is probably the least valuable macromolecule and it is mostly related to high volume/low value biofuel production.
- On the other hand, hemicellulose is composed of different carbohydrate monomers, whose structure and composition may differ depending on the lignocellulosic biomass source.
- Lignin possesses a complex aromatic structure that can be further depolymerized to obtain value-added aromatic compounds.

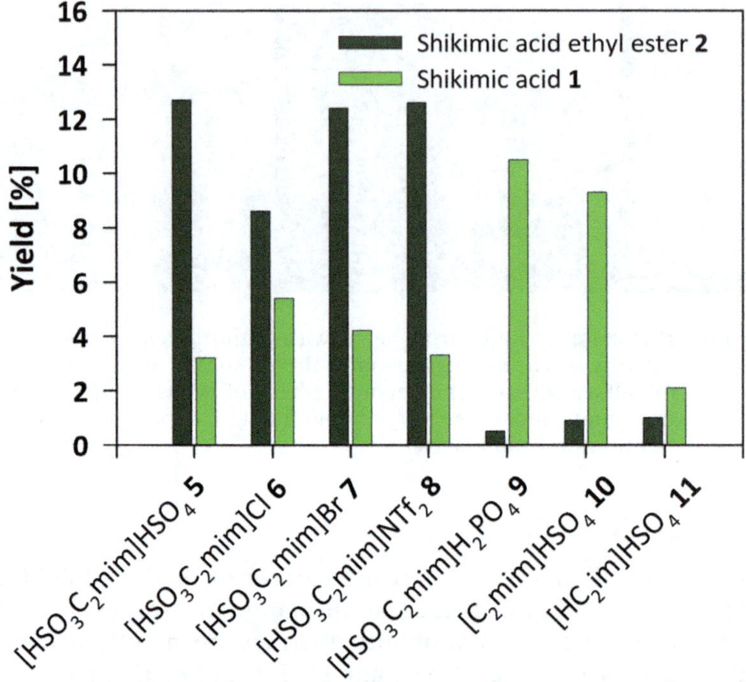

Figure 5.7   Ability of ILs to simultaneously extract and convert shikimic acid from star anise seeds. Ref. 58 with permission from The Royal Society of Chemistry.

Thus, the selective extraction of lignin[64,65] and hemicellulose[66] from lignocellulosic biomass using ILs has been studied. Besides the interesting valorization of the main lignocellulosic fractions, other minority and valuable compounds, such as polyphenols[7,67,68] and terpenes,[69] could also be extracted from this type of biomass using ILs. The main results of selective extraction of biomolecules from lignocellulosic materials with ILs are depicted in Table 5.3.

5.4.1   Lignin

Sugar cane bagasse was used by Tan *et al.*[64] for the first study focusing on the selective extraction of lignin from lignocellulosic biomass with ILs. A specially designed IL based on [emim] cation with a mixture of alkylbenzenesulfonate and xylenesulfonate anions ([emim][ABS]) was synthesized for that purpose. The treatment of biomass with [emim][ABS] was performed at high temperatures (170 and 190 °C). Then, NaOH aqueous solution was added to aid lignin extraction from the liquid stream and to precipitate the polysaccharide fraction of sugar cane, simultaneously. Above 93% extraction yield was achieved with this process and lignin could be easily recovered by

Table 5.3 Lignocellulosic feedstocks used for biomolecule extraction with ionic liquids.

| Biomass | Extraction technology | Target biomolecules | | Ionic liquids | Extraction yield | Ref. |
		Class	Name			
Barley hull	ILPFE	Phenolic compounds	Phenolics	[MHEA][OAc]	189.1[a]	68
		Polysaccharides	Polysaccharides		54.2[a]	
Birch bark	ILMAE	Terpenes	Betulin	[amim][Cl]; [bmpyr][Cl]; [bpy][Cl]; [C$_n$mim][Cl] (n = 2–14); [emim][OAc]; [emim][BF$_4$]; [emim][PF$_6$]; [emim][NTf$_2$]; [mim][Cl]; [mmim][OAc]	31.7[b]/[emim][OAc]	69
Birch outer bark	SLE	—	Suberin	[Ch][Hex]	48.4[b]	81
Bamboo biomass	SLE	—	Lignin	[C$_2$CNaim][Cl]; [C$_2$CNbim][Cl]; [C$_2$CNbzim][Cl]; [C$_2$CNeim][Cl]	66[b]/[C$_2$CN-Bzim][Cl]	73
Bamboo biomass	SLE		Lignin	[emim][Gly]	85.3[b]	74
Cork	SLE	—	Suberin	[emim][Hex]; [Ch][Hex]; [Ch][Oct]; [Ch][Dec]	67.2[b]/[Ch][Hex]	80
Cork	SLE	—	Suberin	[Ch][Hex]	57.9[b]	81
Corn stover	SLE	—	Lignin	[mim][OAc]; [pyr][OAc]; [pyrr][OAc]	72.81[b]/[pyrr][OAc]	10
Apple tree prunings	ILMAE	—	Lignin	[bmim][MeSO$_4$]	18.8[c]	76
Poplar wood	SLE	—	Lignin	[emim][OAc]	5.8[c]	70
Sugar cane bagasse	SLE	—	Lignin	[emim][ABS]	93[b]	64
Sugar cane (leaf, tip, stem and root)	ILUAE	Phenolic compounds	Flavonoids; anthocyanins	[bmim][PF$_6$]	≈14[c]	67
Spruce wood chips	SLE	Polysaccharides	Hemicellulose	Switchable ILs (DBU + butanol + CO$_2$)	38[b]	66
Triticale straw	SLE	—	Lignin	[bmim][Cl]; [emim][OAc]; [DMEA][OAc]; [DMEA][FA]; [DMEA][GA]; [DMEA][SA]	52.7[b]/[emim][OAc]	65
Wood flour	SLE	—	Lignin	[emim][OAc]	86[b]	71

[a]Yield (mg of biomolecule/g of biomass).

[b]% (mass of biomolecule/initial mass content of biomolecule in the original biomass × 100).

[c]% (g of biomolecule/100 g of biomass); ILPFE – ionic liquid-based pressurised fluid extraction; ILMAE – ionic liquid-based microwave-assisted extraction; SLE – solid–liquid extraction; ILUAE – ionic liquid-based ultrasound-assisted extraction.

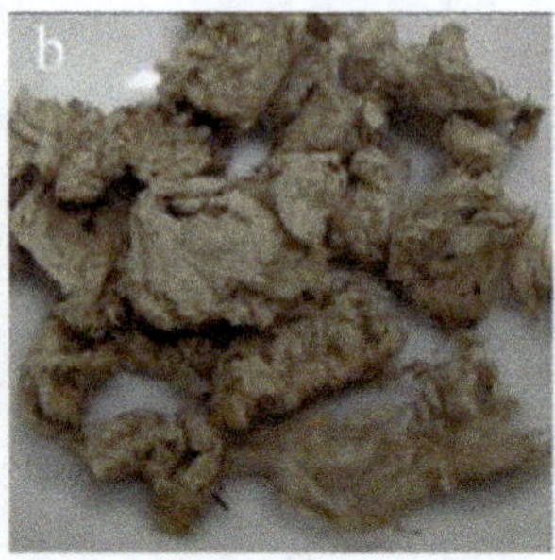

Figure 5.8 Extraction of lignin from sugar cane bagasse with [emim][ABS] IL. (a) Untreated sugar cane bagasse; (b) treated sugar cane bagasse; (c) extracted lignin. Ref. 64 with permission from The Royal Society of Chemistry.

acidifying the extracting solution with HCl (Figure 5.8). However, the possible formation of reactant species between lignin and xylenesulfonate at such high temperatures is a disadvantage of using [emim][ABS]. Additionally, some inconvenient extraction of hemicellulose was also observed.[64]

Several types of biomass, such as agriculture residues and wood materials, was used for lignin extraction[65,70,71] using a more benign and cleaner IL, such as [emim][OAc], which efficiently dissolved lignocellulosic biomass at lower temperatures.[72] Approximately 40% extraction yield was obtained with [emim][OAc] from wood flour, after 5 hours at 90 °C. The extraction yield of lignin more than doubled (86%) after 70 h treatment.[71] Furthermore, Fu *et al.*[65] examined the pre-treatment of triticale straw and extraction of lignin with [emim][OAc] and other ILs, such as [bmim][Cl] and ammonium-based ILs, at 90 °C for 24 h under a nitrogen atmosphere.[65] The extraction of lignin with [emim][OAc] was 30.3% of the original content in triticale straw, which was 2 and 10-fold higher than for [bmim][Cl] and [DMEA][FA] lignin extractions, respectively. Wheat straw and flax shives were also pre-treated with [emim][OAc] reaching 29.6% and 14.0% lignin extraction yield, correspondingly. Nevertheless, when temperature was increased to 150 °C, 52.7% extraction yield was attained in only 1.5 hours for the treatment of triticale straw with [emim][OAc].[65] It is important to discriminate that the type of biomass strongly affects the extraction of lignin, due to the differences in composition and structural interactions occurring. In a different study, Kim *et al.*[70] further investigated the structure features of [emim][OAc] extracted lignin, such as decomposition, functional groups, average molecular weight and polydispersity index.[70] The capacity of IL to extract lignin from wood was related to physical and/or chemical interactions of nitrogen from imidazolium cation with electron-rich oxygen (oxygen atoms involved in β-O-4 and α-O-4, and β–β linkages) present in the lignin structure. The extracted lignin consisted of high levels of free-phenolic moieties resulting from possible partial depolymerization of lignin during extraction. Furthermore, guaiacyl-type lignin was more extractable than syringyl-type. However, the authors showed that [emim][OAc]-extracted lignin decomposes faster than milled wood lignin. Although a lower average molecular weight was observed, the extracted lignin with IL demonstrated to be uniform (polidispersity index close to 1).[70]

Bamboo biomass was also studied for lignin extraction using nitrile-based ILs.[73] Rafiq and co-workers[73] found that unsaturated bonds present in [C_2CNbzim][Cl] and [C_2CNaim][Cl] favoured lignin extraction (53% and 47%, respectively) than without them (38% using [bmim][Cl] IL). Furthermore, the increased lignin extraction is possibly due to more π–π interaction between nitrile-based ILs and the lignin π system. Therefore, a clear difference in cation structure affects lignin extraction. The highest lignin extraction yield (66%) was obtained using [C_2CNbzim][Cl] at 160 °C for 4 h. The [C_2CNbzim][Cl] IL was recycled and reused up to three times without losing efficiency (53%, 54.2% and 52.5% for repeated recycling cycles 1, 2 and 3, correspondingly).[73] The same authors presented an IL composed of amino acid-based anion ([emim][Gly]), which demonstrated enhanced ability to extract lignin.[74] A significant lignin removal of 85.3% of total content in bamboo biomass was accomplished after pre-treatment at 120 °C for 8 h (5% w/w). In this case, using acetone/water mixture (7/3 vol/vol) as the anti-solvent was found to be advantageous to obtain such high lignin extraction yield.[74] Following the same line of research, environmental friendly cholinium based-ILs also composed of amino acid-based anions were screened in the extraction of lignin from rice straw.[75] From the obtained data, [Ch][Gly], [Ch][Phe] and [Ch][Lys] were found to favour the lignin extraction at 90 °C, where maximum 60.4% lignin extraction yield was attained with [Ch][Lys].[75]

Corn stover and apple tree prunings were also subject to lignin extraction using other ILs, such as [pyrr][OAc][10] and [bmim][MeSO$_4$],[76] respectively.

5.4.2 Hemicellulose

Spruce was used for a selective extraction of hemicellulose by processing with switchable ionic liquids (SILs), a very attractive technology that has been lately disclosed for biomass processing.[77,78] The formation of the mentioned SILs occurs *in situ* by adding an equimolar mixture of 1,8-diazabicyclo-[5.4.0]-undec-7-ene (DBU) with alcohol (hexanol or butanol) under CO_2 atmosphere (also acting as a reagent), ambient pressure and at room temperature. The molecular liquid mixture is exothermically transformed into an IL composing of alkyl-carbonate cation (CO_2 plus alcohol) and anion (DBU base). After removal of CO_2, the molecular liquid mixture is formed again. The extraction of hemicellulose was processed at 55 °C for 5 days. The obtained results showed higher extraction efficiency using butanol SIL (36 wt.%) than hexanol SIL (14 wt.%). Furthermore, lignin extraction was practically negligible (around 2% extraction from native biomass) and cellulose was maintained intact after processing with both SILs. A successful selective extraction of hemicellulose from wood was achieved using an innovative and low energy required process (performed at 55 °C), although for a long period. Furthermore, the possibility of switching the IL back to its molecular components turns the process green and safe, as well leading to the achievement of desired recycling yields near 100% and to economical reuse.[66]

5.4.3 Suberin

A different type of lignocellulosic biomass with huge commercial value is cork, which has been experimented for dissolution and selective extraction of the main biopolymer suberin through IL utilization.[79–81] Cholinium alkanoates were demonstrated to be efficient ILs in the dissolution of cork. In addition, these ILs have been classified as biodegradable, biocompatible and of low toxicity. Pereira and co-workers[79–81] verified that the extraction efficiency increases with the length of the anion alkyl chain and progressively with its basicity. The highest alkyl chain of IL anion tested ([Hex]) allowed the most efficient dissolution of cork, especially suberin biopolymer. Furthermore, it was mentioned that hydrogen bond basicity is the key factor to dissolve cork, where [Hex] anion presented the best results.[79] However, this is also a key factor for cellulose dissolution, but small anions, such as [Cl] and [OAc], are preferable. Therefore, due to high porosity of cork, larger and high basicity anions, like hexanoate, provide better molecular interactions with cork leading to high dissolution yield. This also explains the specific extraction of suberin instead of the structural cellulose present in cork.[80,81] Also, the cholinium cation seemed to display an important role in suberin extraction. Using [Ch][Hex] a suberin extraction yield of 67.2% was reached, while 1-[emim][Hex] extracted less than a half of that value (30.6% yield). Moreover, 99% recovery of cholinium IL was accomplished and, subsequently, it showed no loss of extraction efficiency when reused. In the end, the extracted material showed suberin typical features, as demonstrated in Figure 5.9, with an aliphatic and esterified nature along with high thermal resistance.[80,81]

Furthermore, only partial hydrolysis of suberin was detected, indicating that the extracted material presented similar properties as the *in situ* suberin from original materials, like cork or birch outer bark.[81]

5.4.4 Phenolic Compounds

Lignocellulosic residues have also been used for the extraction of phenolic compounds mediated by ILs.[7,67,68] For instance, an ILUAE process was recently applied for the extraction of flavonoids and anthocyanins from different parts of sugar cane, including leaves, tips, stems and roots.[67] ILs composing different anions ([BF$_4$], [PF$_6$],[Br] and [MeSO$_3$]) and cations ([emim], [bmim] and [hmim]) were examined for this purpose. Among them, [bmim][PF$_6$] was the most efficient in the extraction of these compounds (14 g/100 g biomass). Each examined part of sugar cane was extracted with [bmim][PF$_6$] and the extract samples were screened by the DPPH method to estimate bioactivity. As depicted in Figure 5.10, the highest bioactivity yield was obtained for sugar cane tip extract, which was composed of luteolin-8-*C*-rhamnosylgucoside,

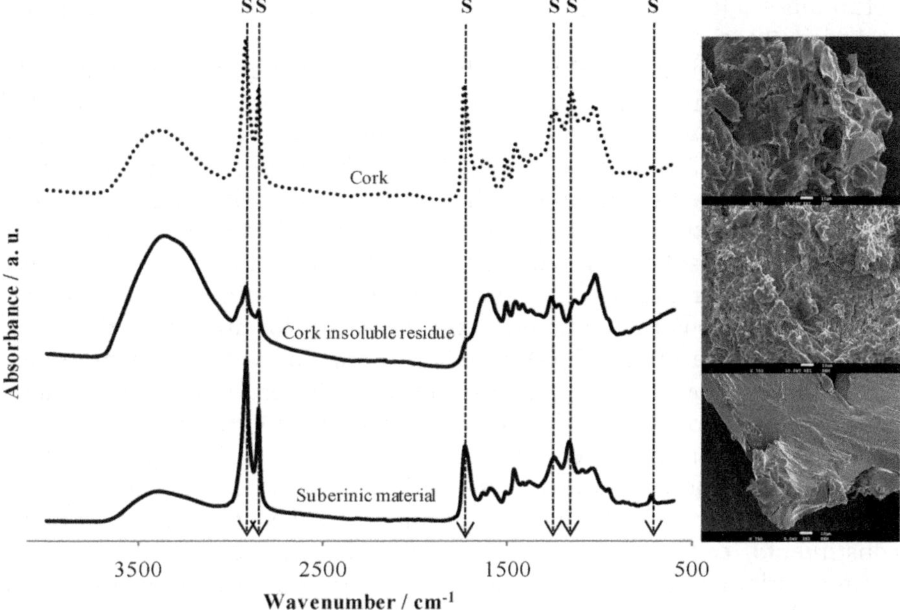

Figure 5.9 ATR-FTIR spectra and respective scanning electron microscope images of untreated cork, resulting cork insoluble residue and extracted suberin with cholinium hexanoate. Vertical lines stand for major characteristic absorption bands of suberin (S). Reproduced from ref. 80 with permission from the Centre National de la Recherche Scientifique (CNRS) and The Royal Society of Chemistry.

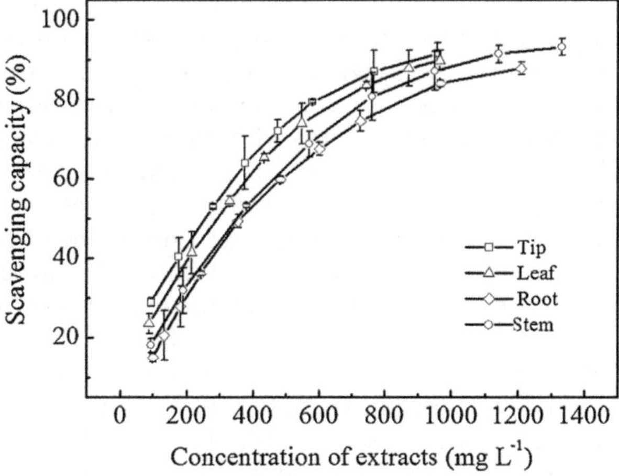

Figure 5.10 DPPH radical scavenging capacities of phenolic extract samples from different parts of sugar cane (extracted with [bmim][PF$_6$] IL). Reprinted from ref. 67 with permission from Elsevier.

petunidin-3-*O*-(6″-succinyl)-rhamnoside, diosmin, cyanidin-3-*O*-glucoside and tricin-7-*O*-rhamnosylgalacturonide.[67]

Barley hull was also used in another study for the extraction of phenolic compounds by pressurized aqueous solution containing [MHEA][OAc] IL.[68] The technology was said to have the advantage of short extraction times, which avoid the formation of aldehydes and browning extracts and degradation of extracted compounds. The results showed that higher phenolic extraction was attained using pressurized [MHEA][OAc] aqueous solution instead regular solid/liquid extraction with water/ethanol or even both pressurized water and water/ethanol mixture. The stability of the ionic species of the aqueous IL was noted as the key factor for high extraction yield, although a major effect of [OAc]⁻ anion interaction with target molecules was highlighted. Among the extracted compounds, ferulic and genistic acids were detected mostly after the extraction process.[68] Furthermore, the ability of [OAc]⁻ anion to extract phenolic compounds was also reported after a fractionation process of wheat straw with [emim][OAc].[7] After a successful application of [emim][OAc] in the fractionation of wheat straw into its major constituents, cellulose, hemicellulose and lignin, the IL was recovered. From an original light yellow colour the recovered IL turned to dark brown. A solid-phase extraction was performed on the recovered IL and surprisingly phenolic compounds, such as vanillin, ferulic and *p*-coumaric acids, were identified.[7] Therefore, an integrated process was developed where several separated fractions, including cellulose, hemicellulose, lignin and phenolic compounds, could be valorized from a low or practically zero value lignocellulosic residue, such as wheat straw.

5.4.5   Terpenes

Birch bark biomass was used for the extraction of betulin, an interesting triterpene for pharmaceutical applications, by ILs.[69] Given the high lignocellulosic biomass solubility recognized for [emim][OAc],[72] this IL was used initially. The process included whole biomass dissolution followed by precipitation of macromolecular compounds with ethanol. Subsequently, a selective precipitation of betulin with water was performed. Up to 31.0% betulin was recovered with [emim][OAc] after 24 h at 100 °C process. Interestingly, an ILMAE process was performed with same IL and similar betulin yield (31.7%) was attained in only 15 min. Furthermore, different ILs were examined to understand the effect of anion and cation on the extraction of betulin. Since the strategy was based on the dissolution of whole biomass and selective precipitation of betulin, 1-alkyl-3-methylimidazolium cations with a short alkyl chain favoured the efficient extraction of betulin. An ILMAE scale-up process was successfully demonstrated, with [emim][OAc] producing a 94% purity betulin fraction and a corresponding 24% extraction yield. According to the authors, IL was easily recovered in an azeotropic distillation of ethanol/water reaching 86–92% IL recovery yield. IL was reused up to four times and a slightly reduction of extraction efficiency was observed.[69]

5.5   Food Waste

The food processing industries generate large quantities of organic waste. This waste is a source of biological compounds, such as polysaccharides, proteins, polyphenols, terpenes and essential oils, among others. Therefore, food waste could also be used as a zero-value raw material to produce several high-value chemicals and biological compounds. The advantage of using ILs for dissolving food waste and extracting its biomolecules is the less harsh processing, which leads to higher quality final products.[2,14,82,83] Table 5.4 summarizes the main data related to food waste processing for extraction of biomolecules with ILs.

5.5.1   Polysaccharides

Chitin, a very useful modified polysaccharide for several medical and industrial purposes, was successfully extracted from shrimp and crab shells using ILs as solvents.[82,84] In the case of shrimp shells, [emim][OAc], [emim][Cl] and [bmim][Cl] ILs, which are known to efficiently dissolve biomass, were analysed. The most efficient IL in dissolving chitin from shrimp shells was [emim][OAc]. The results were very well correlated with the ability of ILs to dissolve cellulose. The higher basicity of the [OAc] anion is the crucial factor that mostly governs the dissolution of polysaccharides, such as chitin, in ILs. Furthermore, ILMAE was also applied to the dissolution process, resulting in more than 100% enhanced performance (around 94% initial chitin was dissolved). Similar to cellulose, water can act as the anti-solvent regenerating the dissolved chitin. The regenerated sample presented higher purity and higher molecular weight than available commercial chitin also used in this work.[82] On the other hand, Setoguchi *et al.*[84] verified that using [amim][Br] in the processing of crab shells not only allows the extraction of chitin, but also can dissolve minerals can be dissolved. Therefore, after the extraction process with IL a following addition of citric acid solution allowed the simultaneous regeneration of the dissolved chitin and the removal of $CaCO_3$ (the main mineral present in crab shells) in the liquid stream. The chelating power of citric acid to $CaCO_3$ allowed higher purification of chitin. XRD analysis gave similar profiles between the extracted chitin and commercial samples, indicating complete removal of $CaCO_3$, as the mineral directly affects the X-ray diffraction pattern of chitin. At the end of the process, [amim][Br] IL could be easily removed just by water evaporation, inducing the precipitation of mineral and chelating agent. Nevertheless, impurities were still identified in the recovered IL, demonstrating that further improvements are needed. Extraction at 120 °C for 24 h gave the highest yield (12.6 g/100 g crab shells), but the isolated chitin presented a high deacetylation degree value, which means that the quality of the obtained chitin was reduced.[84]

Lemon peel was used for extraction of pectin, another polysaccharide, by the ILMAE process.[85] Several ILs, including [bmim][Br], [bmim][Cl], [emim][Br], [bmim][BF$_4$], [emim][BF$_4$] and [amim][Cl], were preliminarily examined.

Table 5.4 Food waste used for biomolecule extraction with ionic liquids.

Biomass	Extraction technology	Target biomolecules		Ionic liquids	Extraction yield	Ref.
		Class	Name			
Chicken feathers	SLE	Proteins	Keratin	$[OHemim][NTf_2]$	21.75^a	83
Duck feathers	SLE	Proteins	Keratin	$[bmim][Cl]$	75.1^a	15
Crab shells	SLE	Polysaccharides	Chitin	$[amim][Br]$	12.6^a	84
Lemon peels	ILMAE	Polysaccharides	Pectin	$[amim][Cl]$; $[bmim][BF_4]$; $[bmim][Br]$; $[bmim][Cl]$; $[emim][BF_4]$; $[emim][Br]$	24.68^b/$[bmim][Cl]$	85
Orange peels	SLE	Essential oils	Limonene	$[amim][Cl]$; $[bmim][Cl]$; $[emim][OAc]$	0.74^a/$[emim][OAc]$	87
Peanut shells	SLE	Phenolic compounds	Luteolin	$[bmim][BF_4]$; $[bmim][NO_3]$; $[bmim][NTf_2]$; $[bmim][PF_6]$; $[hmim][BF_4]$; $[hmim][NO_3]$; $[hmim][PF_6]$	79.8^b/$[bmim][NO_3]$	2
Shrimp shells	ILMAE	Polysaccharides	Chitin	$[emim][OAc]$	94.0^b	82
Shrimp waste	ILUAE	Terpenes	Astaxanthin	$[bmim][BF_4]$; $[bmim][Br]$; $[bmim][Cl]$; $[bmim][MeSO_3]$; $[emim][BF_4]$; $[hmim][BF_4]$; $[NH_2pmim][Br]$	0.093^c/$[NH_2pmim][Br]$	14

[a]% (g of biomolecule/100 g of biomass).

[b]% (mass of biomolecule/initial mass content of biomolecule in the original biomass × 100).

[c]Yield (mg of biomolecule/g of biomass); SLE – solid–liquid extraction; ILMAE – ionic liquid-based microwave-assisted extraction; ILUAE – ionic liquid-based ultrasound-assisted extraction.

The most efficient IL was [bmim][Cl], which allowed 19.91% pectin extraction yield. The increased solubility of pectin in [bmim][Cl] was explained by multiple interactions between both, but mainly hydrogen bonds. Actually, the key dissolution mechanism is expected to be similar to cellulose dissolution in [bmim][Cl], where intermolecular hydrogen bonds between the Cl^- anion and OH groups of polysaccharide are easily established.[86]

5.5.2 Proteins

Chicken and duck feathers were the animal biomass successfully used for the extraction of keratin, using ILs by two different strategies.[15,83] Feathers are insoluble in water, thus a first strategy was based on the application of a hydrophobic IL, such as [OHemim][NTf$_2$], to dissolve feathers and to extract the keratin from them.[83] Then, a solution of $NaHSO_3$ was added to the feather/IL mixture system to reduce the disulfide covalent bonds between cysteines present in the protein structure, which enhanced the extraction of keratin to the IL. An increased yield of extracted keratin was observed when more $NaHSO_3$ was added to the solution mixture. However, an excess of $NaHSO_3$ did not influence the extraction efficiency, as the disulfide bonds were totally reduced. Without $NaHSO_3$, the extraction yield was dramatically reduced. Although $NaHSO_3$ plays an important role in keratin solubilization, the authors explained the importance of the [OHemim][NTf$_2$] as the extractant. The extraction mechanism was explained in two steps: (i) physical interactions occur between IL and feathers, where the IL solvent penetrates into the feather structure; (ii) along with disulfide bond cleavage, chains of feather keratin unfold from aggregation due to solvation effects of IL. Out of the hydrophobic feather structure, the extracted keratin in IL could be further solubilized in water, where IL is immiscible. These contrasting properties allowed the extraction of keratin from IL, and a further purification of the protein was performed by simple dialysis and precipitation with ethanol. Approximately 95% IL was recycled and reused for five times without losing efficiency. The precipitated keratin presented high uniformity, with weight averaged and number-averaged molecular weights of 10 240 and 10 000, respectively.[83] A second strategy was performed for keratin extraction from duck feathers.[15] Among several examined ILs, [amim][Cl] and [bmim][Cl] were demonstrated to be the most efficient in dissolving feathers and, consequently, could extract keratin in only 60 min, as demonstrated in Figure 5.11. However, Ji *et al.*[15] verified that after dissolving duck feathers in those ILs, keratin could be precipitated by adding water (Figure 5.11). This is contradictory to the last revised work,[83] which mentioned that keratin could be easily solubilized by water. Ji *et al.*[15] explained that high amounts of water interferes with IL–keratin interactions, allowing precipitation of keratin, similarly to cellulose regeneration. Furthermore, a similar effect to $NaHSO_3$ was observed when Na_2SO_3 was added to the IL/feathers mixture. After keratin recovery, water was evaporated and the recovered IL containing Na_2SO_3 was directly reused up to three times without losing efficiency.[15]

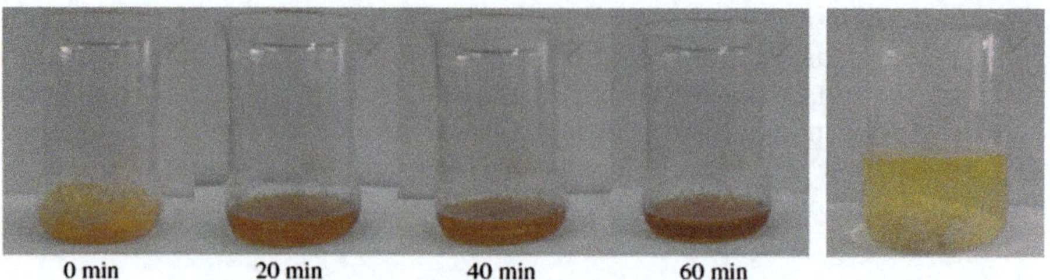

Figure 5.11 Dissolution process of feathers in IL during time and subsequent regeneration (extraction) of keratin after the addition of water to IL solution. Adapted from ref. 15 with permission from Elsevier.

5.5.3 Phenolic Compounds

Peanut shells were other type of food waste in which ILs were applied, in order to extract luteolin, a bioactive flavone found in different dietary sources.[2] A preliminary study of luteolin dissolution in different imidazolium-based ILs was performed. The obtained data showed that luteolin dissolution is anion dependent. Among the examined ILs, the hydrophilic kind demonstrated higher dissolution ability with the IL [bmim][NO$_3$] being the best. Furthermore, this IL demonstrated higher ability for luteolin dissolution than common organic solvents, such as methanol, ethanol, 1-propanol, 2-propanol, 1-butanol, acetone and hexane. Nevertheless, the viscosity of [bmim][NO$_3$] was thought to be inconvenient in the luteolin extraction from peanut shells. Therefore, the IL was slightly diluted with water and it showed higher efficiency in luteolin dissolution. Maximum 79.8% luteolin yield was attained using optimal conditions, but it was also observed that, at the same conditions, other biological phenolic compounds, namely eriodictyol and 5,7-dihydroxychromone, could be simultaneously extracted with [bmim][NO$_3$] aqueous solution, reaching 56.8% and 69.2% yields, respectively.[2]

5.5.4 Terpenes

Shrimp waste was used for the extraction of astaxanthin, a carotenoid and natural dietary component with antioxidant activity, by applying the ILUAE methodology.[14] In this work, ethanol was used instead the common addition of water as co-solvent. ILs composing 1-n-butyl-3-methylimidazolium cation and the anions [Br], [Cl], [BF$_4$] and [MeSO$_4$] were examined. Among them, [bmim][Br] was the most efficient IL for astaxanthin extraction assisted with ultrasound, which was associated with better miscibility of IL with ethanol. The hydrophobicity of the cation seems to be an important factor for the enhanced interaction with astaxanthin. Therefore, the extraction through ILUAE exhibited a higher efficiency than the conventional UAE because of the π–π and π–n, hydrophobic and hydrogen bond interactions.

5.5.5   Essential Oils

Orange peels, as a source of essential oils such as limonene, were subjected to IL processing.[87] The [amim][Cl], [bmim][Cl] and [emim][OAc] ILs were chosen for the extraction, and 20% (w/w) biomass/IL mixture was prepared and magnetically stirred. A complete dissolution of orange peels was only attained with [emim][OAc] after 3 h. The ILs [amim][Cl] and [bmim][Cl] were not efficient to dissolve orange peels completely, even after 24 h dissolution. Subsequently, the solution mixture was distilled and a two-phase distillate consisting of limonene and water from the orange peel was obtained, allowing for an easy separation of limonene through decantation. A limonene maximum extraction yield of 0.74 g/100 g biomass was achieved with [emim][OAc]. Lower extraction yields were obtained with [amim][Cl] and [bmim][Cl]. Therefore, the researchers concluded that complete dissolution must be accomplished for maximum extraction yield of essential oils. After distillation of the orange essential oil, the remaining biomass can be easily precipitated by addition of water. The biomass residue can be further valorized to produce fuels, chemicals and other value-added compounds.[87]

5.6   Microbial Biomass

Biotechnological fermentation is a green and powerful technology that uses microorganisms for the production of desired biomolecules. Interestingly, the grown microorganisms themselves could also be a source of compounds, which mainly correspond to lipids and proteins. Therefore, the opportunity to utilize ILs as extracting agents specifically for those target compounds arose. Although the research on this type of approach is still in its infancy, promising results were reported (Table 5.5), which makes ILs very versatile extracting solvents.

5.6.1   Lipids

Lipids from *Chlorella vulgaris* and *Aurantiochytrium* sp. microalgaes were extracted using blended ILs or mixtures of molten salts/IL, following two different strategies successfully approached by Park and co-workers.[88–90] First, single solution ILs were screened for lipid extraction from *Chlorella vulgaris*[90] and the highest extraction yield was obtained with [emim][DEP] (250.0 mg g^{-1} microalgae) as demonstrated in Figure 5.12. ILs, such as [emim][Cl], [emim][OAc] and [emim][BF$_4$], had the next highest extraction yields, respectively, 235.1, 223.7, and 219.7 mg g^{-1} microalgae. On the other hand, low extraction yields were obtained with [emim][HSO$_4$], [emim][EtSO$_4$], [emim][SCN], [emim][CH$_3$SO$_3$], [amim][Cl] and [bmim][Cl].[90] The authors correlated the obtained data with the lipid extraction mechanism described by Young *et al.*[13] It was suggested that, although examined ILs are hydrophilic, the presence of hydrophobic side groups attached to the nitrogen of the imidazolium ring creates hydrophobic regions within an

Table 5.5　Microbial biomass used for biomolecule extraction with ionic liquids.

Biomass		Extraction	Target biomolecules				
Name	Type	technology	Class	Name	Ionic liquids	Extraction yield	Ref.
Aurantiochytrium (sp. KRS101)	Algae	SLE	Lipids	Docosahexaenoic acid; oleic acid; palmitic acid	$[emim][OAc]^{a}$	$\approx 100^{b}$	85
Blue-green algae	Cyanobacteria	SLE	Lipids	Hexadecanoic acid; octadecanoic acid	$[bmim][Cl]$; $[bmim][PF_6]$	22.0^{c}/$[bmim][Cl]$	90
Chlorella	Algae	SLE	Lipids	Lipids	$[emim][MeSO_4]$	38.0^{d}	47
Chlorella vulgaris	Algae	ILUAE	Lipids	C_{14} to C_{24} (dominant: C16:0, C16:1, C18:2, and C18:3)	$[bmim][MeSO_4]$	60^{b}	88
Chlorella vulgaris	Algae	SLE	Lipids	C_{14} to C_{24} (dominant: C16:0, C16:1, C18:2, and C18:3)	$[bmim][BF_4]$; $[bmim][Br]$; $[bmim][CF_3SO_3]$; $[bmim][MeSO_3]$; $[bmim][Cl]$; $[bmim][MeSO_4]$; $[bmim][PF_6]$; $[bmim][NTf_2]$; $[emim][Cl]$; $[emim][MeSO_4]$; $[emim][OAc]$	19^{d}/$[bmim][CF_3SO_3]$	87
Chlorella vulgaris	Algae	SLE	Lipids	Linoleic acid; linolenic acid; oleic acid; palmitic acid; stearic acid	$[amim][Cl]$; $[bmim][Cl]$; $[emim][AlCl_4]$; $[emim][BF_4]$; $[emim][Cl]$; $[emim][NTf_2]$; $[emim][CH_3SO_3]$; $[emim][DEP]$; $[emim][EtSO_4]$; $[emim][HSO_4]$; $[emim][OAc]$; $[emim][SCN]$	87.5^{b}/$[emim][OAc]$ + $[emim][(CF_3SO_2)_2N]$ mix	83

Chlorella vulgaris	Algae	SLE	Lipids	Linoleic acid; linolenic acid; oleic acid; palmitic acid; stearic acid	[emim][NTf$_2$]; [emim][DEP]; [emim][HSO$_4$]; [emim][OAc]; [emim][SCN]	227.6^c/[emim][OAc]^a	86
Duniella	Algae	SLE	Lipids	Lipids	[emim][MeSO$_4$]	8.6^d	47
Pertusaria pseudocorallina	Lichen	ILMAE ILHE	Depsidones	Norstictic acid	[bmim][NTf$_2$]; [emim][EtSO$_4$]; [emimOH][NTf$_2$]; [mmim][MeSO$_4$]; [pmim][NTf$_2$]	3.2^d/[mmim][MeSO$_4$] 3.8^d/[emim][EtSO$_4$]	89
Rhodosporidium toruloides	Yeasts	SLE	Lipids Monosaccharides	Lipids Glucose; mannose	[emim][OAc]	60.5^d 11.3^d	91
—	Yeasts	SLE	Proteins	Proteins	[DMAPA][FA]		84

^{a}Mixed with molten salt.

^{b}% (mass of biomolecule/initial mass content of biomolecule in the original biomass × 100).

^{c}Yield (mg of biomolecule/g of biomass).

^{d}% (g of biomolecule/100 g of biomass); SLE – solid–liquid extraction; ILUAE – ionic liquid-based ultrasound-assisted extraction; ILMAE – ionic liquid-based microwave-assisted extraction; ILHE – ionic liquid-based heating extraction.

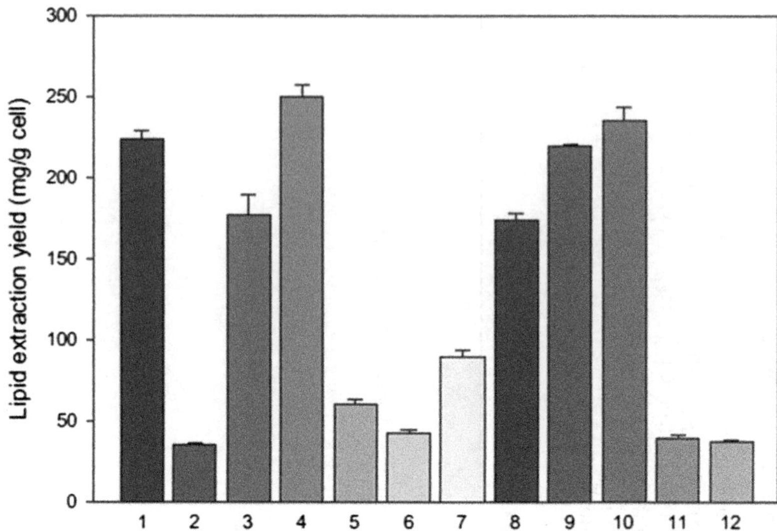

Figure 5.12 Yields of lipid extraction from *Chlorella vulgaris* obtained with different ILs after 120 °C for 2 h. (1) [emim][OAc]; (2) [emim][HSO$_4$]; (3) [emim][AlCl$_4$]; (4) [emim][DEP]; (5) [emim][EtSO$_4$]; (6) [emim][SCN]; (7) [emim][CH$_3$SO$_3$]; (8) [emim][(CF$_3$SO$_2$)$_2$N]; (9) [emim][BF$_4$]; (10) [emim][Cl]; (11) [amim][Cl]; and (12) [bmim][Cl]. Reprinted from ref. 90 with permission from Elsevier.

amphiphilic solution. In this way, IL allows a fast diffusion of lipid molecules through 'hydrophobic tunnels' created by these side chains.[13,90] Nevertheless, when different ILs were mixed, a synergetic effect was observed and lipid extraction yields were enhanced even for the least efficient ILs. The highest lipid extraction was surprisingly attained with [emim][OAc]/[emim][NTf$_2$] mixture, reaching 255.7 mg g^{-1} microalgae yield. In another strategy, the price of ILs was taken into account and molten salts were mixed with ILs to decrease the amount of IL needed for an economically efficient lipid extraction.[89] Using a mixture of FeCl$_3$·6H$_2$O/[emim][OAc] (5:1), the lipid yield was similar to that obtained using only [emim][OAc]. The results showed that replacing part of IL with a molten salt could make the process more viable than using only IL.[89] Furthermore, this strategy was also applied for the extraction of lipids from *Aurantiochytrium* sp., a source of the bioactive docosahexaenoic acid (DHA), and similar results were successfully achieved.[88]

Chlorella and *Duniella* microalgaes were used for the extraction of lipids with ILs[13] and the effect of co-solvent in the extraction was studied. Several organic solvents were examined, along with [emim][MeSO$_4$] IL. Methanol and isopropyl alcohol produced the highest extraction yields, while dimethyl sulfoxide and acetic acid resulted in lower lipid yields. For instance, [emim][MeSO$_4$]/methanol mixture extracted 8.2% from the wet *Chlorella*, resulting in 75% lipid extraction efficiency. Herein, high extraction efficiency was attained without drying biomass, to avoid an intense and expensive water removal from algae that is generally required prior to lipid extraction.[13]

In a different study, an isolation of lipids from the microalgae *Chlorella vulgaris* assisted by ILs was performed within an original strategy.[12] ILs were used for the dissolution of microalgae biomass leaving lipids as insoluble products, which were separated by centrifugation. Methanol was used as the co-solvent to decrease the viscosity of ILs and to act as reactant as well (transesterification). The results demonstrated that lipid recovery was anion-dependent and the efficiency was according to the following order: $[CF_3SO_3]$ > $[MeSO_4]$ > $[CH_3SO_3]$ > $[BF_4]$ > $[PF_6]$ > $[NTf_2]$ > $[Cl]$. It was clear that hydrophobic and water-immiscible ILs, containing $[PF_6]$ and $[NTf_2]$ anions, did not dissolve microalgae biomass. Conversely, hydrophilic and water-miscible ILs, such as [bmim][CF_3SO_3], [bmim][$MeSO_4$] and [emim][$MeSO_4$], demonstrated the highest efficiencies. Nevertheless, considering the methodology used, the high polar ILs with high hydrogen basicity, such as [bmim][Cl] and [emim][OAc], did not allowed high lipid recoveries. This is justified by the high solubility power of those ILs, which are capable of dissolving all microalgae biomass constituents, including lipids. Therefore, for this specific process, dipolarity/polarizability ($\pi*$) and hydrogen bond acidity (α) of ILs are more important than their hydrogen bond basicity (β), looking at the Kamlet–Taft parameters that were scrutinized for each tested IL. The [bmim][CF_3SO_3] was demonstrated to be the most efficient IL for this purpose, reaching 19.0% maximum lipid recovery.

5.6.2 Proteins

Yeast biomass was successfully used in a first attempt for the extraction of proteins.[91] Among the ILs tested different cations were used and the authors ordered the ability to extract proteins as [DMAPA] ~ [DMED] > [DEA] > [TEA] > [TPA] > [TBA] ~ [DIPE]. Therefore, the extraction efficiency increases with decreasing cation hydrophobicity. For anions, enhanced extraction efficiency was observed for high basicity ones, such as trifluoroacetic, formic and acetic acids. Using these anions, the ability to produce a stable hydrogen bond network with proteins is enhanced. Also, the ability of ILs to dissolve the polysaccharides present in the cell walls of yeasts allows the exposure of proteins for extraction. Proteins covering different sizes were extracted using [DMAPA][TFA] and [DMAPA][FA] ILs. [DMAPA][FA] was preferable to be used on protein extraction, due to the possibility of recovering it under vacuum. The mild alkali medium of this IL (pH = 9) aided by the anion basicity allows it to damage both β- and α-D-glucans, resulting in partial exposure of the cell membrane to the IL solution. This allows for an efficient permeation of the IL solution into the cells. Furthermore, the proteins present in yeasts are highly *O*-glycosylated, and are attached to other cell wall components mediated by their O-linked saccharides. Under alkaline conditions, O-chains tend to be cleaved off in a process called beta-elimination, releasing the proteins into the IL medium. Western blotting technique was also performed to extract the proteins and the results showed that immunoreactivity and biological functions were maintained.[91]

5.6.3 Depsidones

Lichens were also used as a source of valuable biomolecules, particularly for the extraction of the depsidone norstictic acid through an ILMAE process.[92] In this study, 18 ILs were examined in preliminary extraction trials and combinations, including the cations 1-alkyl-3-methylimidazolium, 1-alkyloxy-3-methylimidazolium, ammonium and phosphonium as well as the anions chloride, hexafluorophosphate, bis(trifluoromethylsulfonyl) imide and alkylsulfate. Exactly 9 ILs were verified to extract norstictic acid, but the more efficient ones presenting suitable extraction yields and higher selectivity for the acid were [emim][EtSO$_4$], [mmim][MeSO$_4$] and [OHpmim][NTf$_2$]. The [emim][EtSO$_4$] and [mmim][MeSO$_4$] ILs presented the highest extractions, 3.8% (HRE process) and 3.2% (ILMAE process), respectively. Using ILMAE with [emim][EtSO$_4$] the extraction yield was only 0.5%. This low extraction yield was explained by the authors to be a consequence of susceptible lichen degradation to microwave, due to the high ability of [emim][EtSO$_4$] to absorb and transfer energy from this type of radiation. Interestingly, this phenomenon did not occur during microwave extraction with [mmim][MeSO$_4$]. In contrast to ILMAE, a microwave process using tetrahydrofuran (THF) dramatically enhanced the extraction of norstictic acid. However, much less selectivity was obtained compared to ILMAE. The recovery of product and IL was also studied. A primary addition of diethyl ether with water as co-solvent was selected as the best extracting mixture. The choice for diethyl ether depended on its immiscibility with the ILs being tested, which allowed the recovery of organic soluble compounds, such as terpenes, with very low contamination with IL. On the other hand, the aqueous IL phase containing norstictic acid was recovered and acetone was added to precipitate the desired compound. Water/acetone was easily removed by evaporation and the IL could be further reutilized.[92]

5.7 The Effect of Process Parameters in Extraction with ILs

Regardless of the biomass and process used, there are certain factors inherent to the extraction that were evaluated and optimized in order to achieve maximal extraction yields of the target biomolecules with ILs. Among several parameters, temperature, residence time, IL concentration and liquid/solid ratio were the most approached and discussed.

Temperature is probably one of the most important parameters for consideration, as in general, the maximal extraction yield is significantly dependent on it. Furthermore, the efficiency of the process is also considered with regard to the energy required, which in turn is directly associated to the temperature. It was observed that temperature significantly influences the extraction yields of target biomolecules. In fact, increasing temperature enhances IL solubility for biomolecules and decreases IL viscosity, allowing better diffusion in the inner part of the biomass to extract target compounds.[22] Furthermore, at higher temperatures, the biomass structure is disrupted favouring mass transfer of biomolecules into the IL phase.[46] Thus, increasing temperature allows enhanced extraction yields, as reported in several studies.[24,57,64,65,71,73,90] However, too high a temperature also favours thermal decomposition of biomolecules.[15,17,19,22,46,83] For instance, some bioactive compounds, such as quercetin and *trans*-resveratrol, are temperature sensitive, therefore, low temperatures must be used in these cases.[17] Moreover, structural changes on target compounds could also be promoted by high temperatures. For instance, isolated chitin from crab shells presented a higher degree of deacetylation with increasing temperature.[84] Also, for essential oil extraction, very high temperatures results in volatilization of low-boiling oils, influencing the extraction yield.[47] Still, the possibility to perform the extraction at room temperature, so saving energy and making the process economically efficient, was demonstrated.[14,27,28,38,41,42] In fact, in case of ABS processes, a temperature higher than room temperature resulted in lack of efficiency.[93,94] This phenomenon was explained by the transition of water from the IL-rich phase to the salt-rich phase, which decreases the salting-out effect in the salt-rich phase. This occurrence destabilizes the selectivity of both phases for the target biomolecules, decreasing extraction efficiency.[94]

Extraction time is also a parameter studied for optimization. It is not fundamental from the point of view of laboratory scale, but is crucial for the industrial application, when looking at the effectiveness of the process. For the majority of the developed processes, the yield of extracted biomolecules with ILs increases over time, reaching a plateau in which equilibrium of biomolecule concentration is attained.[37,38,41,44] Nevertheless, there are some reports mentioning the degradation of target compounds after a certain time of extraction.[23,28,46] This specially happens when samples are exposed to energy sources (for instance, ultrasound or microwave) for prolonged times.[23,46]

Additionally, two different approaches were observed regarding the IL concentration in the extracting solvent. For instance, pure IL without any dissolution was generally used for the extraction of lignin from lignocellulosic biomass,[64,65,71] or for the extraction of chitin from crustacean shells[82,84] and also for keratin from poultry feathers.[15,83] In such cases, the solubility power and physicochemical properties of IL are crucial for the extraction of those biomacromolecules from the recalcitrant materials. The presence of water or organic solvents could compromise the extraction, decreasing the efficiency of ILs. However, when small biomolecules are the target compounds, maximum extraction efficiency could be attained with diluted ILs in water or organic solvents (*e.g.* methanol). For the majority of cases, increasing the IL concentration allows improved extraction yields, but only to a certain point, which varies from study to study. When increasing the IL concentration, both the solubility and the extracting capacity of the solvent are enhanced.[22] Still, depending on the properties of the target biomolecules, this tendency could change. For instance, the extraction yields of quercetin increased more rapidly than that of caffeic acid after increasing [bmim][Br] concentration.[23] However, no significant effect on essential oil extraction was observed with increasing [omim][Br] concentration.[49] This is related to the natural solubility of each compound in the respective ILs. Nevertheless, it was generally observed that with further increase in IL concentration the extraction efficiency substantially decreases.[19,22,27,44,95] It was stated that the excessive amount of IL increases the viscosity of the extracting solution, compromising the diffusion of IL into biomass sample and mass transference during the extraction.[20,38,95]

Regarding the liquid/solid ratio, similar trends were observed as for IL concentration. In this case, larger volumes should be avoided for the economics of the process, while smaller amounts could lead to incomplete extraction. Therefore, optimization is also needed. In general, larger quantities of extracting solvent allow higher extraction yields, but once again at a certain ratio. Further increases of liquid/solid ratio leads to diminished extraction yields of target biomolecules, as reported elsewhere.[19,23,27,49] For instance, a specific work based on ILNPCAE process, which uses cavitation as the key factor for an efficient extraction, could be compromised by using high liquid/solid ratios, as high quantities of extracting solvent can consume the cavitation energy.[19] However, other studies stated that the increase of liquid/solid ratio simply did not change the extraction efficiency, maintaining the maximum yield of extracted compounds.[37,38,42,57] On the other hand, when the liquid/solid ratio is too low, the extraction is inefficient, decreasing the yields of target biomolecules.[44,73,83] Microwave or ultrasound irradiation energy could be an important factor in assisting the extraction by ILs, for increasing the efficiency of the process. In both cases, an increase of irradiation power enhances the extraction yields of target biomolecules.[38,49,56] Microwave irradiation energy allows the easier penetration of IL into the matrix. In turn, IL absorbs and delivers the energy provided by microwave inside the sample, allowing the disruption of intricate materials and subsequent dissolution of compounds to be extracted in IL. This results in a faster and more efficient extraction than a process without microwave irradiation.[49] In the case of ultrasound, its mechanical effect on biomass samples is preponderant for greater diffusion of ILs, improving mass transfer which subsequently enhances the extraction.[44] However, when exceeding a certain microwave or ultrasound irradiation power, no significant variation occurs in the compound extraction yield.[23,42,44,95]

Although no degradation of target biomolecules was visible, a further increase of irradiation power is not economically efficient. There are also studies that reported a reduced or insignificant effect of microwave on maximal extraction yield of target compounds.[37,46,58] Besides the irradiation power, the pulsing cycle could influence the extraction efficiency. In case of ultrasound, pulsing the energy slows the rate of temperature increase during the extraction. This prevents the clustering that favours the clarification of the cavitation zone to maximize the extraction yield.[44]

Other conditions such as particle size, type and soaking time of the biomass as well as the pH of the extracting solution could also influence the extraction of biomolecules.[16,17,38,57,95]

5.8   Conclusions

Following on the selection of the correct method and appropriate biomass feedstocks, the task-specific ILs can be designed. Hence, ILs proved their great utility as 'designed solvents' and definitively have a position in phytochemical extraction, especially considering the efforts in applying the principles of green chemistry for new health products derived from biomass.

Acknowledgements

This work was supported by the Fundação para a Ciência e a Tecnologia (FCT, Portugal) through Bilateral Cooperation project FCT/CAPES 2014/ 2015 (FCT/1909/27/2/2014/S) and grants SFRH/BD/90282/2012 (AMdCL), IF/00424/2013 (RBL), and CAPES (Brazil) supported the project Pesquisador Visitante Especial 155/2012. Special thanks goes to the European Commission for financing the COST Action TD1203.

References

1. M. D. A. Saldana and C. S. Valdivieso-Ramirez, *J. Supercrit. Fluids*, 2015, **96**, 228–244.
2. L. Ge, F. Xia, Y. Song, K. D. Yang, Z. Z. Qin and L. S. Li, *Sep. Purif. Technol.*, 2014, **135**, 223–228.
3. J. P. Hallett and T. Welton, *Chem. Rev.*, 2011, **111**, 3508–3576.
4. A. Garcia, M. G. Alriols, R. Llano-Ponte and J. Labidi, *Biomass Bioenergy*, 2011, **35**, 516–525.
5. S. Behera, R. Arora, N. Nandhagopal and S. Kumar, *Renewable Sustainable Energy Rev.*, 2014, **36**, 91–106.
6. W. Lan, C. F. Liu and R. C. Sun, *J. Agric. Food Chem.*, 2011, **59**, 8691–8701.
7. S. P. Magalhães da Silva, A. M. da Costa Lopes, L. B. Roseiro and R. Bogel-Lukasik, *RSC Adv.*, 2013, **3**, 16040–16050.
8. I. Kilpelainen, H. Xie, A. King, M. Granstrom, S. Heikkinen and D. S. Argyropoulos, *J. Agric. Food Chem.*, 2007, **55**, 9142–9148.
9. S. Hyvarinen, J. P. Mikkola, D. Y. Murzin, M. Vaher, M. Kaljurand and M. Koel, *Catal. Today*, 2014, **223**, 18–24.
10. E. C. Achinivu, R. M. Howard, G. Li, H. Gracz and W. A. Henderson, *Green Chem.*, 2014, **16**, 1114–1119.
11. P. Varanasi, P. Singh, M. Auer, P. Adams, B. Simmons and S. Singh, *Biotechnol. Biofuels*, 2013, **6**, 14.

12. Y. H. Kim, Y. K. Choi, J. Park, S. Lee, Y. H. Yang, H. J. Kim, T. J. Park, Y. H. Kim and S. H. Lee, *Bioresour. Technol.*, 2012, **109**, 312–315.

13. G. Young, F. Nippgen, S. Titterbrandt and M. J. Cooney, *Sep. Purif. Technol.*, 2010, **72**, 118–121.

14. W. Bi, M. Tian, J. Zhou and K. H. Row, *J. Chromatogr. B*, 2010, **878**, 2243–2248.

15. Y. M. Ji, J. Y. Chen, J. X. Lv, Z. L. Li, L. Y. Xing and S. Y. Ding, *Sep. Purif. Technol.*, 2014, **132**, 577–583.

16. F. Y. Du, X. H. Mao and G. K. Li, *J. Chromatogr. A*, 2007, **1140**, 56–62.

17. F. Y. Du, X. H. Xiao, X. J. Luo and G. K. Li, *Talanta*, 2009, **78**, 1177–1184.

18. J. L. Anderson, J. Ding, T. Welton and D. W. Armstrong, *J. Am. Chem. Soc.*, 2002, **124**, 14247–14254.

19. M. H. Duan, M. Luo, C. J. Zhao, W. Wang, Y. G. Zu, D. Y. Zhang, X. H. Yao and Y. J. Fu, *Sep. Purif. Technol.*, 2013, **107**, 26–36.

20. L. J. Zhang and X. Wang, *J. Sep. Sci.*, 2010, **33**, 2035–2038.

21. L. Crowhurst, P. R. Mawdsley, J. M. Perez-Arlandis, P. A. Salter and T. Welton, *Phys. Chem. Chem. Phys.*, 2003, **5**, 2790–2794.

22. H. A. Zeng, Y. Z. Wang, J. H. Kong, C. Nie and Y. Yuan, *Talanta*, 2010, **83**, 582–590.

23. Z. X. Lou, H. X. Wang, S. Zhu, S. W. Chen, M. Zhang and Z. P. Wang, *Anal. Chim. Acta*, 2012, **716**, 28–33.

24. B. Tang, Y. J. Lee, Y. R. Lee and K. H. Row, *J. Chromatogr. B*, 2013, **933**, 8–14.

25. D. Y. Zhang, Y. G. Zu, Y. J. Fu, M. Luo, C. B. Gu, W. Wang and X. H. Yao, *Sep. Purif. Technol.*, 2011, **83**, 91–99.

26. W. Liu, Y. J. Fu, Y. G. Zu, Y. Kong, L. Zhang, B. S. Zu and T. Efferth, *J. Chromatogr. A*, 2009, **1216**, 3841–3850.

27. Y. S. Sun, W. Li and J. H. Wang, *J. Chromatogr. B*, 2011, **879**, 975–980.

28. S. A. Chowdhury, R. Vijayaraghavan and D. R. MacFarlane, *Green Chem.*, 2010, **12**, 1023–1028.

29. C. X. Lu, H. X. Wang, W. P. Lv, C. Y. Ma, Z. X. Lou, J. Xie and B. Liu, *Nat. Prod. Res.*, 2012, **26**, 1842–1847.

30. R. H. Jin, L. Fan and X. N. An, *Sep. Purif. Technol.*, 2011, **83**, 45–49.

31. Z. F. Wei, Y. G. Zu, Y. J. Fu, W. Wang, M. Luo, C. J. Zhao and Y. Z. Pan, *Sep. Purif. Technol.*, 2013, **102**, 75–81.

32. W. Xu, K. D. Chu, H. Li, Y. Q. Zhang, H. Y. Zheng, R. L. Chen and L. D. Chen, *Molecules*, 2012, **17**, 14323–14335.

33. Y. F. Zhang, Z. Liu, Y. L. Li and R. Chi, *Sep. Purif. Technol.*, 2014, **129**, 71–79.

34. J. P. Fan, J. Cao, X. H. Zhang, J. Z. Huang, T. Kong, S. Tong, Z. Y. Tian, Y. L. Xie, R. Xu and J. H. Zhu, *Food Chem.*, 2012, **135**, 2299–2306.

35. X. Q. Li, R. L. Guo, X. P. Zhang and X. Y. Li, *Sep. Purif. Technol.*, 2012, **88**, 146–150.

36. L. Yang, Y. Liu, Y. G. Zu, C. J. Zhao, L. Zhang, X. Q. Chen and Z. H. Zhang, *Chem. Eng. J.*, 2011, **175**, 539–547.

37. W. Y. Ma, Y. B. Lu, R. L. Hu, J. H. Chen, Z. Z. Zhang and Y. J. Pan, *Talanta*, 2010, **80**, 1292–1297.
38. L. Yang, H. Wang, Y. G. Zu, C. J. Zhao, L. Zhang, X. Q. Chen and Z. H. Zhang, *Chem. Eng. J.*, 2011, **172**, 705–712.
39. M. G. Bogdanov and I. Svinyarov, *Sep. Purif. Technol.*, 2013, **103**, 279–288.
40. M. G. Bogdanov, I. Svinyarov, R. Keremedchieva and A. Sidjimov, *Sep. Purif. Technol.*, 2012, **97**, 221–227.
41. F. Liu, D. Wang, W. Liu, X. Wang, A. Y. Bai and L. Q. Huang, *Sep. Purif. Technol.*, 2013, **110**, 86–92.
42. W. Bi, M. Tian and K. H. Row, *Talanta*, 2011, **85**, 701–706.
43. S. Q. Zhang, H. M. Bi and C. J. Liu, *Sep. Purif. Technol.*, 2007, **57**, 277–282.
44. S. M. Harde, S. L. Lonkar, M. S. Degani and R. S. Singhal, *Ind. Crops Prod.*, 2014, **61**, 258–264.
45. W. T. Bi, M. L. Tian and K. H. Row, *J. Chromatogr. B*, 2013, **913**, 61–68.
46. Y. S. Chi, Z. D. Zhang, C. P. Li, Q. S. Liu, P. F. Yan and U. Welz-Biermann, *Green Chem.*, 2011, **13**, 666–670.
47. J. Jiao, Q. Y. Gai, Y. J. Fu, Y. G. Zu, M. Luo, W. Wang and C. J. Zhao, *J. Food Eng.*, 2013, **117**, 477–485.
48. T. V. Doherty, M. Mora-Pale, S. E. Foley, R. J. Linhardt and J. S. Dordick, *Green Chem.*, 2010, **12**, 1967–1975.
49. T. T. Liu, X. Y. Sui, R. R. Zhang, L. Yang, Y. G. Zu, L. Zhang, Y. Zhang and Z. H. Zhang, *J. Chromatogr. A*, 2011, **1218**, 8480–8489.
50. Y. Xiao, Y. Wang, S. Q. Gao, R. Zhang, R. B. Ren, N. Li and H. Q. Zhang, *J. Chromatogr. B*, 2011, **879**, 1833–1838.
51. P. Wang, C. Y. Ma, S. W. Chen, S. Zhu, Z. X. Lou and H. X. Wang, *Trop. J. Pharm. Res.*, 2014, **13**, 1339–1345.
52. B. D. Ribeiro, M. A. Z. Coelho, L. P. N. Rebelo and I. M. Marrucho, *Ind. Eng. Chem. Res.*, 2013, **52**, 12146–12153.
53. Y. S. Sun, Z. B. Liu, J. H. Wang, S. F. Yang, B. Q. Li and N. Xu, *Ultrason. Sonochem.*, 2013, **20**, 180–186.
54. T. Usuki, N. Yasuda, M. Yoshizawa-Fujita and M. Rikukawa, *Chem. Commun.*, 2011, **47**, 10560–10562.
55. Y. B. Lu, W. Y. Ma, R. L. Hu, X. J. Dai and Y. J. Pan, *J. Chromatogr. A*, 2008, **1208**, 42–46.
56. X. J. Cao, X. M. Ye, Y. B. Lu, Y. Yu and W. M. Mo, *Anal. Chim. Acta*, 2009, **640**, 47–51.
57. A. F. M. Claudio, A. M. Ferreira, M. G. Freire and J. A. P. Coutinho, *Green Chem.*, 2013, **15**, 2002–2010.
58. A. K. Ressmann, P. Gaertner and K. Bica, *Green Chem.*, 2011, **13**, 1442–1447.
59. A. Al-Maaieh and D. R. Flanagan, *J. Pharm. Sci.*, 2002, **91**, 1000–1008.
60. C. H. Ma, S. Y. Wang, L. Yang, Y. G. Zu, F. J. Yang, C. J. Zhao, L. Zhang and Z. H. Zhang, *Chem. Eng. Process.*, 2012, **57–58**, 59–64.
61. S. Y. Wang, L. Yang, Y. G. Zu, C. J. Zhao, X. W. Sun, L. Zhang and Z. H. Zhang, *Ind. Eng. Chem. Res.*, 2011, **50**, 13620–13627.

62. J. Jiao, Q. Y. Gai, Y. J. Fu, Y. G. Zu, M. Luo, C. J. Zhao and C. Y. Li, *Sep. Purif. Technol.*, 2013, **107**, 228–237.

63. G. Severa, G. Kumar, M. Troung, G. Young and M. J. Cooney, *Sep. Purif. Technol.*, 2013, **116**, 265–270.

64. S. S. Y. Tan, D. R. MacFarlane, J. Upfal, L. A. Edye, W. O. S. Doherty, A. F. Patti, J. M. Pringle and J. L. Scott, *Green Chem.*, 2009, **11**, 339–345.

65. D. Fu, G. Mazza and Y. Tamaki, *J. Agric. Food Chem.*, 2010, **58**, 2915–2922.

66. I. Anugwom, P. Maki-Arvela, P. Virtanen, S. Willfor, R. Sjoholm and J. P. Mikkola, *Carbohydr. Polym.*, 2012, **87**, 2005–2011.

67. X. Li, Z. Ma and S. Yao, *Food Bioprod. Process.*, 2015, **94**, 547–554.

68. S. Sarkar, V. H. Alvarez and M. D. A. Saldana, *J. Supercrit. Fluids*, 2014, **93**, 27–37.

69. A. K. Ressmann, K. Strassl, P. Gaertner, B. Zhao, L. Greiner and K. Bica, *Green Chem.*, 2012, **14**, 940–944.

70. J. Y. Kim, E. J. Shin, I. Y. Eom, K. Won, Y. H. Kim, D. Choi, I. G. Choi and J. W. Choi, *Bioresour. Technol.*, 2011, **102**, 9020–9025.

71. S. H. Lee, T. V. Doherty, R. J. Linhardt and J. S. Dordick, *Biotechnol. Bioeng.*, 2009, **102**, 1368–1376.

72. N. Sun, M. Rahman, Y. Qin, M. L. Maxim, H. Rodriguez and R. D. Rogers, *Green Chem.*, 2009, **11**, 646–655.

73. N. Muhammad, Z. Man, M. A. Bustam, M. I. A. Mutalib and S. Rafiq, *J. Ind. Eng. Chem.*, 2013, **19**, 207–214.

74. N. Muhammad, Z. Man, M. A. Bustam, M. I. A. Mutalib, C. D. Wilfred and S. Rafiq, *Appl. Biochem. Biotechnol.*, 2011, **165**, 998–1009.

75. X. D. Hou, T. J. Smith, N. Li and M. H. Zong, *Biotechnol. Bioeng.*, 2012, **109**, 2484–2493.

76. R. Prado, X. Erdocia and J. Labidi, *J. Chem. Technol. Biotechnol.*, 2013, **88**, 1248–1257.

77. P. D. de Maria, *J. Chem. Technol. Biotechnol.*, 2014, **89**, 11–18.

78. I. Anugwom, V. Eta, P. Virtanen, P. Maki-Arvela, M. Hedenstrom, M. Hummel, H. Sixta and J. P. Mikkola, *ChemSusChem*, 2014, **7**, 1170–1176.

79. H. Garcia, R. Ferreira, M. Petkovic, J. L. Ferguson, M. C. Leitao, H. Q. N. Gunaratne, K. R. Seddon, L. P. N. Rebelo and C. S. Pereira, *Green Chem.*, 2010, **12**, 367–369.

80. R. Ferreira, H. Garcia, A. F. Sousa, M. Petkovic, P. Lamosa, C. S. R. Freire, A. J. D. Silvestre, L. P. N. Rebelo and C. S. Pereira, *New J. Chem.*, 2012, **36**, 2014–2024.

81. R. Ferreira, H. Garcia, A. F. Sousa, C. S. R. Freire, A. J. D. Silvestre, L. P. N. Rebelo and C. S. Pereira, *Ind. Crops Prod.*, 2013, **44**, 520–527.

82. Y. Qin, X. M. Lu, N. Sun and R. D. Rogers, *Green Chem.*, 2010, **12**, 968–971.

83. Y. X. Wang and X. J. Cao, *Process Biochem.*, 2012, **47**, 896–899.

84. T. Setoguchi, T. Kato, K. Yamamoto and J. Kadokawa, *Int. J. Biol. Macromol.*, 2012, **50**, 861–864.

85. G. L. Huang, S. Jeffrey, K. Zhang and X. L. Huang, *J. Anal. Methods Chem.*, 2012, 302059.

86. R. C. Remsing, R. P. Swatloski, R. D. Rogers and G. Moyna, *Chem. Commun.*, 2006, 1271–1273.

87. K. Bica, P. Gaertner and R. D. Rogers, *Green Chem.*, 2011, **13**, 1997–1999.
88. S. A. Choi, J. Y. Jung, K. Kim, J. H. Kwon, J. S. Lee, S. W. Kim, J. Y. Park and J. W. Yang, *Bioprocess Biosyst. Eng.*, 2014, **37**, 2199–2204.
89. S. A. Choi, J. S. Lee, Y. K. Oh, M. J. Jeong, S. Kim and J. Y. Park, *Algal Res.*, 2014, **3**, 44–48.
90. S. A. Choi, Y. K. Oh, M. J. Jeong, S. W. Kim, J. S. Lee and J. Y. Park, *Renewable Energy*, 2014, **65**, 169–174.
91. L. Y. Ge, X. T. Wang, S. N. Tan, H. H. Tsai, J. W. H. Yong and L. Hua, *Talanta*, 2010, **81**, 1861–1864.
92. S. Bonny, L. Paquin, D. Carrie, J. Boustie and S. Tomasi, *Anal. Chim. Acta*, 2011, **707**, 69–75.
93. Z. J. Tan, F. F. Li, X. L. Xu and J. M. Xing, *Desalination*, 2012, **286**, 389–393.
94. J. K. Yan, H. L. Ma, J. J. Pei, Z. B. Wang and J. Y. Wu, *Sep. Purif. Technol.*, 2014, **135**, 278–284.
95. L. J. Zhang, Y. L. Geng, W. J. Duan, D. J. Wang, M. R. Fu and X. Wang, *J. Sep. Sci.*, 2009, **32**, 3550–3554.
96. M. Rajabi, H. Ghanbari, B. Barfi, A. Asghari and S. Haji-Esfandiari, *Food Res. Int.*, 2014, **62**, 761–770.
97. H. M. Lin, Y. G. Zhang, M. Han and L. M. Yang, *Ultrason. Sonochem.*, 2013, **20**, 680–684.

Toxicity and Bio-Acceptability in the Context of Biological Processes in Ionic Liquid Media

HANNAH PRYDDERCH[a], ANDREAS HEISE[a] AND NICHOLAS GATHERGOOD*[b]

[a]School of Chemical Sciences, Dublin City University, Ireland;
[b]Department of Chemistry, Tallinn University of Technology, Estonia
*E-mail: Nicholas.Gathergood@ttu.ee

6.1 Introduction

The physical properties that the majority of ionic liquids (ILs) possess make them an attractive candidate as an alternative solvent for use in a biorefinery. These properties include, but are not limited to, low vapour pressure, low flammability and high thermostability. A solvent with these properties would have a clear advantage over volatile organic chemicals (VOCs) currently employed in (bio)refineries and could thus be looked upon as a 'green' alternative and suitable for introduction into a biorefinery, which aims to reduce the use and release of toxic chemicals. Despite research showing that not all ILs possess these properties[1,2] there is an even bigger factor that needs to be considered when evaluating the 'greenness' of an IL. These factors fall under the umbrella of the environmental properties of ILs.

Despite great care taken to prevent the release of an IL utilized in a biorefinery into the environment, it has to be considered that it is indeed a

RSC Green Chemistry No. 36
Ionic Liquids in the Biorefinery Concept: Challenges and Perspectives
Edited by Rafal Bogel-Lukasik

Published by the Royal Society of Chemistry, www.rsc.org

possibility that some of the IL may reach the environment. Potential causes for this could be due to product contamination, release through wastewater streams or release through solvent disposal. Even small amounts of IL released will interact with the environment, thus it is highly necessary to study and understand how each individual IL will interact with the environment and the potential impact on the ecosystems affected. Thus for an IL to be classified as a 'green' alternative to VOCs, environmental factors such as biocompatibility, bioaccumulation, toxicity and biodegradability need to be considered.[3] It is also pertinent to consider the biodegradation pathways that an IL follows during its break down.[4-7] Metabolites generated in the biodegradation process may be more toxic and bioaccumulate to a greater extent in the environment than the parent IL.[8] Thus an assessment of metabolite formation, toxicity and stability is highly recommended.[9-11]

Enzyme catalysis is a useful tool for biomass conversion to value added products. Conversion of biomass by enzymes is challenging due to the insolubility of the biomass in a solvent where the enzyme remains stable, highly active and selective. Highly polar organic solvents necessary for solubilization of biomass are usually not compatible with the enzymes utilized.[12] Thus, a wide range of ILs have been employed as a solvents for enzyme catalysis and have been shown to give high catalytic conversions where the enzymes are stable, selective, easily separated from the products and can be recycled.[13,14]

This chapter will firstly focus on the toxicity and biodegradation of ILs by providing an overview of the current research in the area of IL toxicity and biodegradability. It is important to understand how to design an IL for increased biodegradability and low toxicity, and this is discussed from a structure–activity perspective. Enzymes utilized in an IL biorefinery will be in contact with ILs, thus it is necessary to understand how ILs effect and interact with enzymes. IL toxicity towards enzymes has been reviewed using the acetylcholinesterase (AChE) assay, providing an insight into how IL structures influence toxicity towards enzymes. It is hoped that this information can be utilized to subsequently design ILs that will be less toxic towards enzymes. IL compatibility with enzymes is also vital to the success of an IL biorefinery and has been presented from the perspective of enzyme stability, selectivity and activity in ILs. Finally, to address the idea of a fully integrated biorefinery, enzymatic polymerization reactions carried out in ILs are presented. This is an area that truly has the potential to provide the economy with renewable materials necessary for a sustainable future.

6.2 Toxicity and Biodegradation of ILs

Initial publications in the field of ILs focused on their suitability for a wide range of applications and promoted how their isoteric physical properties made them 'green' alternatives to VOCs. It was not until the early 2000s that the environmental impact of ILs was investigated with the first toxicity tests being carried out by Jastorff *et al.* in 2002.[15] In this work the environmental risk potential of ILs were evaluated and theoretical toxicity evaluations of the commonly used C_4 and C_{10} [BF$_4$] ILs were made. A high 'uncertainty' factor was identified in the multidimensional risk analysis carried out due to the lack of information available on the IL decomposition residues, bioaccumulation, biodegradation and biological activity (Figure 6.1). This highlighted not only the potential overlooked toxicity of the second generation of ILs (see below) but also the crucial need for the evaluation of IL toxicity and biodegradation. This work was also able to demonstrate the use of quantitative structure–activity relationships (QSAR) as a methodology for predicting IL toxicity.

Since this work, numerous toxicity and biodegradation studies have been carried out on the second generation of ILs, highlighting the need for the re-design of ILs for lower toxicity and increased biodegradation.[16–19] It is now also known that the widely employed anions [BF$_4$] and [PF$_6$] readily hydrolyse to highly corrosive hydrogen fluoride (HF), especially in the presence of water and heat, with [PF$_6$] additionally releasing the toxic and corrosive hydrolysis products HPO_2F_2, H_2PO_3F and POF_3.[20,21] With this knowledge, there has been speculation that some of the early results of [BF$_4$] and [PF$_6$] IL catalysis may have been no more than catalysis by HF produced from IL breakdown.[22] However, the main problem highlighted here is that the additional toxic and corrosive hydrolysis products generated from these ILs, which are often overlooked, clearly present a barrier to their widespread application in industrial processes.

The series of ILs that have found the most applications to date are often referred to as the second generation of ILs.[23] These ILs are oxygen stable and water stable, with weakly coordinating anions such as [BF$_4$], [PF$_6$] or halides ([Cl], [Br] and [I]). The most common cations found in the second generation of ILs are imidazolium, pyridinium, ammonium and phosphonium. These ILs are designed to remain inert and are also the most studied and applied in the field of biocatalysts.[24] Modifications to the structure of the second generation of ILs resulted in the production of a third generation of task specific

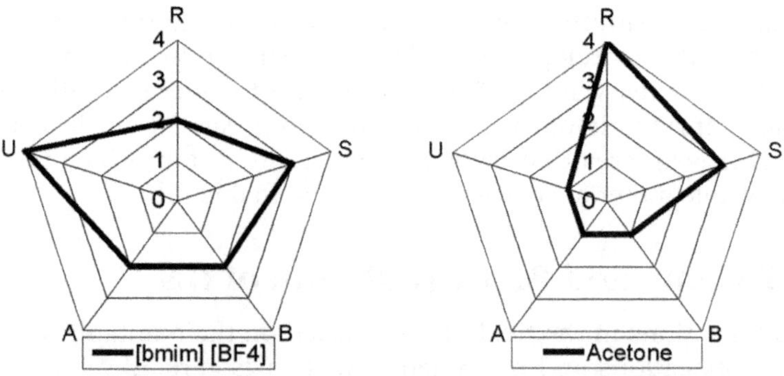

Figure 6.1   5-Dimensional risk analysis comparison of [BMIM][BF$_4$] with acetone. Very high risk = 4 and very low risk = 1. Release, S = Spatiotemporal Range, B = Bioaccumulation, A = Biological activity, U = Uncertainty. (Adapted from ref. 15.)

ILs (TSILs). These were designed with an anion or cation that would actively take part in chemical reactions. An example is the imidazolium ILs synthesized and utilized by Myles *et al.* that act as Brønsted acidic catalysts in the presence of protic additives.[25,26]

However, research into the biodegradability and toxicity of the second generation ILs placed uncertainty over their 'greenness'. Progress has been made towards increasing the biodegradability of the second generation of ILs by introducing functionalities such as ester groups and ether linkages into the cation side chain, and by replacing the common halogenated anions with the biodegradable low toxicity counter ion octylsulfate.[27–29] IL toxicity has also been reduced by the addition of ether linkages and the avoidance of long alkyl chains ($>C_6$).[30] However, the imidazolium ring has been shown to be highly resistant to biodegradation and therefore inappropriate for designing biodegradable ILs.[31]

This has led to the recent introduction of a fourth generation of ILs based on more biodegradable, readily available, renewable and less toxic cations and/or anions. More hydrophobic stable anions such as sugars, amino acids, organic acids, and the cation choline have thus been employed. However, despite the introduction of these types of moieties into ILs results still show that the ILs designed in this way are not always readily biodegradable ($>60\%$ biodegradation over 28 days).[32,33] It appears that incorporating both high biodegradation and low toxicity into an IL is still a challenge today and is an area that has a lot of scope to still be explored.

A class of solvents with similar properties to ILs are the deep eutectic solvents (DES). These are mixtures of salts (*e.g.* choline chloride) and uncharged hydrogen bond donors (*e.g.* urea, glycerol).[34] They have also been proposed as green alternatives to the second and third generations of ILs due to the non-toxic, renewable and biodegradable nature of their synthesis materials.[35] They are also likely to be less expensive than previous generations of ILs with no purification required after the formation of the DES.

Toxicity and biodegradation tests on ILs are often carried out independently of each other, yet it is important not to consider these tests as two entirely separate entities.[36] IL toxicity can have a significant bearing on the biodegradation test results as the IL being tested may be toxic to the microorganisms necessary for biodegradation.[37,38] The microorganisms will not be given the opportunity to biodegrade the IL and a low level of biodegradation will be observed in the test, even if the IL has many accessible breakdown pathways. To prevent erroneous biodegradation results IL concentrations should not exceed, if possible, levels that are toxic to the microorganisms used in the test. However to promote IL breakdown in the environment, the avoidance of high toxicity needs to be considered when designing an IL for enhanced biodegradability.

A compound that breaks down too readily will not be useful for a wide range of applications with varying conditions, such as high temperatures and a range of pH valves. When considering carrying out biological process

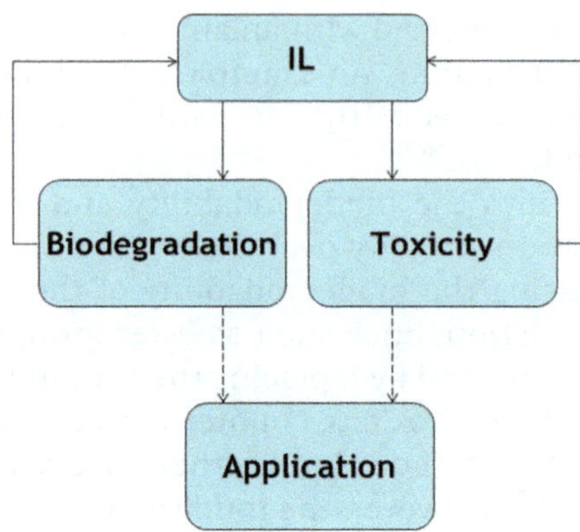

Figure 6.2   Design strategy for a 'green' IL.

in IL media it is critical to establish that the IL will not be broken down by the reaction conditions or the enzymes used in the process. The resulting degradation products of the IL may hinder the reaction, react in a way that damages the process, or contaminate the final product.[39] The IL would also no longer be able to be recycled in the process and additional costs would be incurred as a result.[40] These are likely to be for separation of contaminants from the product, purchasing of fresh IL and the disposal of the unusable degraded IL. Thus, although counterintuitive, it is possible to design a compound that is potentially 'too biodegradable'.

The research groups directly related to the synthesis and study of 'green' ILs will likely have a different design approach to those synthesizing task specific ILs where 'green' credentials of ILs may be secondary to the IL application. The approach of a green chemist would be to design for biodegradation and low toxicity first and to select an appropriate application afterwards. If no suitable application can be found the toxicity and biodegradation data can be fed back into the design of the next class of 'green' ILs (Figure 6.2). This is where one has to think about 'greenness' *versus* application and decide which is more important and should be considered first during the IL design. Preferably, fulfilling all three credentials of biodegradability, low toxicity and task specificity would lead to an improved, potentially 'ideal' IL (Figure 6.3).

6.2.1   Toxicity of ILs

IL toxicity has been studied using bacteria,[41–44] fungi,[41] algae,[45–51] enzymes[52–54] and multicellular organisms including *Daphnia magna*,[55] *Danio rerio*,[56] soil invertebrates,[42] mammals[57] and mammalian[42,43,58–60] and fish cells.[61] A generalized overview of the current data available on the assessment of the second generation ILs (for example ILs with cations [BMIM], [MMIM], [MPY]) shows that they have toxicity, in some cases, equal to or even two to four orders of magnitude higher than conventional organic solvents such as benzene,

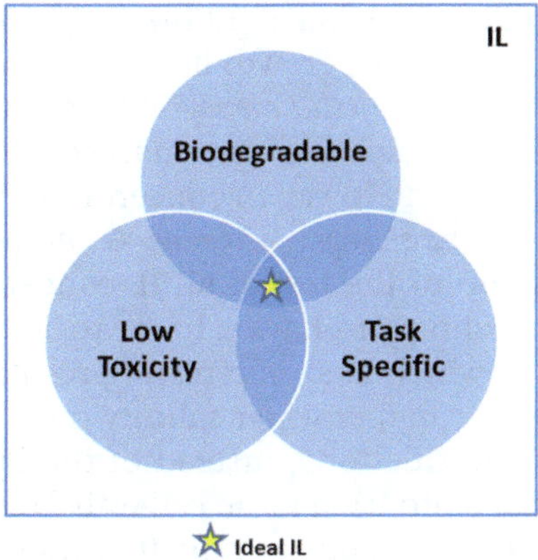

Figure 6.3   Credentials of an 'ideal' IL.

methanol, dimethylformamide or propan-2-ol.[18,45,51,55] The exact results are dependent on the structural nature of the IL (cation and anion pair) and the test system selected.

The toxicity of ILs is known to be much more affected by the structure of the cation than the anion.[60,62] The cation alkyl chain length has been shown to be very important, with the longer alkyl chains resulting in higher toxicity.[62] This is directly related to the longer alkyl chains increasing the hydrophobicity of the IL and thus lipophilicity. The increased lipophilicity increases the likelihood of an interaction with the phospholipid bilayers of cell membranes and membrane protein hydrophobic domains. This interaction causes the cell membranes to rupture, leading ultimately to cell death.[63] Polar groups, such as ether groups, introduced on the cation side chain have also been shown to reduce the toxicity of ILs.[30,58]

Below, individual test systems and areas of IL toxicity are discussed. These include aquatic toxicity, bacterial and fungal toxicity, cytotoxicity, antimicrobial resistance and acetylcholinesterase toxicity.

6.2.1.1   Aquatic Toxicity

When the environmental effects of ILs were first considered around 2005 the aquatic toxicity of ILs was the first area to be investigated.[50] This can be rationalized by the fact that if ILs were ever accidently released into the environment it is likely that their first point of contact would be with the aquatic ecosystem. The majority of ILs have high water solubilities and thus they are likely to move swiftly through the environment, interacting with a large range of aquatic organisms.

Algae are primary producers of organic matter necessary to sustain the lives of many species in the freshwater food chain. They also have short life cycles so they respond quickly to environmental change.[17] Both these factors have led to the use of algae to assess the aquatic toxicity of ILs.

In 2005 Latała *et al.* studied the effect of [EMIM][BF$_4$], [BMIM][BF$_4$], [HEXMIM][BF$_4$] and [BZMIM][BF$_4$] ILs on the growth inhibition of the green alga *Oocystis submarina* and the diatom *Cyclotella meneghiniana* from the southern Baltic Sea.[50] The responses of both species were markedly different. *O. submarina* appeared to acclimatize to the lower IL concentrations and recovered their growing ability after *ca.* five days, whereas *C. meneghiniana* growth was inhibited for the duration of the test at all IL concentrations. [BMIM][BF$_4$] and [BZMIM][BF$_4$] were shown to have a lower toxic effect on *O. submarina* at higher water salinity, which is an interesting result as the salinity of the Baltic Sea varies greatly. At higher water salinity the likelihood of the IL cations pairing with the chloride ions, instead of the hydroxyl groups (green algae) or silanol groups (diatom) in algae cell wall functionalities, increases. This in turn reduces the permeability of the IL cations through the algal cell walls making them less toxic to the algae.

Since 2005 algae assays have been widely used to evaluate the aquatic toxicity of ILs. Further work has been conducted by Latała *et al.* studying the toxicity of 1-alkyl-3-methylimidazolium ILs towards planktonic green algae (*Chlorella vulgaris* and *O. submarina*), diatoms (*Cyclotella meneghiniana*, *Skeletonema marinoi* and *Bacillaria paxillifer*), and blue-green algae (*Geitlerinema amphibium*).[47–49] A pronounced alkyl chain effect was observed with all organisms studied and indicated that sensitivity towards ILs decreases in the order blue-green algae > diatoms > green algae. The main reason for this observed trend is the difference in algae cell wall structures. Blue-green algae have peptidoglycan cell walls, diatoms have silica-based cell walls and green algae have cellulose-based cell walls. [BF$_4$] and [CF$_3$SO$_3$] anions were shown to increase IL toxicity due to fluoride formation from [BF$_4$] hydrolysis and increased cell wall penetration with highly lipophilic [CF$_3$SO$_3$]. Cell size is also significant in determining the level of toxicity, with a 10 times difference in cell size leading to a 100% more sensitive reaction to ILs in green algae and diatoms.[48] 1-Alkyl-3-methylimidazolium chloride ILs also showed the same previously observed trend of lower toxicity at higher water salinities.[47]

Additional studies on algae toxicity all followed the same observed alkyl chain trend of increasing IL toxicity with increasing chain length.[42,46,64] Kulacki and Lamberti also stated that the EC$_{50}$ values they reported suggested that the ILs tested were more toxic, or just as toxic, to the algae than many of the solvents they are intended to replace.[46] Cho and Pham *et al.* also showed that imidazolium ILs comprising the anion [SbF$_6$] were the most toxic of anions tested and found pyridinium ILs were more toxic than their imidazolium counterparts.[45,51]

Daphnia magna are a species of cladoceran freshwater water flea widely used in the laboratory environment to examine the toxic effect a compound has on reproduction and life expectancy. The first IL toxicity studies on *D. magna* undertaken in 2005 by Bernot *et al.* investigated the acute and chronic toxic effects of imidazolium ILs with various anions such as [Cl], [Br], [BF$_4$] and [PF$_6$].[55] Median lethal toxicity (LC$_{50}$) values were more than an order of magnitude higher than the corresponding sodium salts, showing that the

IL toxicity was related to the imidazolium anion and not the cation. Toxicities were comparable to that of common industrial solvents of the time such as phenol and ammonia; however, they were two to three orders of magnitude more toxic than methanol or acetonitrile.[55] Since this work, numerous studies on the toxicity of ILs towards *D. magna* have been undertaken.[38,65,66] The overriding conclusion from this work is that imidazolium, pyridinium and quaternary ammonium ILs with increasing alkyl chain length have an increased toxicity towards *D. magna*. Again the nature of the anion was shown to have a smaller effect than the cation. Couling *et al.* used QSAR modelling to help identify the specific part of the IL responsible for the toxic effects towards *D. magna*.[66] It was determined that ILs having longer alkyl residues attached to aromatic nitrogen atoms, and a higher number of aromatic nitrogen atoms, had increased toxicity. Methylating the aromatic carbons was shown to reduce the IL toxicity.

The model vertebrate zebrafish (*Danio rerio*) were used by Pretti *et al.* in 2006 to study the toxic effect of ammonium, imidazolium, pyridinium and pyrrolidinium ILs.[56] The imidazolium, pyridinium and pyrrolidinium ILs gave LC_{50} values >100 mg L^{-1} and were thus regarded as non-highly lethal towards zebrafish. In stark contrast the ammonium ILs AMMOENG™ 100 and 130 were much more toxic with LC_{50} values 5.9 and 5.2 mg L^{-1} respectively. These LC_{50} values were remarkably lower than those for common organic solvents and tertiary amines. It was thus concluded that ILs may have a different effect on fish related to the IL chemical structure. However, it is important to note that the ammonium ILs used in the study were AMMOENG™ 100, 110, 112 and 130. These acyclic ammonium salts are used industrially as cationic surfactants and contain PEG chains of various lengths on the cation.[67] These PEG chains are much longer than the ethyl and butyl chains on the imidazolium and pyridinium ILs used in the study. Long alkyl chain length has been shown to increase IL toxicity and this should be taken into account when concluding that ammonium ILs were more toxic to fish than the imidazolium or pyridinium ILs.

6.2.1.2 Bacterial and Fungal Toxicity

The short generation times of bacteria make them an ideal starting point for toxicity studies. Preliminary toxicity studies showed that compounds containing quaternary ammonium and pyridinium moieties have clinical inhibitory effects on a variety of bacteria and fungi.[17]

Pernak *et al.* published a paper in 2003 on the antimicrobial properties of 3-alkoxymethyl-1-methylimidazolium ILs with [Cl], [BF$_4$] and [PF$_6$] anions. Here they showed an increase in alkyl chain length increased the toxicity towards rods, cocci and fungi.[62] ILs with 10, 11, 12 and 14 carbon atoms in the alkyl chain were shown to have the highest toxicity towards the screened bacteria. Additional work by Pernak *et al.* corroborated this trend with pyridinium salts.[68,69] Pernak *et al.* also found that the stereochemistry of the anion in lactate ILs affected their toxicity towards bacteria. In general the L-lactate ILs gave lower MIC values than the racemates.[68]

In 2005 Docherty and Kulpa examined the antimicrobial effects of various alkyl chain length imidazolium and pyridinium ILs at a concentration of 1000 ppm on the growth of *Escherichia coli, Staphylococcus aureus, Bacillus subtilis, Pseudomonas fluorescens* and *Saccharomyces cerevisiae*.[41] Increased antimicrobial activity was observed at increased alkyl chain lengths. Imidazolium and pyridinium bromide ILs with hexyl and octyl chains showed considerable antimicrobial activity towards pure cultures of the microorganisms, with ILs [OMIM][Br], [OPY][Br] and [HEXMIM][Br] having log EC_{50} values 0.07, 0.25 and 0.81 respectively.

In general, studies on the microbial toxicity of imidazolium ILs have shown that the anion has a slight effect on activity.[41,62,65,68] Whereas for alkyltrihexyl phosphonium ILs both the anion type and cation structure have been shown to have an effect on the IL microbial toxicity.[69]

The rod-shaped bacterium *Vibrio fischeri* is often used to measure IL toxicity using the acute bioluminescence assay (DIN EN ISO 11348).[42,64,66] Gathergood *et al.* also used *Photobacterium phosphoreum* for the same purpose.[65] For both bacteria increasing alkyl chain length was again shown to increase the toxicity of the ILs. With *V. fischeri* the same trend was observed by Docherty *et al.* for imidazolium and pyridinium IL using the Microtox assay.[41] Ranke *et al.* showed IL toxicity was mainly due to the cationic portion of the IL; however, ILs containing the anion [PF_6] were slightly more toxic. At concentrations below the ILs inhibitory concentration, slight hormesis was observed.[43] Couling *et al.* found that the same trend of increasing alkyl chain length leading to increased IL toxicity was apparent for the toxicity of ILs towards *D. magna*, with quaternary ammonium ILs being the least toxic of those studied.[66] The same QSAR model and trends found for the toxicity of the ILs towards *V. fischeri* also applied to *D. magna*.

Stolte *et al.* studied the effect of head groups and side chains on the aquatic toxicity of ILs and showed that short functionalized side chains reduce the lipophilicity and thus toxicity (six to seven orders of magnitude).[64] The morpholinium head group was recommended for use and the dimethylaminopyridinium recommended for avoidance due to its effect on the aquatic organisms in the study. It was found that the halide anions ([Cl] and [Br]) did not exhibit an intrinsic toxic effect; however, ILs containing the [NTf_2] anion were more toxic for *V. fischeri*. This is in contrast to the work by Matzke *et al.* where [NTf_2] showed no intrinsic toxicity to *V. fischeri*.[42]

In 2006 Docherty *et al.* carried out the Ames test with *Salmonella typhimurium* strains TA98 and TA100 to assess the mutagenicity of ten ILs containing imidazolium, pyridinium and quaternary ammonium cations.[70] The level that the ILs caused mutation in the bacteria was measured and related to higher organism carcinogenicity. None of the ILs tested fell under the US EPA classification for mutagenicity.[71] However, imidazolium ILs [BMIM][Br] and [OMIM][Br] indicated the potential for mutagenicity at high doses (20 mg per plate for [BMIM][Br] and 1.0 mg per plate for [OMIM][Br]), suggesting that further testing is required before the mutagenic potential of ILs is fully understood. Based on this study alone, designing ILs based on pyridinium and quaternary ammonium cations instead of imidazolium cations is encouraged to reduce the potential for mutagenicity.

6.2.1.3 Cytotoxicity

Stolte *et al.* studied the cytotoxicity of ILs on the IPC-81 rat leukaemia cell line using the WST-1 cell variability assay.[58,72] 100 ILs containing various anions, cations and side chains were tested and showed a low cytotoxicity compared to previous ILs tested. Low cytotoxicity was linked to the polar ether, hydroxyl and nitrile functional groups in the IL side chains decreasing the IL lipophilicity and consequently cell uptake.[58] The head group was shown to have little effect on cytotoxicity with the [NTf$_2$] anion having a marked effect on cytotoxicity. The correlation between IL cation lipophilicity and cytotoxicity was also confirmed using a HPLC derived lipophilicity parameter. Further work by Ranke *et al.* on 74 ILs with comparatively small anions ([Cl], [Br], [BF$_4$] and [PF$_6$]) demonstrated that the IL side chain has a greater influence on IL cytotoxicity than the IL head group or the anion, reinforcing the link between increased lipophilicity and increased IL cytotoxicity.[60]

6.2.1.4 Antimicrobial Resistance

It is important to consider what happens to ILs if they are released into the environment where they are likely to come into contact with a plethora of different microorganisms. A major problem that has recently been highlighted in the media is that of antimicrobial resistance. This is described by the World Health Organization (WHO) as an increasingly serious threat to global public health that requires action across all government sectors and society.[73]

In 2014 Luo *et al.* investigated the ability of the IL [BMIM][PF$_6$] to promote the proliferation and dissemination of antibiotic resistance genes in environmental bacteria.[74] [BMIM][PF$_6$] was found to increase the abundance of a *sulI* gene resistance in freshwater microcosms by 500 times when compared to untreated controls. The IL also significantly increased the abundance of class I integrons, which are particularly adapted to transfer and disseminate antibiotic resistance through horizontal gene transfer between different bacterial strains. Cell membranes in bacteria are hugely significant in preventing horizontal gene transfer. Results suggested that *sulI* propagation mediated by class I integrons was up to 88 times higher in the presence of the IL. This was concluded to be due to the presence of the IL causing an increase in cell membrane permeability. Flow cytometry results showed that in the presence of just 0.5 g L^{-1} of the IL the bacterial cell membrane permeability increased by over 230% when compared to untreated cells.

This work by Luo *et al.* is the first to report that ILs can facilitate the proliferation of antibiotic resistant genes in environmental bacteria and highlights ILs as a potential public health risk. Despite the clear merits of this work it should be highlighted that only one IL has been evaluated in this study. The

IL [BMIM][PF$_6$] is also a second generation IL and various work in the past has already shown that this IL has high levels of toxicity to certain bacteria and aquatic organisms. This IL also presents a high usage and release hazard especially through the formation of degradation products. Thus a wider range of ILs needs to be evaluated, including non-toxic and biodegradable ILs, before a clear picture can be built up as to how concerned we should be over the ability of ILs to promote antimicrobial resistance.

6.2.1.5 Acetylcholinesterase Inhibition

A widely used test to study the molecular toxicity of ILs is the acetylcholinesterase (AChE) assay where the purified enzyme from the electric eel (*Electrophorus electricus*) is most often used. This enzyme has a vital role in the nervous system of almost all higher organisms, including humans, where it plays a vital role as a neurotransmitter. Inhibition of AChE has been linked to various conditions such as muscular weakness, fatigue and heart disease.[53]

Stock *et al.* in 2004 were the first to use the AChE assay to evaluate the toxicity of ILs.[52] They give many reasons for their selection of the AChE assay, including that the active site of AChE has been widely studied and was the main target during extensive research into the design of insecticides based on phosphoric acid esters and carbamates.[75–77] This work would help to facilitate a detailed QSAR study linking IL structural features to enzyme inhibition. Previous work on pharmaceuticals showed that AChE inhibition was linked to a positively charged nitrogen, an electron-deficient aromatic system, and a certain lipophilicity, all common features of ILs.[78] The AChE active site has also been shown to be comparable between organisms, thus any QSAR conclusions drawn would be applicable to organisms other than those tested.[52]

Imidazolium, pyridinium and phosphonium ILs were chosen for AChE inhibition assays in particular because the chemical structures of the IL cations and acetylcholine (substrate for acetylcholinesterase enzyme) are similar.[52] Both are cationic and the positive charge is masked by the lipophilic residues.[79] Inhibition was shown to be dependent on the cationic structure and alkyl side chain with 1-butyl-2-methylpyridinium, [OMIM] and [DECMIM] ILs all showing pronounced AChE inhibition only 3–10 times weaker than a potent inhibiting insecticide aldicarb.[21] ILs with short alkyl chains gave less inhibition, correlating to results in *V. fischeri* and leukaemia cells. When considering only AChE inhibition phosphonium < pyridinium < imidazolium thus phosphonium IL use is preferred.[52] The influence of the anion was shown to not dominate the toxicity of the IL.

A further study by Arning *et al.* in 2007 identified a positively charged nitrogen atom, a widely delocalized aromatic system and the lipophilicity of the cation alkyl chain as key structural elements of IL binding to the AChE active site.[53] The diethylaminopyridinium, quinolinium and pyridinium head groups were listed as having a very strong inhibitory potential, whereas the polar and non-aromatic morpholinium head group was only weakly inhibiting. The inhibitory potential was shown to be lowered when ether, nitrile or polar hydroxyl groups were introduced into the alkyl side chain. A wide range of commonly used anions were tested with most showing no inhibitory potential. Only fluorine and fluoride containing anions which readily undergo hydrolytic cleavage showed any inhibitory potential. The 'side chain' effect was observed in this study with increased lipophilicity of the cation alkyl side chain increasing enzyme inhibition.

A review in 2007 by Ranke *et al.* listed the AChE inhibition (log EC_{50}) values for 292 compounds, including a large number of ILs and closely related salts.[21] The review compiled into one table the largest selection of ILs tested in a single biological test to date. The AChE inhibitory concentrations spanned more than three orders of magnitude and followed the trends of potent IL AChE inhibitors having a positively charged nitrogen centre, a broadly delocalized aromatic ring system and a certain lipophilicity.

The results obtained in the aforementioned AChE inhibition studies were often rationalized using qualitative and quantitative structure–activity relationships (QSAR).[52,53] QSAR analysis aims to relate the chemical structures of molecules to their biological activity, thus enabling a prediction about the toxicity of novel molecules to be made. QSAR modelling is currently being used more readily in the area of predicting IL toxicity, although it is still rare for the prediction of AChE inhibition.[80,81] Most recently work has been published by Das *et al.* on the *in silico* modelling of ILs towards inhibition of AChE.[82] In this study the chemical attributes of a broad range of 292 ILs were investigated using predictive regression and classification-based quantitative mathematical models. Topological features and electronic parameters of the ILs were considered in the model. The results showed that the main contributors to IL toxicity were positively charged nitrogen species, branching, hydrophobicity and hydrophilicity, as summarized in Figure 6.4. These conclusions were additionally backed up by molecular docking studies, carried out for the first time in this area, which investigated the structural features responsible for IL binding to the AChE cavity.

A recent paper by Hou *et al.* evaluated a series of cholinium amino acid ILs, composed entirely of renewable biomaterials, for acetylcholinesterase inhibition.[83] The cholinium ILs had much lower toxicity towards AChE, with inhibition potentials approximately an order of magnitude lower than [BMIM][BF$_4$]. They also displayed low toxicity towards the bacteria tested against. For several ILs, lengthening the amino acid side chain led to an increase in the AChE inhibition. This is due to an increased anion lipophilicity resulting in stronger hydrophobic interactions with the lipophilic amino acid residues in the AChE enzyme cavity. However, for amino acids with hydrophobic aliphatic side chains, increasing the size of the side chain beyond valine resulted in lower AChE inhibition. This is likely due to steric hindrance preventing the amino acid entering the active site of the enzyme. The introduction of hydrophilic groups such as hydroxyl, carboxyl and amino into the amino acid generally decreased the IL AChE inhibition. Aromatic systems introduced into the cation side chains also lead to greater AChE inhibition.

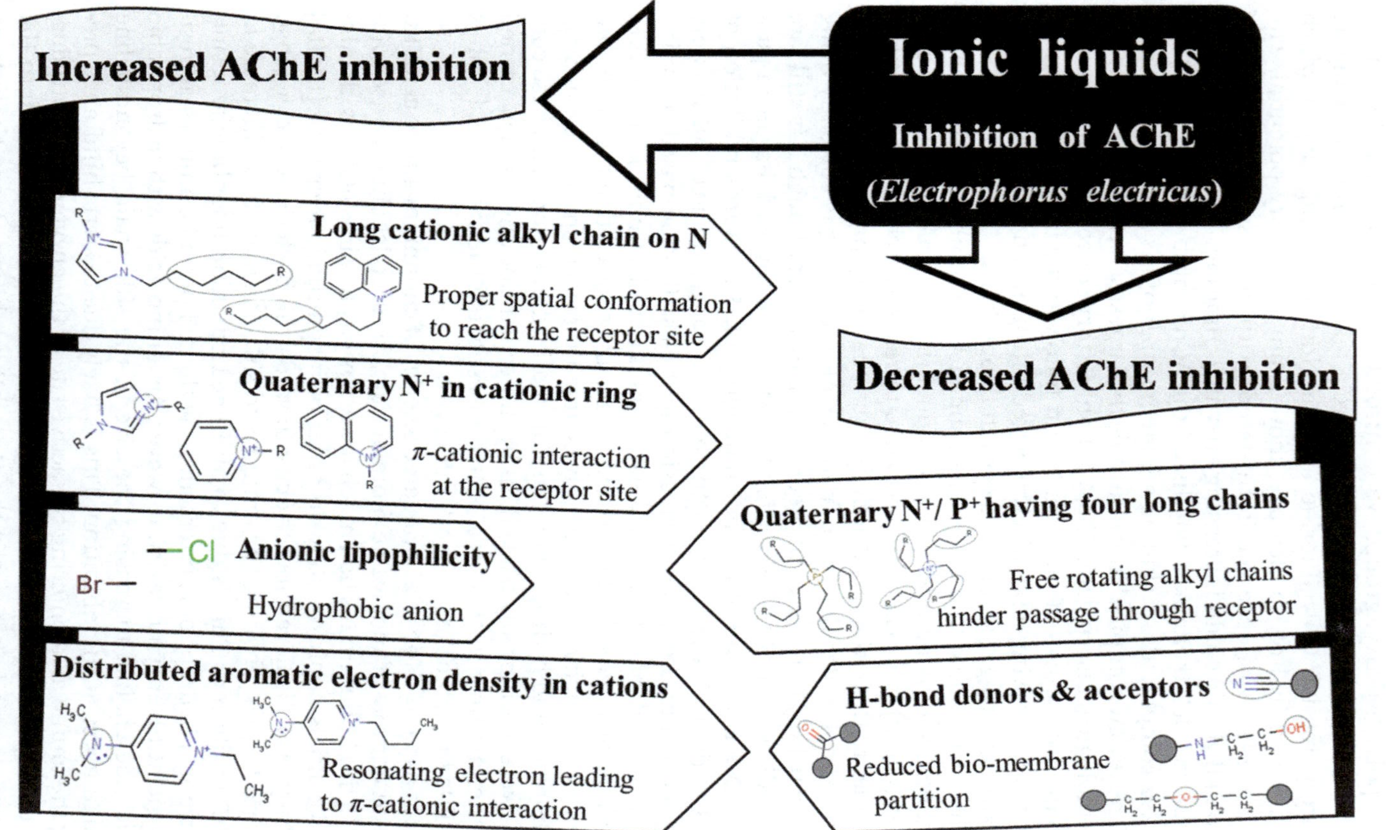

Figure 6.4 Features of ILs for toxicity towards *E. electricus* derived using discernment, regression and docking analysis carried out by Das *et al.*[82] Reprinted with permission from R. N. Das and K. Roy, *Ind. Eng. Chem. Res.*, 2014, 53, 1020–1032. Copyright 2014 American Chemical Society.

6.2.2 Biodegradation of ILs

Biodegradation is classified as the breakdown of a compound by microorganisms *via* enzymatic processes. This breakdown pathway is preferred over chemical degradation especially if the release of the compound into the environment is a possibility. The ultimate biodegradation of a compound (total breakdown of the test compound by microorganisms) can be studied by several methods. The most widely used today are those carried out following OECD guidelines for chemical testing and include the dissolved oxygen content (DOC) die-away test (OECD 301 A),[84] the CO_2 evolution/modified Sturm test (OECD 301 B),[84] the closed bottle test (OECD 301 D),[84] the manometric respirometry test (OECD 301 F),[84] and the CO_2 headspace test (OECD 310/ISO 14593).[85,86]

There are many classifications for evaluating the level of biodegradation of a compound. The term most often used when evaluating the ultimate biodegradation of an IL is readily biodegradable. If an IL passes a specified level of degradation in a certain amount of time, as determined by the particular ultimate biodegradation test used, the IL can be classed as readily biodegradable. For the OECD tests 301 A, B, D and F a removal of at least 70% dissolved organic carbon or a production of 60% theoretical oxygen demand/theoretical carbon dioxide, during the 28 day test period, classifies a compound as readily biodegradable.[84] There is also a 10-day window for biodegradation following attainment of 10% biodegradation, in which the compound must biodegrade to the specified level to be classified as readily biodegradable. For the CO_2 headspace test (OECD 310) production of >60% theoretical inorganic carbon within the 10-day window demonstrates that the test substance is readily biodegradable under aerobic conditions.[85]

The biodegradation of ILs was first reviewed by Coleman and Gathergood in 2010 and subsequently by Stolte *et al.* in 2011.[7,16] Both reviews provide a detailed explanation of IL biodegradation assays, and the links found between IL structure and biodegradation are discussed. The next section of this book contains an overview as to how to best design a biodegradable IL based on the research contained in the reviews. Recent examples of interest in the literature will be highlighted.

6.2.2.1 Designing for Biodegradation

Gathergood and Scammells were the first to study the biodegradation of ILs in 2002.[87] They introduced functional groups susceptible to enzymatic hydrolysis to increase IL biodegradation. Incorporating ester groups into the IL side chain significantly improved IL biodegradation; however, ILs with additional amide groups were poorly biodegradable.[27] Further work by Gathergood *et al.* focused on the effect of the anion on IL biodegradation, leading to a series of results where imidazolium ester ILs combined with octylsulfate anions (first used in ILs by Wasserscheid *et al.*)[29] showed high levels of biodegradation.[65] Readily biodegradable ILs of this type were identified using the CO_2 headspace test.[28] ILs prepared from pyridine or nicotinic acid with an ester side chain moiety by Harjani *et al.* in 2009 also showed high levels of biodegradation in the CO_2 headspace test and were classed as readily biodegradable.[88] In 2009 Morrissey *et al.* incorporated ether and poly ether side chains into imidazolium ester ILs.[30] Six of the ILs were classed as readily biodegradable by the CO_2 headspace test, all containing the octylsulfate anion, with amide ILs again shown to be less biodegradable than ester ILs. The introduction of ether and poly ether side chains was also shown to reduce IL bacterial toxicity compared to long alkyl chain derivatives.

Further work in the area of biodegradable ILs has led to research being conducted on investigating the biodegradation pathways of ILs and the resulting metabolites.[7] It is of paramount importance to understand the resultant metabolites in IL biodegradation, especially if a compound is not ultimately biodegradable, and they have been predominantly studied using HPLC-MS. It is vital to know at what point a compound stops being biodegraded so further generations of ILs can be designed to overcome this. Analysis of the toxicity of metabolites is also necessary to evaluate if more toxic entities are remaining after the IL has undergone the maximum biodegradation possible in the environment.

Recent work on the biodegradation of ILs has included the evaluation of cholinium amino acid ILs based on renewable resources.[83] The choline cation was paired with 18 different amino acids as anions and each IL was evaluated for aerobic biodegradability *via* the closed bottle and CO_2 headspace test. The ILs were compared to choline acetate and the reference sodium benzoate. Biodegradability was >60% after 28 days for all ILs in both tests so all can be classed as readily biodegradable. The high level of biodegradation for the cholinium amino acid ILs was attributed to the cholinium cation and the presence of the carboxyl and primary amino groups in the anion. The addition of branched side chains decreased the levels of biodegradation, and the addition of hydroxyl groups increased IL biodegradation. The ILs with carboxylic acid and primary amide groups gave very high levels of biodegradability (>86%). Particularly interesting to note from this work is that previous problems with designing ILs for both low toxicity and biodegradation were not present in this study. With relatively few exceptions, the cholinium amino acid ILs with lower toxicity had a higher level of biodegradability.

The toxicity of an IL can have an effect on biodegradation test results. Recent work by Markiewicz *et al.* showed this to be the case for ILs with highly hydrophobic cations and anions.[37] It was recommended that for such ILs, lower than normal IL test concentrations should be used to avoid false negative biodegradation results. This work also showed that the type of inoculum used in the biodegradation test had a strong influence on the test result. However when considering toxicity, no significant differences in sensitivity/tolerance of the activated sludge could be determined.

Neumann *et al.* recently published a paper on the biodegradability of 27 pyrrolidinium, morpholinium, piperidinium, imidazolium and pyridinium

IL cations under aerobic conditions.[31] The ILs contained different alkyl chains and halide counter ions and were assessed for ultimate biodegradability using the manometric respirometry test (OECD 301 F). ILs classed as readily biodegradable were assessed for full mineralization through BOD measurements. For all five head groups ILs that were readily or inherently (20–70%) biodegradable were identified, with biodegradation being solely attributed to the cation. The head group influenced the biodegradation level, with pyrrolidinium and pyridinium cations having higher levels of biodegradation than piperidinium and morpholinium cations. The imidazolium cation was classified as the least biodegradable head group and, as in previous work, was presented as recalcitrant. Side chain type also influenced biodegradation levels, with hydroxyl substitution and longer chain length both increasing biodegradation. This work also corroborated previous work stating that the level of biodegradation was dependant on the inoculum used and the conditions of the test.

When thinking about the structural attributes of an IL that influence biodegradability, increased biodegradability has been observed in pyridinium, imidazolium and 4-(dimethylamino)pyridinium ILs with elongated alkyl side chains.[7] However, this design strategy is directly opposed to creating a low toxicity IL where long alkyl side chains have been shown to increase IL toxicity. It has also been shown that introducing ester groups into the IL side chains can increase their biodegradation.

With regards to the IL head group, pyridinium ILs generally give a higher degree of biodegradation than imidazolium and 4-(dimethylamino)pyridinium ILs. The pyridinium head group often presents as readily biodegradable even when linked to short alkyl chains and tested under stringent readily biodegradable analysis conditions. Imidazolium ILs only show degradation of the alkyl chain under the same readily biodegradable test conditions, with the imidazolium core resistant to biodegradation. However, the use of industrial inocula have indicated that degradation of the imidazolium core is possible under certain conditions.[7]

Anions that have been used to increase the biodegradation of ILs include alkylsulfates (such as methylsulfate or octylsulfate), linear alkylsulfonates (such as methylsulfonate), linear alkylbenzylsulfonates (such as *p*-toluenesulfonate) and salts of organic acids (such as acetate or lactate). These anions are also recommended for use from when designing from a low toxicity point of view. Fluorine-containing cations (such as $[NTf_2]$ and $[(C_2F_5)_3PF_3]$) should be avoided as they are likely to be recalcitrant in the environment and, due to their high hydrophobicity, accumulate in the tissues of living organisms.

Modelling and predicting the biodegradation of ILs is important. It can be costly and time consuming to have to reconsider the use of a particular IL far down the line of development if it is found to be unsuitable for use due to its hazardous nature. Thus, it would be a major benefit to have QSAR models that were used widely and highly reliable for the prediction of an IL's biodegradation. Unfortunately we are not there yet and a review from 2012 by Rücker and Kümmerer details some of the reasons why.[89] A particular problem highlighted was the slow progress of QSAR model development due to the poor reproducibility of experimental data. A large part of this is, however, unavoidable due to the variation in bacterial populations. Also noted was the dearth of high-quality experimental data, often remaining confidential in industry, and the need for it to be made publically available. The conclusion was that models need to be published in detail, including all the data a model is built on and any corresponding software, to facilitate the wider use and enhancement of such models for predicting biodegradation.

Adding biodegradable functionalities to IL cations and pairing cations with biodegradable anions may help ILs to reach readily biodegradable pass levels, but will not lead to the production of ultimately biodegradable ILs. At the end of the day if the core of the IL is not biodegradable it will be left in the environment unable to be biodegrade when the biodegradable additions have been metabolized. Thus we should be using the information gathered from these previous studies to help design ultimately biodegradable ILs from the ground up. In this way we can prevent the build-up of recalcitrant IL components in the environment should their release ever occur.

6.3   Enzyme Catalysis in IL

Enzymatic biocatalysis in the presence of ILs has been successfully reported for a number of reactions including esterification, transesterification, alcoholysis, hydrolysis, perhydrolysis, aminolysis, ammonolysis and polymerizations.[14,90] A wide variety of enzymes has also been employed, such as lipases, cellulases and oxidoreductases,[13] in a wide range of ILs and have been shown to give comparable or higher catalytic activities in ILs than conventional solvents.[63] In many cases higher reaction rates, higher conversions and improved regioselectivity and enantioselectivity have been reported.[90]

It is necessary to understand how an IL affects enzyme stability, selectivity and activity before a suitable IL can be selected for a particular enzymatic reaction. Thus, these factors will be discussed in the following section, with particular emphasis on how the structural features of an IL impact on its enzyme compatibility. The structural features present in green ILs, such as medium length alkyl chains and hydrolysable bonds, will be considered in this analysis.

This section will focus on the use of an IL as the sole solvent for enzyme catalysis, thus only non-aqueous biocatalysis will be discussed. The use of IL as an additive, as or with a co-solvent, or in a biphasic system will not be considered. In this way it is clearly the nature of the IL that is affecting enzyme stability, activity and selectivity and will enable more reliable conclusions between enzyme activity and IL structure and toxicity to be drawn.

6.3.1 Green ILs in Enzyme Catalysis

There are many factors that need to be considered when designing a green IL for enzyme catalysis. These include:

(1) Will any hydrolysable bonds in the IL be broken down by the reaction enzyme?
(2) How will a longer alkyl chain on the cation affect its enzyme compatibility?
(3) How temperature stable is the IL to the reaction conditions?
(4) How does the toxicity of the IL affect its enzyme compatibility?
(5) How does the selection of cations and anions affect enzyme compatibility?

The introduction of hydrolysable bonds is important for designing increased biodegradation into an IL. However, an enzyme that would break the hydrolysable bonds in the IL would not be compatible with the IL for catalysis reactions. The IL would be broken down and not only would the solvent be changed, there would be IL degradation products contaminating the final product and the remaining IL. The contaminants would need to be separated from the final product and the IL could no longer be recycled in the process, which would lead to additional costs for product isolation, decontamination, IL disposal and new IL purchase. There would also be a cost associated with redesigning the process. Thus careful thought must be given to the enzyme utilized in the reaction and the structure of the IL chosen. One must accept, however, that it would likely be easier to redesign the process to have a less hydrolysable IL as a solvent than to replace the enzyme which has a specific active site necessary for the given reaction. Also a more enzyme-compatible and reaction condition-stable IL is likely to be one that is more resistant to biodegradation.

The biodegradability of an IL can be increased by having a longer length of alkyl chain; however, data on the toxicity of ILs towards AChE showed that longer cation IL alkyl chains were more toxic to AChE due to an increase in lipophilicity.[53] This conclusion is consistent with the IL alkyl chain length and compatibility with enzyme catalysis. Thus the QSAR rules outlined by Das *et al.* for IL structures and AChE inhibition may be useful to think about when beginning to design an IL for biocatalysis.[82]

Yamamoto *et al.* reported that increasing the alkyl chain length of *N*-alkylpyridinium and *N*-alkyl-*N*-methylpyrrolidinium chlorides destabilized native lysozyme due to an increase in IL hydrophobicity at longer alkyl chain lengths.[91] 500 mM of IL decreased the melting temperature of lysozyme, indicating a decrease in protein thermal stability. More hydrophobic ILs [OPY][Cl], [DODECPY][Cl] and [OPYRR][Cl] even denatured lysozyme at room temperature. Attri *et al.* found that the ILs with more hydrophobic cations and a longer alkyl chain ([BZMIM][Cl], [BZMIM][BF$_4$] and tetrabutylphosphonium bromide) were less stabilizing towards α-chymotrypsin than ILs with shorter alkyl chains (tetraethylammonium acetate and tetraethylammonium phosphate).[92] The activity of α-chymotrypsin in the ILs also decreased with increased IL hydrophobicity and increased alkyl chain length. It has also been reported that IL anions with long alkyl substitution such as hexylsulfate were destabilizing and unable to promote enzyme refolding.[93] Decyl chains on imidazolium IL cations have also been shown to destabilize bovine serum albumin more than shorter chain ILs due to the hydrophobic interactions of the IL with the protein leading to disruption of the secondary structure and deactivation.[94] However, some contradictory results reported that an increase in IL hydrophobicity, by increasing alkyl chain length from [BMIM] to [OMIM], led to an increase in enzyme activity, enantioselectivity and thermostability.[95] This effect though may also be linked to IL viscosity, which greatly increases when the alkyl chain length is extended from butyl to octyl.[96]

Research has shown that terminal hydroxylation of alkyl chains leads to a more enzyme-compatible IL with increased enzyme stability. Ou *et al.* obtained high activity and stability of *Candida antarctica* Lipase B (CALB) when [BMIM] was replaced with $[C_2OHMIM]$ in ILs with $[BF_4]$ and $[NO_3]$ anions, in the presence of an IL phosphorus buffer.[97] AChE inhibition has also been reduced by ILs with shorter alkyl chains and those with additional hydroxyl functionality.[53,83] The addition of hydroxyl groups has also resulted in increased biodegradability of ILs.[31] Thus, introducing hydroxyl function-ality into ILs may be a positive design strategy towards producing ILs with high enzyme compatibility, low toxicity and increased biodegradability.

Inorganic fluoride-containing ILs are best avoided when using an IL as a co-solvent in aqueous biocatalyst due to the hydrolysis of $[PF_6]$ and $[BF_4]$ to hydrofluoric acid which can cause enzyme inactivation and denaturing.[34]

The scale most often used to quantify overall IL hydrophobicity is the log *P* scale, which is the logarithm of the IL partition coefficient between octanol and water. When thinking about the overall hydrophobicity of an IL, it has been shown that more hydrophobic ILs are more compatible with enzymes than hydrophilic ILs. Enzymes have shown higher enantioselectivity, stabil-ity and activity in hydrophobic ILs over hydrophilic ILs.[96] However, as previ-ously mentioned, some contradictory reports do show high enzyme stability and activity in hydrophilic ILs.[91]

The idea of increased IL hydrophobicity leading to increased enzyme stability differs from the research showing that the avoidance of long alkyl chains in IL cations, which increase IL hydrophobicity, increases enzyme stability in ILs. It is possible that the reason the long alkyl chain cation is undesired is due to the resultant increase in IL toxicity and not necessarily the increase in IL hydrophobicity. An overall IL hydrophobicity may be better for the enzyme catalysis, but due to toxicity factors it appears to be better to design this into the anion of the IL rather than the cation.

Another factor that needs to be considered is the temperature of the reaction. If the enzymatic reaction needs to be carried out at a temperature where the hydrolysable bonds in the biodegradable IL will break down, then it is clearly not suitable for the reaction and a more stable and likely less biodegradable

IL is necessary for the process. Therefore it is necessary to know the thermal stability of the ILs, the enzyme in the IL and the necessary temperature required for the reaction, before a suitable IL/enzyme combination can be chosen. Enzymes have thus far been shown in many reports to have high thermostability in ILs, with enzyme immobilization having been shown to increase thermal and operational stability compared to the free enzyme.[13,98]

6.3.2 Enzyme Compatibility

Lipases, in particular *Candida antarctica* Lipase B (CALB), are the most studied enzyme in ILs, with CALB being the first lipase reported active in [BMIM][BF$_4$] and [BMIM][PF$_6$].[13] There have been a number of studies detailing the activity of lipases in ILs to be comparable or higher than in conventional organic solvents.[63] The stability of CALB has been shown to be higher in ILs with [BF$_4$], [PF$_6$] and [NTf$_2$] anions when compared to ILs with [SbF$_6$] or [CF$_3$SO$_3$] anions, where it has lower activity.[34] A more delocalized negative charge in the anion of the IL has been shown to lead to higher enzyme activity. The reactions with lipases are most often carried out with low or no addition of water and thus hydrophobic ILs are used.

The most important features of an IL to consider when thinking about enzyme compatibility were listed in a review by Zhao in 2010 to be IL polarity, hydrogen bond basicity/anion nucleophilicity, IL network, ion kosmotropicity, viscosity, hydrophobicity, enzyme dissolution and the surfactant effect.[99]

There have been many studies to try and increase enzyme compatibility in ILs and these include enzyme immobilization, physical or covalent attachment to PEG, solvent rinsing, water-in-IL microemulsions and modification of the IL structure.[99]

6.3.3 Enzyme Stability

Enzyme activity is much higher in organic solvents, especially at low water content, than in aqueous solutions.[63] Thus it is reasonable to expect ILs to have the same stabilizing effects on enzymes as organic solvents. It is widely accepted that a hydrophobic reaction medium enables the water molecules surrounding the protein core to remain intact, which in turn reduces the protein–ion contact and increases enzyme stability.

The stabilizing effect that ILs have on enzymes has been further enhanced by incubation of the enzyme in the presence of the substrate, where significant increases in enzyme half-life have been reported.[63] The substrate may be associating with the enzyme in a way that activates it by causing a change in the enzyme conformation. This active confirmation is well retained whilst the substrate is present.

Enzymes have been shown to have high stability (operational and thermal) in ILs. Work on the thermostability of enzymes in ILs has shown that ILs offer a significant thermodynamic stabilization effect to the enzymes. This is due to alteration in the protein hydration levels and structural compaction.[63]

Further work has shown that enzyme stabilization in ILs may be due to structural changes in the protein. The melting temperature and heat capacity of the enzymes was enhanced in ILs and fluorescence spectroscopy has shown that the enzymes compact to their native structural conformations in ILs which prevents thermal unfolding of the enzyme. Circular dichroism studies have also shown that the secondary structure of the protein changes in an IL, further indicating that IL media are able to stabilize enzymes.[100]

6.3.4 Enzyme Activity

When enzymes are active in ILs they do not usually dissolve and so remain suspended as a powder or immobilized by a solid support.[63] When in ILs, enzymes have been shown to have catalytic activity comparable or higher than those observed in conventional organic solvents. It has also been observed that pre-coating lactases with a second hydrophobic IL prior to dispersion in the primary IL solvent resulted in higher enzyme activity. This proposed the idea that the pre-treat IL was acting as a stabilizing microenvironment for the enzyme. It has been found that the enzyme activity can be increased by adjusting the solvent parameters of the ILs such as their hydrophobicity, polarity, viscosity, water content, fillers, pH and impurities.[63]

Enzymes that dissolve in ILs have been found to be inactive. This is thought to be due to denaturing of the protein structure on enzyme dissolution. The cation or anion may be interacting with charged groups on the enzyme's active site or exterior, causing a change in the enzyme's structure. Enzyme denaturing by dissolution in ILs is, however, often reversible by the addition of water and precipitation of the enzyme from the IL.[14]

For successful biocatalysis, the IL should have minimal interactions with the enzyme, the substrates and the reaction products. The interactions between proteins and ILs are often said to follow the Hofmeister series. This series has been used to evaluate the enzyme activity in aqueous solutions of ILs, where kosmotropic (increased water–water interactions) anions and chaotropic (decreased water–water interactions) cations are classed as protein stabilizing (Figure 6.5).[101] Kosmotropes stabilize intramolecular protein

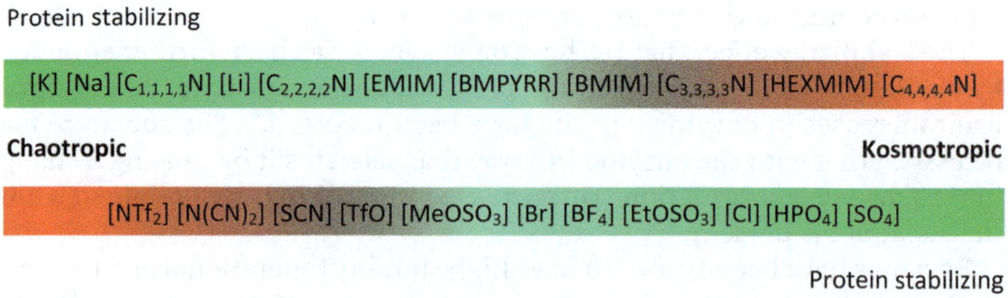

Figure 6.5 The Hofmeister series, depicting protein stabilizing ability of IL anions and cations. Data taken from literature.[131]

interactions, preventing unfolding and denaturing of the enzyme, and chaotropic agents disrupt the hydrogen bonding network, affecting the enzyme stability and inducing enzyme denaturing.

However, the Hofmeister series shows poor correlation for enzyme stability in pure ILs or ILs with trace water. The chaotropic anion $[PF_6]$ should be enzyme destabilizing as described by the Hofmeister series, but ILs containing $[PF_6]$ are typically found to be enzyme stabilizing. They are hydrophobic and thus have poor water solubility and ion dissociation in water. Thus, the Hofmeister series has been noted to apply only to ILs in aqueous solutions and hydrophilic ILs and not to hydrophobic or anhydrous hydrophilic ILs.[13]

The anions of ILs have been found to have more of an impact on the enzyme activity than IL cations, with enzymes being active in IL containing $[BF_4]$, $[PF_6]$, $[NTf_2]$ but inactive in ILs containing $[CF_3CO_2]$ and $[CF_3SO_3]$ anions.[63] The enzyme-compatible anions have a lower hydrogen bond basicity which minimizes interference with the internal hydrogen bonds of the enzymes. Less interference prevents structural changes from occurring in the enzymes confirmation and prevents enzyme deactivation and denaturing.[96] It has also been shown that there is less interaction between the charged sites in the enzyme structures with less nucleophilic anions. However, contradictory work implied higher enzyme activity of immobilized CALB in ILs with highly nucleophilic anions, highlighting again the complex nature of the how an IL affects enzyme activity and stability.[102]

A rule of thumb that is also useful when considering the compatibility of IL anions with enzymes is that anions that spread their charge over multiple atoms are more stabilizing than those with a negative charge on a single atom.[34] In this way ILs with an $[BF_4]$, $[PF_6]$ and $[NTf_2]$ anion are more stabilizing than those with halide anions such as $[Cl]$ and $[Br]$ or acetate anions. Several groups have reported the presence of halide and acetate anions to have a significant effect on the stability and activity of enzyme with a decrease in enzyme compatibility at higher halide concentrations observed.[96] ILs containing halide impurities also reduced enzyme compatibility. Strong hydrogen bonds can form between the anions and the enzymes promoting enzyme unfolding, irreversible aggregation and precipitation of the enzyme from the IL.[34] This result correlates with the increased inhibition of AChE observed for ILs containing halide anions.

Some researchers have shown that an increase in IL polarity leads to an increase in enzyme activity.[103] This is contradictory to the statement that more hydrophobic ILs increase enzyme stability. The observation of an increase in IL activity at higher polarity may actually be due to a decrease in IL viscosity, as IL polarity and viscosity are linked.[63,90] As the alky chain on the IL cation gets longer the polarity decreases and viscosity increases. A longer alkyl chain length on the cation of an IL will result in a slight decrease in polarity but a large increase in viscosity. The rate of the reaction may actually be increased due to the lower viscosity of the IL and not the increase in IL polarity. Many groups have shown no clear trend between IL polarity and activity and again this is likely due to the multifaceted nature of IL interactions with enzymes.[96]

6.3.5 Enzyme Selectivity

Enzymes have been shown to have increased chemo-, enantio- and regioselectivity in ILs over organic solvents.[63,104–106] Enhanced enantioselectivities of acylation reactions by lipases in ILs have been reported.[95,107] The enhanced solubility of substrates can increase the rate of reactions and often increases the regio- or enantioselectivity.[103]

6.4 Polymer Synthesis from Renewable Building Blocks

At the end of 2013 fossil fuel reserves-to-production ratios predicted the exhaustion of fossil fuels within less than two centuries.[108] Current chemical production is unsustainable due to the heavy reliance on the use of petrochemicals. Consequently, the utilization of biomass for the production of speciality, consumer and commodity chemicals is envisaged to have a positive impact on both the economy and the environment.

The definition of a sustainable material is one that has a minimal social, economic and environmental impact.[109] Thus a sustainable material is likely to be less hazardous, will be cost effective, will generate minimal waste, will require minimum energy input and will have a suitable product lifetime. The cost of material processing and feedstock are vitally important in sustainable chemistry. High volume feedstocks and those that yield monomers in the fewest number of steps are the most economically viable.[110] Minimal use of toxic, petrochemical-based, persistent and volatile chemicals is necessary in a sustainable and green chemistry process.

Sheldon stated in a 2014 critical review that 'a so-called bio-based economy is envisaged in which biofuels, commodity chemicals and novel materials such as bioplastics will be produced in integrated biorefineries'.[111] Integrated biorefineries that produce platform chemicals and materials alongside biofuels will have an increased economic viability than those solely producing biofuels. Materials produced in this way, such as bio-plastics, will be more sustainable with a lower environmental footprint, minimal toxicity and enhanced biodegradability.

Feedstock sources that have been considered viable for renewable materials are those that are not in direct or indirect competition with food. These include lignocellulosic biomass and inedible seed crops.[109] A sustainable method to obtain the feedstock is the valorization of waste biomass from production of crops such as rice, sugar cane and corn.[112] It is envisaged that food supply chain waste (FSCW) can be a valuable feedstock for the production of platform chemicals in biorefineries through waste utilization and valorization using green and sustainable chemistry.[111]

It is necessary to consider all aspects that contribute to the synthesis of a material before the material or the process can be classed as sustainable. A polymer consisting of monomers from renewable resources will be preferred over a polymer made from petrochemical building blocks; however, the former may not necessarily be sustainable. It is important to consider the solvents, catalysts and conditions used to undertake the synthesis. The resulting product from the sustainable process also needs to be of comparable quality to its petrochemical-based equivalent for it to be a viable replacement. Another consideration is that the overall manufacturing processes needs to remain cost effective.

A wide range of natural monomers can be prepared from renewable resources such as terpenes, rosin, sugars, glycerol, vegetable oils, furans, tannins, lignin, suberin, citric acid and tartaric acids.[113] For example, the widely used polymer building blocks caprolactone and caprolactam can be synthesized from the platform chemical 5-hydroxymethylfurfural, which can be obtained in high yield through the conversation of D-glucose found in lignocellulosic biomass.[114] Polymerization of renewable monomers can be carried out using a number of techniques which include ring-opening polymerization (ROP), anionic polymerization and radical polymerization.[110,115]

The use of greener solvents (ILs, ScCO$_2$) and greener catalysts (enzymes instead of metal catalysts) should be considered in the design of a sustainable process. Production costs can be reduced by using solvents that can be recycled and catalysts that are highly efficient, selective, long-lived and work in ambient reaction conditions. Hazards can also be reduced by eliminating the use of toxic solvents and toxic catalysts. Thus the following section will focus on the area of enzyme-catalysed polymer synthesis in ILs with the hope of paving the way for sustainable polymer chemistry.

6.4.1 Polymerization Reactions in ILs

Polymerization reactions were included in the first investigations into the use of ILs as reaction solvents.[116] In recent years homo- and co-polymerization reactions in ILs utilizing a wide range of monomers have been investigated.[117] Polymerization reactions have been shown to give very different monomer conversions, polymer molecular weights (MW) and polydispersity (PD) when carried out in an IL instead of a typical organic solvent. The main reason for this difference is often attributed to polymer solubility in the IL media and thus the interactions between the polymer and IL.[116,118]

Polymerization reactions are often carried out in organic solvents to try and overcome the problems with high viscosity encountered in bulk polymerizations. In this way a large amount of organic solvent is used. However, organic solvents are not always effective at dissolving the highly polar monomers of the reaction. Also, even if the monomers are soluble in the organic solvent as the polymer grows and the MW increases, there is likely a point that the polymer is no longer soluble in the organic solvent. Due to this the polymer will precipitate out and will no longer be in the same phase as the monomers. Chain extension now needs to occur across a phase boundary and because of this only low MW polymer is produced. Highly polar aprotic solvents, such as DMSO and DMF, are able to dissolve high MW polymer fractions but enzyme activity in these solvents is often very low.[12] ILs have been employed to overcome the problems of low monomer solubility, polymer precipitation at low MWs and low enzyme activities in highly polar solvents. Substrates that are highly polar or hydrophilic, such as amino acids and carbohydrates, have been successfully solubilized for biotransformation in ILs.[116]

Numerous reports in the literature conclude that the degree of polymerization (DP) obtained from polymerization in an IL is controlled by separation of the polymer from the IL. As an example Ma *et al.* showed that a well-controlled reverse atom-transfer radical polymerization (ATRP) of methyl methacrylate (MMA) was possible in [DODECMIM][BF$_4$], but not in [BMIM][BF$_4$] due to the poor solubility of poly(methyl methacrylate) (PMMA).[119] The paper states that the catalyst was easily isolated and the IL was recycled and successfully reused after a simple purification. However, it is clearly apparent that to isolate the polymer from the IL a large amount of solvent (in this case THF and methanol) is used. This is often the case with polymer reactions carried out in ILs. It is clear that an alternative isolation method is necessary to make sure that the use of green ILs in the place of organic solvents as a reaction medium for polymerizations is not negated by the large amount of organic solvent used to isolate the polymer from the IL. Alternative strategies that could be employed to maintain the green credentials of the reaction include the use of supercritical carbon dioxide (ScCO$_2$) and membrane technologies such as pervaporation.[120]

Other polymerizations in ILs, especially free-radical polymerizations, have shown that the IL plays a crucial role in the resultant polymer MW. Higher MWs of polymers obtained by reactions in ILs imply that there is an increase in the rate of propagation and/or a decrease in the rate of termination. One case reported both to be happening simultaneously in the polymerization of MMA in [BMIM][PF$_6$]. Here propagation increased with increased polarity and termination decreased with increased viscosity.[121]

Strehmel *et al.* carried out the free-radical polymerization of *n*-butyl methacrylate in 29 ILs.[122] Significantly higher DP was observed for reactions in ILs than in toluene, with increases also observed when compared to a bulk reaction. Imidazolium ILs gave the highest DP when compared to pyridinium and ammonium ILs, with an increased DP observed at increased viscosity. The polymers had higher glass transition temperatures and thermal stabilities when synthesized in ILs instead of toluene. However, additional efforts are needed to purify the polymer and when using highly viscous ILs extraction of the polymer from the IL with large amounts of organic solvent are necessary. The use of organic solvents is once again a clear deviation from the implementation of a green process and alternative strategies for product recovery from the IL could potentially be employed to overcome this.

In 2014 Peng *et al.* used Brønsted acidic ILs for the synthesis of biodegradable co-polymers of L-lactic acid and ε-caprolactone.[123] The ILs functioned as both solvent and catalyst and consisted of various anions and cations. Co-polymers were synthesized with different monomer ratios and it was observed that the anions of the ILs had the largest influence over the polymerization. The ILs could be recovered by phase separation and recycled up to four times without a decrease in efficiency. This application of ILs with bi-functionality as both solvent and catalyst is encouraging from a sustainable chemistry point of view. There is a decrease in the use of reagents and less waste will be produced in a reaction where the components can be recovered and recycled.

6.4.1.1 Enzymatic Polymerization Reactions

The use of enzymes for polymerization reactions is receiving great interest as a way to replace the toxic catalysts currently in use, such as tin(II) chloride. Enzymatic catalysts are also advantageous as they are able to proceed in mild reaction conditions.

6.4.1.2 Enzymatic Synthesis of Polyesters

Polyesters represent a useful class of polymers that contain hydrolysable ester bonds in their backbone. These bonds promote the biodegradation of polyesters making them significant candidates for biopolymers.[124] To date the enzymatic synthesis of polyesters in ILs has been carried out by condensation polymerization and ROP.[118,125] Condensation polymerization reactions (Scheme 6.1) are step-growth polymerizations that only reach high MWs at high conversions. Small amounts of water associated with the enzyme, and the water or alcohol released during the condensation reaction, prevent the reaction from reaching high conversion rates and as a result product MWs are lower.[34] ROP reactions (Scheme 6.2) are a type of chain-growth polymerization where high MWs can be reached even at incomplete conversions without the release of water, thus ROP is preferred for the enzymatic synthesis of polyesters.

The enzymatic polymerization of polyesters was first conducted in ILs by Uyama *et al.* using lipase from CALB in the ROP of ε-caprolactone in

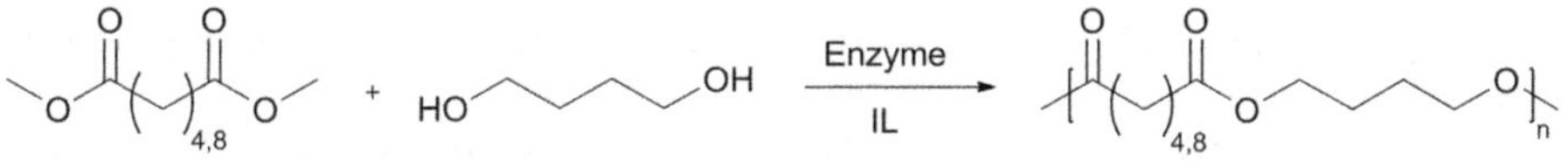

Scheme 6.1   Polycondensation of 1,4-butanediol with dimethyl adipate/dimethyl sebacate.

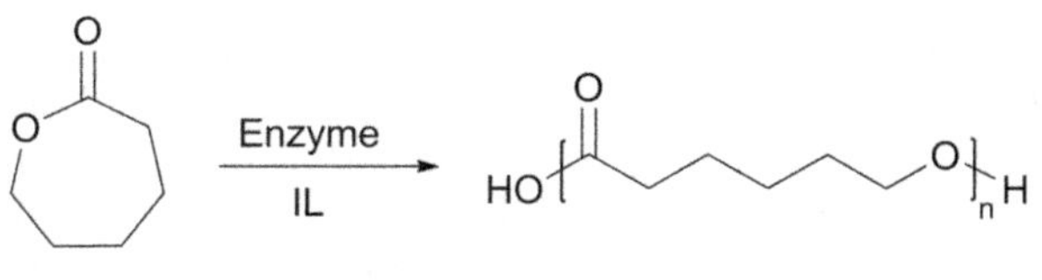

Scheme 6.2   Ring opening polymerization of ε-caprolactone.

[BMIM][BF$_4$] and [BMIM][PF$_6$] at 60 °C.[125] The reaction took one week to produce polyesters but showed that lactones of high number average molecular weight (Mn) (up to ~4200 g mol^{-1}) could be produced with enzymes in ILs. Condensation polymerization of dicarboxylic acid diesters and 1,4-butanediol was also carried out under the same conditions but conversion and polymer MWs were low (Mn up to ~1500 g mol^{-1}).

Nara *et al.* studied the lipase-catalysed polycondensation of diethyl octane-1,8-dicarboxylate and 1,4-butanediol at room temperature and at 60 °C in [BMIM][PF$_6$] with lipase PS-C.[126] Relatively high MW polymers were obtained at 60 °C (Mn up to ~4300 g mol^{-1}). This work has since been deemed to show that polymer solubility in the IL limits the MW of the resultant polymer, with a higher polymer MW obtained at 60 °C where the polymer is more soluble in the IL. Low PD values of the polymers were believed to be potentially due to insolubility of the polymer in the IL at high MWs; however, values were measured after precipitation of the polymer from the IL with methanol which leaves behind the lower MW oligomers in the IL.[34]

Marcilla *et al.* carried out the enzymatic ROP of ε-caprolactone and the polycondensation of dimethyl adipate and dimethyl sebacate, respectively, with 1,4-butanol in [BMIM][NTf$_2$], [BMIM][PF$_6$] and [BMIM][BF$_4$].[118] All reactions were undertaken using CALB immobilized on an acrylic resin (Novozyme 435). MWs (Mn) of 7000–9500 g mol^{-1} were obtained for the ROP but were not as high as that from the reaction in toluene with MWs (Mn) of 13 000 g mol^{-1}. For the polycondensation reaction MWs (Mn) of up to 5400 g mol^{-1} were produced and the reactions could be carried out in open reaction vessels, close to the boiling point of the condensation by-product.

Gorke *et al.* produced poly(hydroxyalkanoates) from five lactones using immobilized CALB (Novozyme 435) with both water-miscible ILs and water-immiscible ILs.[127] In the pre-screening reaction, transesterification of ethyl valerate to methyl valerate, higher conversions were obtained in water-immiscible ILs such as [BMIM][NTf$_2$] and [BMIM][PF$_6$] as opposed to water-miscible ILs such as [BMIM][DCA]. This result is likely due to lipase deactivation with the water-miscible ILs because of their high polarity. Ring-opening polymerization of β-propiolactone and ε-caprolactone in [BMIM][NTf$_2$] gave MWs (Mn) of ~10 000 g mol^{-1}, but other lactones gave only low MW oligomers.

The lipase-catalysed (CALB) ROP of L-lactide was carried out by Fujita *et al.* in ILs at room temperature in 2008.[128] They reported higher MWs compared to the same reactions carried out in bulk and in toluene (Mn <44 000 g mol^{-1}); however, lower conversions were observed. The anion of the IL showed a significant effect on the conversion, MW and yield, suggesting that the anion has a molecular interaction, likely with the enzyme, during the reaction. In [BMIM][BF$_4$] a MW (Mn) of 54 600 g mol^{-1} was reached at 110 °C with a 24.3% yield. It was postulated by the authors that lower polymer yields were obtained in ILs due to the high solubility of the polymer in the IL. It proved difficult to thoroughly extract the polymer even after multiple extractions with toluene and isolation of the polymer was obtained by precipitation in methanol. Again the excessive use of organic solvents is clearly not ideal for a green process and an alternative greener method of extraction (*e.g.* ScCO$_2$) is recommended and may facilitate product isolation.

The synthesis of poly-L-lactide (PLLA) and poly-L-lactide-co-glycolide (PLLGA) in [HEXMIM][PF$_6$], mediated by immobilized CALB (Novozyme 435), was reported by Chanfreau *et al.* in 2010.[129] PLLA was obtained in at 63% yield after a 7 day reaction at 90 °C (Mn = 37 800 g mol^{-1}, PD = 1.3) with 10 wt% catalyst. Highly crystalline polymers (up to 85%) were produced in this way. The reaction was also able to proceed without the presence of the enzyme; however, in an equivalent 7 day reaction at 90 °C a much lower MW was obtained (Mn = 1500 g mol^{-1}, PD = 1.2). Synthesis of PLLA at 65 °C in [HEXMIM][PF$_6$] was only possible in the presence of the enzyme, therefore showing that only the enzymatic mechanism (ROP) was present at 65 °C, with other contributions induced at higher temperatures. PLLGA was additionally synthesized in [HEXMIM][PF$_6$] at 65 °C with Novozyme 435 and the resulting co-polymer contained up to 19% lactyl units.

The main processing challenges for carrying out polymerization reactions in ILs are the solubility of the monomers and polymers, the stability of the enzyme and the isolation of the resultant polymer. There are also environmental factors to consider. It is clear to date that research in the area of polymerization reactions in green ILs is very sparse. Most of the reactions have been carried out in [BMIM][BF$_4$] and [BMIM][PF$_6$], which definitely cannot be considered as green ILs. There needs to be a greater interest in utilizing green ILs for polymerization reactions before we can know if they will be suitable for the enzymatic synthesis of polymers. Only in this way can the utilization of green ILs for the enzymatic synthesis of high MW polymers, and their implementation on a large scale in a biorefinery, be realized.

It is clear that the enzymatic production of polyesters in ILs has the potential to offer an advantage over reactions in organic solvents, with high MWs of polyesters in ILs being reported. However, currently the MWs produced are most often similar or lower than reactions in organic solvents or in bulk, with lower conversions reported. It therefore has to be asked, what can a reaction in an IL provide that one in bulk or in an organic solvent cannot? This is most evidently the solubilization of highly polar monomers for polymerization reactions. However, there is still the problem of extracting the product from the highly solubilizing IL. Thus the main advantage of using an IL may be the replacement of the highly toxic organic solvents. When thinking on the toxicity issue it is unfortunate that the ILs that have thus far found the most applicability to biocatalytic reactions in ILs are the second generation ILs, where most have been evaluated to be of very low biodegradability and of moderate to high toxicity. It is hoped that researchers will prioritize the need for the use of low toxicity biodegradable ILs and their implementation in biocatalytic reactions. One promising example related to this is the use of DES in the ROP of ε-caprolactone by CALB.[130] Here the components of the DES are biodegradable, non-toxic and of low cost. Thus, it is clear that there needs to be a distinct advantage to using ILs in biocatalysts before they will be utilized on a large scale in industry.

6.5   Conclusions and Prospects

It is evident from the literature presented and discussed in this chapter that interactions between ILs and enzymes are of a complex nature. Numerous parameters play a part in the compatibility of enzymes and ILs. As a result, designing an IL for biocatalytic reactions involves multiple considerations, especially when trying to design for low toxicity and biodegradation simultaneously. When one specific feature of an IL is altered it is inevitable that one or more physiochemical properties of the IL in question will be affected. This will inevitably lead to complex changes in the enzyme compatibility and is a clear reason as to why there are a significant amount of contradictory trends in the literature concerning IL compatibility with enzymes.

It is apparent that ILs can be utilized as a real solution to the problems presented by biocatalysts in organic solvents and aqueous media. However, the use of environmentally friendly ILs for biocatalysis is an area of research that is currently lacking attention. The vast majority of enzymatic reactions in ILs have been carried out with second generation ILs. However, this research has clearly provided the academic community with a solid knowledge base on the compatibility of enzymes with ILs and the factors that govern enzyme stability, activity and selectivity. Hopefully this research area can be extended to third and fourth generation ILs in the near future. The tying together of green ILs and biocatalysis is highly necessary, in the authors' opinion, before biocatalysis in ILs can be truly classed as a green alternative to traditional catalysis and be introduced into biorefineries for the production of sustainable chemicals.

Acknowledgements

The authors H. P., A. H. and N. G. have received funding from the European Union's Seventh Framework Programme for research, technological development and demonstration under grant agreement No. 289253 and No. 621364 (TUTIC-Green, N. G.). The authors would like to acknowledge the COST Actions TD1203 (EUBIS) and CM1206 (EXIL) for their role in aiding networking of researchers and facilitating relevant discussions surrounding this work.

References

1. P. Wasserscheid, *Nature*, 2006, **439**, 797.
2. M. Smiglak, W. M. Reichert, J. D. Holbrey, J. S. Wilkes, L. Y. Sun, J. S. Thrasher, K. Kirichenko, S. Singh, A. R. Katritzky and R. D. Rogers, *Chem. Commun.*, 2006, 2554–2556.
3. N. Sun, H. Rodríguez, M. Rahman and R. D. Rogers, *Chem. Commun.*, 2011, **47**, 1405–1421.
4. K. M. Docherty, J. K. Dixon and C. F. Kulpa, Jr, *Biodegradation*, 2007, **18**, 481–493.
5. S. Stolte, S. Abdulkarim, J. Arning, A. K. Blomeyer-Nienstedt, U. Bottin-Weber, M. Matzke, J. Ranke, B. Jastorff and J. Thoming, *Green Chem.*, 2008, **10**, 214–224.

6. T. P. T. Pham, C. W. Cho, C. O. Jeon, Y. J. Chung, M. W. Lee and Y. S. Yun, *Environ. Sci. Technol.*, 2009, **43**, 516–521.
7. S. Stolte, S. Steudte, A. Igartua and P. Stepnowski, *Curr. Org. Chem.*, 2011, **15**, 1946–1973.
8. A. B. A. Boxall, C. J. Sinclair, K. Fenner, D. Kolpin and S. J. Maud, *Environ. Sci. Technol.*, 2004, **38**, 368a–375a.
9. K. Kummerer, *Green Chem.*, 2007, **9**, 899–907.
10. H. Sutterlin, R. Alexy, A. Coker and K. Kummerer, *Chemosphere*, 2008, **72**, 479–484.
11. K. M. Docherty, M. V. Joyce, K. J. Kulacki and C. F. Kulpa, *Green Chem.*, 2010, **12**, 701–712.
12. M. Moniruzzaman, N. Kamiya and M. Goto, *Org. Biomol. Chem.*, 2010, **8**, 2887–2899.
13. A. P. M. Tavares, O. Rodríguez and E. A. Macedo, in *Ionic Liquids – New Aspects for the Future*, ed. J. Kadokawa, InTech, 2013, ch. 20, pp. 537–556.
14. F. van Rantwijk and R. A. Sheldon, *Chem. Rev.*, 2007, **107**, 2757–2785.
15. B. Jastorff, R. Stormann, J. Ranke, K. Molter, F. Stock, B. Oberheitmann, W. Hoffmann, J. Hoffmann, M. Nuchter, B. Ondruschka and J. Filser, *Green Chem.*, 2003, **5**, 136–142.
16. D. Coleman and N. Gathergood, *Chem. Soc. Rev.*, 2010, **39**, 600–637.
17. T. P. T. Pham, C.-W. Cho and Y.-S. Yun, *Water Res.*, 2010, **44**, 352–372.
18. M. C. Bubalo, K. Radosevic, I. R. Redovnikovic, J. Halambek and V. G. Srcek, *Ecotoxicol. Environ. Saf.*, 2014, **99**, 1–12.
19. M. Petkovic, K. R. Seddon, L. P. N. Rebelo and C. S. Pereira, *Chem. Soc. Rev.*, 2011, **40**, 1383–1403.
20. R. P. Swatloski, J. D. Holbrey and R. D. Rogers, *Green Chem.*, 2003, **5**, 361–363.
21. J. Ranke, S. Stolte, R. Stormann, J. Arning and B. Jastorff, *Chem. Rev.*, 2007, **107**, 2183–2206.
22. S. Sowmiah, V. Srinivasadesikan, M. C. Tseng and Y. H. Chu, *Molecules*, 2009, **14**, 3780–3813.
23. A. Pathak, N. Jangid, R. Ameta and P. B. Punjabi, in *Green Chemistry: Fundamentals and Application*, ed. S. C. Ameta and R. Ameta, Apple Academic Press, Inc., Canada, 2013, ch. 5, pp. 109–136.
24. P. Domínguez de María, in *Ionic Liquids in Biotransformations and Organocatalysis*, John Wiley & Sons, Inc., 2012, pp. 1–14.
25. L. Myles, R. G. Gore, M. Spulak, I. Beadham, T. M. Garcia, S. J. Connon and N. Gathergood, *Green Chem.*, 2013, **15**, 2747–2760.
26. L. Myles, R. Gore, M. Spulak, N. Gathergood and S. J. Connon, *Green Chem.*, 2010, **12**, 1157–1162.
27. N. Gathergood, M. T. Garcia and P. J. Scammells, *Green Chem.*, 2004, **6**, 166–175.
28. N. Gathergood, P. J. Scammells and M. T. Garcia, *Green Chem.*, 2006, **8**, 156–160.
29. P. Wasserscheid, R. v. Hal and A. Bosmann, *Green Chem.*, 2002, **4**, 400–404.
30. S. Morrissey, B. Pegot, D. Coleman, M. T. Garcia, D. Ferguson, B. Quilty and N. Gathergood, *Green Chem.*, 2009, **11**, 475–483.

31. J. Neumann, S. Steudte, C. W. Cho, J. Thoming and S. Stolte, *Green Chem.*, 2014, **16**, 2174–2184.
32. N. Ferlin, M. Courty, A. N. V. Nhien, S. Gatard, M. Pour, B. Quilty, M. Ghavre, A. Haiss, K. Kummerer, N. Gathergood and S. Bouquillon, *RSC Adv.*, 2013, **3**, 26241–26251.
33. N. Ferlin, M. Courty, S. Gatard, M. Spulak, B. Quilty, I. Beadham, M. Ghavre, A. Haiss, K. Kummerer, N. Gathergood and S. Bouquillon, *Tetrahedron*, 2013, **69**, 6150–6161.
34. J. Gorke, F. Srienc and R. Kazlauskas, *Biotechnol. Bioprocess Eng.*, 2010, **15**, 40–53.
35. Q. Zhang, K. De Oliveira Vigier, S. Royer and F. Jerome, *Chem. Soc. Rev.*, 2012, **41**, 7108–7146.
36. R. Alexy, T. Kumpel and K. Kummerer, *Chemosphere*, 2004, **57**, 505–512.
37. M. Markiewicz, M. Piszora, N. Caicedo, C. Jungnickel and S. Stolte, *Water Res.*, 2013, **47**, 2921–2928.
38. A. S. Wells and V. T. Coombe, *Org. Process Res. Dev.*, 2006, **10**, 794–798.
39. G. Chatel and D. R. MacFarlane, *Chem. Soc. Rev.*, 2014, **43**, 8132.
40. S. I. Abu-Eishah, in *Ionic Liquids – Classes and Properties*, ed. S. Handy, InTech, Croatia, 2011, ch. 11, pp. 239–241.
41. K. M. Docherty and C. F. Kulpa, *Green Chem.*, 2005, **7**, 185–189.
42. M. Matzke, S. Stolte, K. Thiele, T. Juffernholz, J. Arning, J. Ranke, U. Welz-Biermann and B. Jastorff, *Green Chem.*, 2007, **9**, 1198–1207.
43. J. Ranke, K. Molter, F. Stock, U. Bottin-Weber, J. Poczobutt, J. Hoffmann, B. Ondruschka, J. Filser and B. Jastorff, *Ecotoxicol. Environ. Saf.*, 2004, **58**, 396–404.
44. S. P. M. Ventura, C. S. Marques, A. A. Rosatella, C. A. M. Afonso, F. Goncalves and J. A. P. Coutinho, *Ecotoxicol. Environ. Saf.*, 2012, **76**, 162–168.
45. C. W. Cho, T. P. T. Pham, Y. C. Jeon and Y. S. Yun, *Green Chem.*, 2008, **10**, 67–72.
46. K. J. Kulacki and G. A. Lamberti, *Green Chem.*, 2008, **10**, 104–110.
47. A. Latala, M. Nedzi and P. Stepnowski, *Green Chem.*, 2010, **12**, 60–64.
48. A. Latala, M. Nedzi and P. Stepnowski, *Green Chem.*, 2009, **11**, 580–588.
49. A. Latala, M. Nedzi and P. Stepnowski, *Green Chem.*, 2009, **11**, 1371–1376.
50. A. Latala, P. Stepnowski, M. Nedzi and W. Mrozik, *Aquat. Toxicol.*, 2005, **73**, 91–98.
51. T. P. T. Pham, C. W. Cho, J. Min and Y. S. Yun, *J. Biosci. Bioeng.*, 2008, **105**, 425–428.
52. F. Stock, J. Hoffmann, J. Ranke, R. Stormann, B. Ondruschka and B. Jastorff, *Green Chem.*, 2004, **6**, 286–290.
53. J. Arning, S. Stolte, A. Boschen, F. Stock, W. R. Pitner, U. Welz-Biermann, B. Jastorff and J. Ranke, *Green Chem.*, 2008, **10**, 47–58.
54. A. C. Skladanowski, P. Stepnowski, K. Kleszczynski and B. Dmochowska, *Environ. Toxicol. Pharmacol.*, 2005, **19**, 291–296.
55. R. J. Bernot, M. A. Brueseke, M. A. Evans-White and G. A. Lamberti, *Environ. Toxicol. Chem.*, 2005, **24**, 87–92.
56. C. Pretti, C. Chiappe, D. Pieraccini, M. Gregori, F. Abramo, G. Monni and L. Intorre, *Green Chem.*, 2006, **8**, 238–240.

57. M. Yu, S. M. Li, X. Y. Li, B. J. Zhang and J. J. Wang, *Ecotoxicol. Environ. Saf.*, 2008, **71**, 903–908.
58. S. Stolte, J. Arning, U. Bottin-Weber, A. Muller, W. R. Pitner, U. Welz-Biermann, B. Jastorff and J. Ranke, *Green Chem.*, 2007, **9**, 760–767.
59. X. F. Wang, C. A. Ohlin, Q. H. Lu, Z. F. Fei, J. Hu and P. J. Dyson, *Green Chem.*, 2007, **9**, 1191–1197.
60. J. Ranke, A. Muller, U. Bottin-Weber, F. Stock, S. Stolte, J. Arning, R. Stormann and B. Jastorff, *Ecotoxicol. Environ. Saf.*, 2007, **67**, 430–438.
61. K. Radosevic, M. Cvjetko, N. Kopjar, R. Novak, J. Dumic and V. G. Srcek, *Ecotoxicol. Environ. Saf.*, 2013, **92**, 112–118.
62. J. Pernak, K. Sobaszkiewicz and I. Mirska, *Green Chem.*, 2003, **5**, 52–56.
63. Z. Yang and W. B. Pan, *Enzyme Microb. Technol.*, 2005, **37**, 19–28.
64. S. Stolte, M. Matzke, J. Arning, A. Boschen, W. R. Pitner, U. Welz-Biermann, B. Jastorff and J. Ranke, *Green Chem.*, 2007, **9**, 1170–1179.
65. M. T. Garcia, N. Gathergood and P. J. Scammells, *Green Chem.*, 2005, **7**, 9–14.
66. D. J. Couling, R. J. Bernot, K. M. Docherty, J. K. Dixon and E. J. Maginn, *Green Chem.*, 2006, **8**, 82–90.
67. B. Weyershausen and K. Lehmann, *Green Chem.*, 2005, **7**, 15–19.
68. J. Pernak, I. Goc and I. Mirska, *Green Chem.*, 2004, **6**, 323–329.
69. A. Cieniecka-Roslonkiewicz, J. Pernak, J. Kubis-Feder, A. Ramani, A. J. Robertson and K. R. Seddon, *Green Chem.*, 2005, **7**, 855–862.
70. K. M. Docherty, S. Z. Hebbeler and C. F. Kulpa, *Green Chem.*, 2006, **8**, 560–567.
71. W. J. Stang, *National Enforcement Investigations Center*, United States Environmental Protection Agency, Office of Enforcement, 1980.
72. S. Stolte, J. Arning, U. Bottin-Weber, M. Matzke, F. Stock, K. Thiele, M. Uerdingen, U. Welz-Biermann, B. Jastorff and J. Ranke, *Green Chem.*, 2006, **8**, 621–629.
73. W. H. Organization, Antimicrobial resistance: global report on surveillance, http://www.who.int/mediacentre/factsheets/fs194/en/, accessed September 2014.
74. Y. Luo, Q. Wang, Q. Lu, Q. Mu and D. Mao, *Environ. Sci. Technol. Lett.*, 2014, **1**, 266–270.
75. P. H. Axelsen, M. Harel, I. Silman and J. L. Sussman, *Protein Sci.*, 1994, **3**, 188–197.
76. M. A. Sogorb and E. Vilanova, *Toxicol. Lett.*, 2002, **128**, 215–228.
77. J. Massoulie, L. Pezzementi, S. Bon, E. Krejci and F. M. Vallette, *Prog. Neurobiol.*, 1993, **41**, 31–91.
78. J. Kaur and M. Q. Zhang, *Curr. Med. Chem.*, 2000, **7**, 273–294.
79. B. Jastorff, K. Molter, P. Behrend, U. Bottin-Weber, J. Filser, A. Heimers, B. Ondruschka, J. Ranke, M. Schaefer, H. Schroder, A. Stark, P. Stepnowski, F. Stock, R. Stormann, S. Stolte, U. Welz-Biermann, S. Ziegert and J. Thoming, *Green Chem.*, 2005, **7**, 362–372.
80. F. Y. Yan, S. Q. Xia, Q. Wang and P. S. Ma, *J. Chem. Eng. Data*, 2012, **57**, 2252–2257.

81. J. S. Torrecilla, J. Garcia, E. Rojo and F. Rodriguez, *J. Hazard. Mater.*, 2009, **164**, 182–194.

82. R. N. Das and K. Roy, *Ind. Eng. Chem. Res.*, 2014, **53**, 1020–1032.

83. X. D. Hou, Q. P. Liu, T. J. Smith, N. Li and M. H. Zong, *PLoS One*, 2013, **8**, e59145.

84. Organisation for Economic Co-operation and Development, *OECD Guidelines for the Testing of Chemicals; Ready Biodegradability*, Paris, 1992.

85. Organisation for Economic Co-operation and Development, *OECD Guidlines for the Testing of Chemicals; Ready Biodegradability – CO_2 in sealed vessels (Headspace Test)*, Paris, 2006.

86. International Organization for Standardization, *ISO Standard14593; Water quality – Evaluation of ultimate aerobic biodegradability of organic compounds in aqueous medium – Method by analysis of inorganic carbon in sealed vessels (CO_2 headspace test)*, Geneva, 1999.

87. N. Gathergood and P. J. Scammells, *Aust. J. Chem.*, 2002, **55**, 557–560.

88. J. R. Harjani, R. D. Singer, M. T. Garciac and P. J. Scammells, *Green Chem.*, 2009, **11**, 83–90.

89. C. Rucker and K. Kummerer, *Green Chem.*, 2012, **14**, 875–887.

90. M. Moniruzzaman, K. Nakashima, N. Kamiya and M. Goto, *Biochem. Eng. J.*, 2010, **48**, 295–314.

91. E. Yamamoto, S. Yamaguchi and T. Nagamune, *Appl. Biochem. Biotechnol.*, 2011, **164**, 957–967.

92. P. Attri, P. Venkatesu and A. Kumar, *Phys. Chem. Chem. Phys.*, 2011, **13**, 2788–2796.

93. R. Buchfink, A. Tischer, G. Patil, R. Rudolph and C. Lange, *J. Biotechnol.*, 2010, **150**, 64–72.

94. H. Yan, J. Wu, G. Dai, A. Zhong, H. Chen, J. Yang and D. Han, *J. Lumin.*, 2012, **132**, 622–628.

95. W.-Y. Lou and M.-H. Zong, *Chirality*, 2006, **18**, 814–821.

96. M. Naushad, Z. A. Alothman, A. B. Khan and M. Ali, *Int. J. Biol. Macromol.*, 2012, **51**, 555–560.

97. G. N. Ou, J. Yang, B. Y. He and Y. Z. Yuan, *J. Mol. Catal. B: Enzym.*, 2011, **68**, 66–70.

98. D. Alsafadi and F. Paradisi, *Mol. Biotechnol.*, 2014, **56**, 240–247.

99. H. Zhao, *J. Chem. Technol. Biotechnol.*, 2010, **85**, 891–907.

100. T. De Diego, P. Lozano, S. Gmouh, M. Vaultier and J. L. Iborra, *Biotechnol. Bioeng.*, 2004, **88**, 916–924.

101. Z. Yang, *J. Biotechnol.*, 2009, **144**, 12–22.

102. R. Irimescu and K. Kato, *J. Mol. Catal. B: Enzym.*, 2004, **30**, 189–194.

103. S. Park and R. J. Kazlauskas, *J. Org. Chem.*, 2001, **66**, 8395–8401.

104. S. Kobayashi, *Proc. Jpn. Acad., Ser. B*, 2010, **86**, 338–365.

105. K. W. Kim, B. Song, M. Y. Choi and M. J. Kim, *Org. Lett.*, 2001, **3**, 1507–1509.

106. S. H. Schöfer, N. Kaftzik, P. Wasserscheid and U. Kragl, *Chem. Commun.*, 2001, 425–426.

107. T. Itoh, E. Akasaki, K. Kudo and S. Shirakami, *Chem. Lett.*, 2001, 262–263.
108. *BP Statistical Review of World Energy*, http://www.bp.com/statisticalreview, 2014.
109. J. H. Clark, *J. Chem. Technol. Biotechnol.*, 2007, **82**, 603–609.
110. A. L. Holmberg, K. H. Reno, R. P. Wool and I. I. I. T. H. Epps, *Soft Matter*, 2014, **10**, 7405–7424.
111. R. A. Sheldon, *Green Chem.*, 2014, **16**, 950–963.
112. C. O. Tuck, E. Perez, I. T. Horvath, R. A. Sheldon and M. Poliakoff, *Science*, 2012, **337**, 695–699.
113. A. Gandini, in *Biocatalysis in Polymer Chemistry*, Wiley-VCH Verlag GmbH & Co. KGaA, 2010, pp. 1–33.
114. T. Buntara, S. Noel, P. H. Phua, I. Melián-Cabrera, J. G. de Vries and H. J. Heeres, *Angew. Chem., Int. Ed.*, 2011, **50**, 7083–7087.
115. K. J. Yao and C. B. Tang, *Macromolecules*, 2013, **46**, 1689–1712.
116. N. Winterton, *J. Mater. Chem.*, 2006, **16**, 4281–4293.
117. P. Kubisa, *Prog. Polym. Sci.*, 2009, **34**, 1333–1347.
118. R. Marcilla, M. de Geus, D. Mecerreyes, C. J. Duxbury, C. E. Koning and A. Heise, *Eur. Polym. J.*, 2006, **42**, 1215–1221.
119. H. Y. Ma, X. H. Wan, X. F. Chen and Q. F. Zhou, *Polymer*, 2003, **44**, 5311–5316.
120. P. Lozano, *Green Chem.*, 2010, **12**, 555–569.
121. S. Harrisson, S. R. Mackenzie and D. M. Haddleton, *Macromolecules*, 2003, **36**, 5072–5075.
122. V. Strehmel, A. Laschewsky, H. Wetzel and E. Gornitz, *Macromolecules*, 2006, **39**, 923–930.
123. Q. Peng, K. Mahmood, Y. Wu, L. Wang, Y. Liang, J. Shen and Z. Liu, *Green Chem.*, 2014, **16**, 2234–2241.
124. M. J. L. Tschan, E. Brule, P. Haquette and C. M. Thomas, *Polym. Chem.*, 2012, **3**, 836–851.
125. H. Uyama, T. Takamoto and S. Kobayashi, *Polym. J.*, 2002, **34**, 94–96.
126. S. J. Nara, J. R. Harjani, M. M. Salunkhe, A. T. Mane and P. P. Wadgaonkar, *Tetrahedron Lett.*, 2003, **44**, 1371.
127. J. T. Gorke, K. Okrasa, A. Louwagie, R. J. Kazlauskas and F. Srienc, *J. Biotechnol.*, 2007, **132**, 306–313.
128. M. Yoshizawa-Fujita, C. Saito, Y. Takeoka and M. Rikukawa, *Polym. Adv. Technol.*, 2008, **19**, 1396–1400.
129. S. Chanfreau, M. Mena, J. Porras-Domínguez, M. Ramírez-Gilly, M. Gimeno, P. Roquero, A. Tecante and E. Bárzana, *Bioprocess Biosyst. Eng.*, 2010, **33**, 629–638.
130. T. G. Johnathan, S. Friedrich and J. K. Romas, in *Ionic Liquid Applications: Pharmaceuticals, Therapeutics, and Biotechnology*, American Chemical Society, 2010, vol. 1038, ch. 14, pp. 169–180.
131. D. Constantinescu, H. Weingartner and C. Herrmann, *Angew. Chem., Int. Ed.*, 2007, **46**, 8887–8889.

Synthesis of HMF in Ionic Liquids: Biomass-Derived Products

GUANG-WAY JANG*[a], JINN-JONG WONG[a],
YING-TING HUANG[a], AND CHIA-LING LI[a]

[a]Material and Chemical Research Laboratories, Industrial Technology
Research Institute, 321 Kuang Fu Road, Hsinchu, 30011, Taiwan
*E-mail: billjang@gmail.com

7.1   Introduction

There has been a great interest in 5-hydroxymethylfurfural (HMF) as an intermediate for the production of a wide variety of valuable chemicals, shown in Scheme 7.1, and materials[1–3] through oxidation[4–11] and reduction.[4,12,13] Recent years have seen several reviews on various aspects of HMF synthesis[3,14–17] and applications.[3] HMF and derivatives are considered as important precursors to the production of high-value polymers, plasticizers,[18] biofuels,[19] pharmaceuticals and antifungal agents.[20] Valuable chemicals derived from the platform molecule include dimethylfuran (DMF) for biofuel and 2,5-furandicarboxylic acid (FDCA) for polyesters and polyamides synthesis, as well as 3,5-dihydroxymethylfuran, caprolactone, adipic acid, levulinic acid, *etc.*[1,2] FDCA, obtained from the catalytic oxidation of HMF,[5–11] is a structural analogue of terephthalic acid. It was considered

RSC Green Chemistry No. 36
Ionic Liquids in the Biorefinery Concept: Challenges and Perspectives
Edited by Rafal Bogel-Lukasik

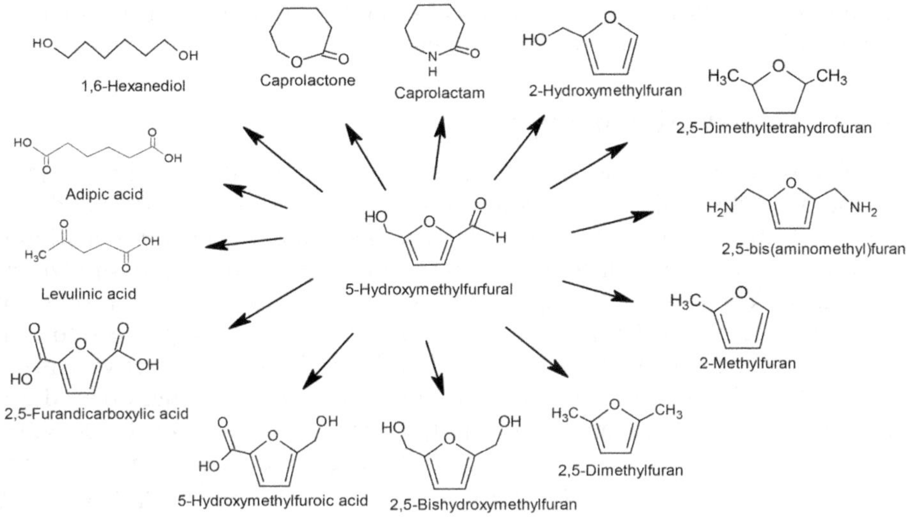

Scheme 7.1 HMF derivatives.[3,25]

as 'the sleeping giant' with biomass origin.[21] FDCA and levulinic acid are among the top twelve value-added building blocks from biomass identified by the DOE in 2004. Multiple functional groups render the furanic molecules with potential to transform into new families of useful chemicals or materials with unique properties. Bioplastic poly(ethylene furanoate) (PEF) has attractive mechanical properties similar to that of poly(ethylene terephthalate) (PET) and a greatly improved barrier performance.[22] These are critical characteristics for packaging uses, the largest potential application for bioplastics at present. A number of furanic polyesters and co-polyesters can be synthesized by polymerization of FDCA with a variety of diols.[23] Economic analysis of FDCA production from HMF using acetic acid as the solvent, Pt/ZrO_2 as the catalyst and air as the oxidant suggested a minimum sale price of $3157 per ton.[5] The FDCA price reduces to $2458 per ton if pure oxygen is used as the oxidant. A cradle-to-grave life cycle assessment (LCA) suggested a reduction of 45–55% greenhouse gas (GHG) emissions for the production of PEF instead of PET. According to HIS forecast, the global demand for PET packaging resins will reach 20 million metric tons (MMTs) in 2014. In an even larger market, polyester fibre demand was roughly 38 MMTs in 2012. Bottling, film, textile and automotive industries are the main application sectors driving the demand for renewable bioplastics.

7.2   Synthesis of HMF

HMF is generally synthesized *via* acid-catalysed carbohydrate dehydration[24] or alternatively from lignocellulosic materials, such as cellulose, with selected solvents and catalyst combinations. A challenge for the commercial production of HMF is that economically viable techniques for its isolation and purification have yet to be developed.

7.2.1 Dehydration of Sugars

Dehydration of fructose to lose three water molecules produces HMF. In many studies, high conversion of sugars with limited HMF yield was observed due to the formation of by-products, such as levulinic acid and humins. Formation of dark brown-coloured humins due to cross-polymerization is a key obstacle to improving HMF production yield. In presence of water molecules, HMF can take up two water molecules and proceed to consecutive reaction to yield levulinic acid and formic acid. A selective synthesis of HMF relies on an adequate combination of catalyst, reaction medium and process. A wide range of acid catalysts have been used for the dehydration of fructose, such as mineral acids,[16,17,25] organic acids,[17,26] ion-exchange resins,[17,27] metal salts,[28] heteropoly acids,[29] zeolites[19] and supported ionic liquids (ILs).[30] Homogeneous catalysts are in general more effective for fructose conversion and moderate HMF yield. However, homogeneous catalysts pose a challenge for the purification of product and/or solvent recycling. On the other hand, heterogeneous acid catalysts possess high HMF selectivity but low fructose conversion. The reaction medium interacts with feedstocks, products and, most importantly, with reaction intermediates. It affects the thermodynamics, kinetics and yield of the HMF production processes.

Although very high conversion and HMF yield were achieved using fructose as a feedstock, it is a limited naturally occurring resource and thus a high-cost raw material. This drives the search for alternatives, including glucose,[31–36] disaccharides,[37–39] polysaccharides,[24,40,41] and even lignocellulosic biomass.[32,42,43] Conversion of glucose to HMF proceeds through competing reaction pathways of non-furan cyclic ether formation and carbon–carbon bond scission. This leads to the formation of by-products. *In situ* isomerization to fructose using a suitable metal halide to achieve high HMF yield was proposed.[36] The isomerization is the rate-determining step for the conversion of glucose to HMF. Study of selective dehydration of glucose in DMF using solid acid alone or combination of solid acid and base catalysts in a one-pot process indicated that isomerization was catalysed by solid base and proceeded to formation of HMF in presence of solid acid while anhydroglucose was produced when only solid acid was used.[44]

Dehydration of sucrose with mineral acids, solid acid, ILs and metallic compounds results in a relatively low HMF yield, 50% or less. This is because only the fructose moiety of sucrose was converted into HMF while the glucose molecules were mostly left untouched. Tong, Zhang and co-workers identified ammonium halides as promoters in combination with $CrCl_3$ to enhance the efficiency of sucrose conversion to HMF.[37] The authors suggested that the halide component of the promoters serves as a ligand for the catalytic metal centre of the intermediate and attacks the fructofuranosyl oxocarbenium ion in the following steps.

7.2.2 One-Pot Synthesis of HMF from Lignocellulosic Biomass

Plants are considered renewable alternatives for the production of energy and chemicals to replace fossil fuel due to concerns of limited crude oil reserves and global warming. Cellulose and hemicellulose, the main components of plants, are constructed from sugar molecules. Cellulose is a linear crystalline polymer chain that is composed of glucose repeat units. On the other hand, hemicellulose is a branched amorphous polymer made of pentose and hexose units. Use of lignocellulosic raw materials could minimize food crises but requires increasingly complex synthetic routes. It is inevitable when considering the conversion of cellulose or oligosaccharides from inedible bioresources to HMF to encounter the complexity of the structure and the existence of the other biomass components, such lignin and protein. The process further includes dissolution of lignocellulose biomass and cellulose depolymerization to produce glucose for the synthesis of HMF. Hydrolysis of disaccharides and polysaccharides is catalysed by a base while dehydration of the resulting sugar monomers favours the acid condition. Thus, the selection of solvent and catalyst are the key challenges for a high-efficiency synthesis of HMF. The solvent has to be a good solvent for feedstocks and intermediates, including lignocellulosic biomass, cellulose, glucose or fructose, and inert to HMF during synthesis, as shown in Scheme 7.2.

7.2.3 Solvent Effects

There are intensive studies on the influence of solvents on the synthesis of HMF from various bioresources. Solvents evaluated for HMF synthesis include water,[26,45] organic solvents,[21,28,46] ILs[17,42,47] and biphasic systems.[15,33,48,49] Other critical factors governing yield and purity are reaction temperature and time. Water is a good solvent for both fructose and HMF. However, the presence of water causes the dehydration equilibrium to be reversed, hydrolysis of HMF and formation of humins. It also suffers from low selectivity and a slow rate of HMF formation in a pure water solvent. Rapid rehydration of HMF with a Brønsted acidic catalyst limits the desired product yield. A very high HMF yield of about 50% in water was reported using Brønsted and Lewis acidic niobic acid as the catalyst. Essayem *et al.*

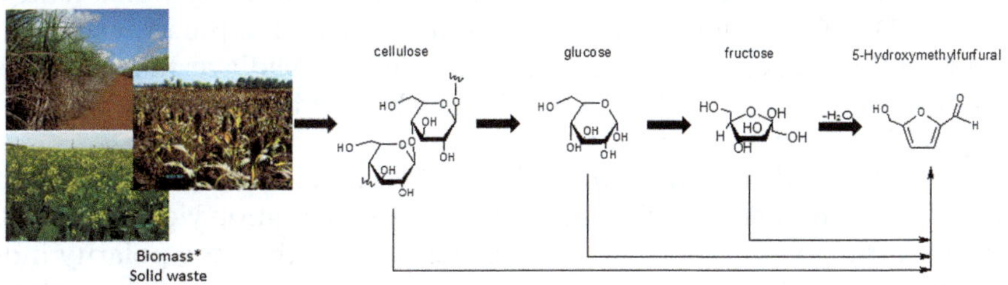

Scheme 7.2   Synthesis of HMF from various bioresources.[32] * Photos of biomass were taken by Mark Tseng.

studied a series of solid catalysts for the conversion of glucose and fructose to HMF in water with very limited yields.[26] They achieved 28% HMF yield in case of glucose/NaOH (1:1 w/w) at 150 °C for 2 h. The study suggested that Brønsted and Lewis sites ratio is a critical factor governing the HMF production efficiency. Alternatively, they were able to achieve much higher HMF yield using concentrated carboxylic acids as reaction media. The highest HMF yield claimed to date from fructose in aqueous media, however, is 64% obtained in a 50 wt.% lactic acid solution at 150 °C for 2 h. Dehydration of fructose in 20 wt.% aqueous acetic and formic acids also led to high HMF yields of 43–53%.

Dumesic *et al.* developed water–organic biphasic systems for the synthesis of HMF to prevent low solubility of sugars in organic solvents and degradation of HMF in a reactive aqueous phase.[48,49] The biphasic systems performed well for the conversion of fructose, glucose, sucrose, inulin, starch, cellobiose and xylan. An unprecedented high HMF yield of 90% from fructose was achieved in a water/2-butanol biphasic system using hydrated tantalum oxide $(Ta_2O_5 \cdot nH_2O)$ at 160 °C.[31] Tantalum oxide catalysts are water-tolerant and can be reused without substantial decrease in HMF selectivity. Direct production of HMF in the biphasic system from polysaccharides, inulin and Jerusalem artichoke juice resulted in HMF yields of 87% and 50%, respectively. A tin-containing, high-silica molecular sieve with the zeolite beta topology (Sn-Beta zeolite) was found to be effective in catalysing glucose to fructose in H_2O/THF/NaCl biphasic systems at low pH.[33] HMF selectivity of 72% at 79% glucose conversion was achieved in the system at 180 °C for 70 min.

It has been demonstrated that dehydration of fructose to HMF occurred in DMSO without addition of a catalyst. DMSO serves as a solvent and reaction promoter in this reaction. High yield of HMF was reported by using DMSO as the solvent and ion-exchange resin as the catalyst. Separation by evaporation of DMSO can be a challenge for the stability of HMF and is an energy costly procedure due to the high boiling point of the solvent (189 °C). The HMF produced was simultaneously extracted with MIBK. A yield of 97% was achieved but the concentration of HMF in MIBK was only 2%. In an acetone–DMSO (70:30 w/w) solution using sulfated zirconia as a solid acid catalyst, an HMF yield of 72.8% was achieved at 180 °C by microwave heating.[46] Reducing the water concentration by using aqueous solvent mixtures as reaction media improves the yield. High HMF production was achieved by water removal from the DMSO reaction mixture by a mild evacuation approach using solid acidic catalysts, such as heteropoly acid, zeolite and acidic resins.[50] The yield was further increased to about 100% by crushed and sieved Amberlyst-15 to the particle size of 0.15–0.05 mm.

It was observed that the HMF yield of 80% in 1,4-dioxane, boiling point ~101.1 °C, is comparable with that of the 81% dehydration yield in DMSO.[27] However, the polarity index of 1,4-dioxane is lower than the polarity index of DMSO and high fructose corn syrup is not fully miscible with the solvent. Nevertheless, the syrup was gradually dissolved as the reaction proceeded. Separation procedures include filtration of Amberlyst-15, evaporation of dioxane, and the dissolving of the crude reaction mixture in ethyl acetate and then washing with water to eliminate by-products. After evaporation of ethyl acetate, 72% of HMF was recovered. Compounds with free α-diol groups destabilize the aldehyde group of HMF while isolated hydroxyl groups attack the hydroxymethyl functionality.[2] Other drawbacks of using organic solvents include cost, health and environmental safety concerns. Caes and Raines applied an industrial organic solvent sulfolane, boiling point 285 °C, for the dehydration of fructose to HMF.[37] They achieved a very high HMF yield of 93% at 100 °C using 1 : 1 w/w LiCl/fructose and 5 wt.% HBr as a catalyst. Extraction of HMF was performed with a H$_2$O/sulfolane 9 : 1 aqueous phase and a MIBK/2-butanol 8 : 2 organic phase. It was suggested that non-aqueous solvents containing a high concentration of chloride ions could be as effective for HMF synthesis from sugars.[32] Conversion of fructose to a 92% HMF yield was achieved in a DMA–LiI or DMA–LiBr solvent.[32] This is because bromide and iodide ions tend to be less ion-paired than fluoride or chloride.

Supercritical fluids were used to minimize rehydration of HMF with 80% fructose conversion and 62% selectivity for HMF. Soluble polymers were the major by-products in this case. Sulfated zirconia was used for the dehydration of fructose with 93.6% conversion and 72.8% HMF yield by microwave heating at 180 °C for 20 min in acetone–dimethylsulfoxide (DMSO) mixed solvent.

7.3 Synthesis HMF in Ionic Liquids

A selected of number of ionic liquids (ILs) with high hydrogen bond basicity were found to be efficient solvents for the depolymerization of cellulose to sugars and subsequently dehydration of the sugars to HMF.[17,47] It was reported that ILs such as hydrogen bond basic media break intra- and inter-molecular hydrogen bonds in crystalline cellulose. The further side reaction of HMF can be suppressed due to the high viscosity of ILs. ILs consisting of 1-ethyl-3-methylimidazolium [EMIM] or 1-butyl-3-methylimidazolium [BMIM] cations were often used for the synthesis of HMF based on patent mapping analysis (Figure 7.1). On the other hand, the most common anion for this application is chloride followed by bromide. Utilization of 1-ethyl-3-methylimidazolium chloride [EMIM][Cl] as a solvent for pre-treatment of lignocellulose prior to the enzymatic hydrolysis process was demonstrated. ILs could act as a solvent and a catalyst in the synthesis of HMF. In recent years, microwave irradiation has been applied in the biorefinery field for the conversion of biomass to chemicals and fuels to take advantage of uniform heating as well as improved reaction rate and selectivity. Microwave-assisted catalytic conversion of carbohydrates and lignocellulosic biomass to value-added materials in ILs reduces the reaction time from hours to minutes. There have been many related review articles with large portions focused on IL-mediated production of HMF in recent years.[17]

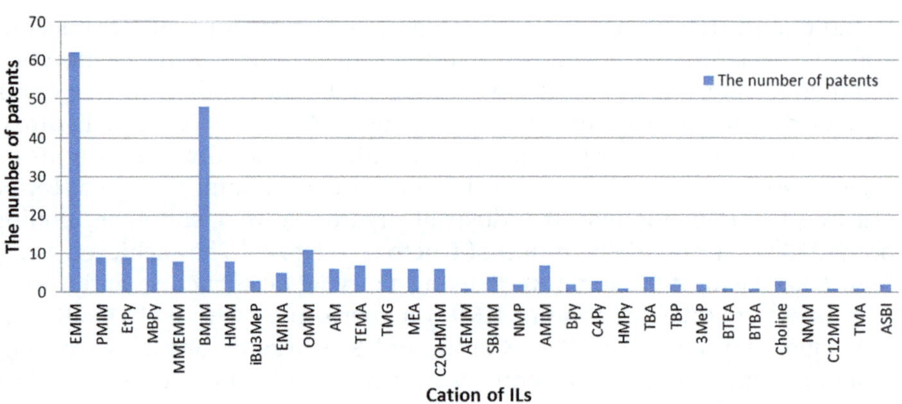

Figure 7.1   Cations of ILs for the synthesis of HMF.

7.3.1   Dehydration of Fructose in Ionic Liquids

It has been successfully demonstrated that acidic ILs may be used for the efficient dehydration of fructose to HMF (Table 7.1). An HMF yield of 88% was reported for a reaction system using 1-ethyl-3-methylimidazolium hydrogensulfate ([EMIM][HSO$_4$]) as a solvent and isobutyl ketone as a co-solvent at 100 °C. To consider large-scale production of HMF, Zhang *et al.* investigated the dehydration process with high initial fructose loading of greater than 10 wt.%.[51] They achieved a very high HMF yield of 97% from 10 wt.% in 1-butyl-3-methylimidazolium chloride ([BMIM][Cl]) catalysed with 9 mol% hydrochloric acid (HCl) at 80 °C in just 8 min. The yield was 51% at 67 wt.% fructose loading. The study suggested water had little effect on the yield if it is kept below 15.4 wt.%.

The acidic IL, 1-(4-sulfonic acid)butyl-3-methylimidazolium hydrogensulfate ([SBMIM][HSO$_4$]) was effective in converting xylose to furfural with 91.5% yield. Functionalization of the cation portion of the IL with a sulfonic acid moiety gave the resulting IL increased acidity. The sulfonic acid-functionalized ILs, 1-butyl-3-(3-sulfopropyl)imidazolium chloride [BSPIM][Cl] and 1-methyl-3-(3-sulfopropyl)imidazolium chloride [MSPIM][Cl] were used as a solvent and a catalyst for the dehydration of fructose to synthesize 5-alkoxymethylfurfural ether in a biphasic system comprising hexane as an extracting solvent. 5-Ethoxymethylfurfural (EMF) is a promising fuel additive due to a comparable energy density with standard gasoline and favourable blending characteristics. The IL–hexane biphasic system minimized the formation of the humin by-product. Yields of 49% and 57% of 5-methoxymethylfurfural (MMF) were obtained at 100 °C in 20 min in the presence of methanol using ILs [BSPIM][Cl] and [MSPIM][Cl] respectively. The slightly lower yield in the case of IL with the butyl group on the 1-position of the imidazolium ring may be attributed to the increased tendency of [BSPIM][Cl] to aggregate as compared to that of [MSPIM][Cl]. It took 80 min for the production of EMF at the same temperature and the yields were about the same,

Table 7.1 Dehydration of fructose in ILs.

Cation	Anion	Solvent/co-solvent	Catalyst	Temp (°C)	Time (min)	HMF Selectivity (%)	HMF Yield (%)	Loading	Ref.
IL-HSO$_4$		DMSO	SILnPs	130	30		63.0	Fructose: 50.0 mg DMSO: 0.5 mL Catalyst: 20.0 mg	30
		ChoCl	pTsOH	100	30		67	Fructose/ChCl = 4:6 Catalyst: 10 mol%	55
[ClC$_2$MIN]$^+$	[Cl]$^-$	H$_2$O	[ClC$_2$MIN]Cl	100	40		75.6	ILs: 4.0 g Fructose: 0.5 g H$_2$O: 1.0 g	52
[BMIM]$^+$	[Cl]$^-$		Amberlyst R-15	80	10		83.3	Catalyst: 18 mol% Amberlyst R-15: 0.05 g	53
[NMP]$^+$	[CH$_3$SO$_3$]$^+$	DMSO		90	120	87.2	72.3	DMSO: 12 mL ILs: 7.5 mol% Fructose: 1.0 g	38
[Tetra-EG(MIN)$_2$]$^+$	[OMs]$_2^-$			120	40	92.3	92.3	1:1:1 Equimolar of ILs, metal halide, fructose	39
[BMIM]$^+$	[Cl]$^-$		H$_2$SO$_4$	120	60		82.3	ILs/fructose/catalyst = 100:10:1	70
[BMIM]$^+$	[Cl]$^-$		P$_2$O$_5$	50	60		81.2	Fructose: 1.0 mmol P$_2$O$_5$: 0.5 mmol [BMIM][Cl]: 10.0 mmol	71
[BMIM]$^+$	[Br]$^-$	DMSO	H$_3$PW$_{12}$O$_{40}$	120	120	94.7	96.7	ILs: 10 g Catalyst: 0.25 g Fructose: 0.5 g	29
[NMM]$^+$	[CH$_3$SO$_3$]$^-$	DMF–LiBr		90	120		74.8	Fructose: 1.0 g ILs: 10 mol% Solvent: 10 mL (DMF–LiBr 70:1)	56
[C$_4$Py]$^+$	[Cl]$^-$		HCl	80	10		97	ILs: 4.0 g Fructose: 0.4 g Catalyst: 0.1 mmol	51

55% and 54% in both ILs. However, there was more ethyl levulinate by-product generated in the case of [BSPIM][Cl]. On the other hand, ethyl levulinate was the main product when 1-methyl-3-(3-sulfopropyl)imidazolium hydrogensulfate [MSPIM][HSO$_4$] was used for the reaction under the same conditions. This may be due to the cumulated and higher acidity of the IL. The HMF yield was 69.5% when the fructose loading increased from 10% to 50%. Additional examples include Brønsted acid 1-*H*-3-methyl imidazolium chloride to achieve a 92% HMF yield and Lewis acid 3-allyl-1-(4-sulfurylchloride butyl)-imidazolium trifluoromethanesulfonate ([ASCBI][OTf]) effectively catalysed the dehydration of fructose under microwave irradiation as well as neutral ILs [BMIM][PF$_6$] and [BMIM][BF$_4$] as reaction media in the presence of a heterogeneous catalyst, such as Amberlyst-15. Li, Argyropoulos and co-workers studied the influence of the functionality of cation and anion-type ILs on the conversion of fructose to HMF.[52] Dissolution of sugars in ILs is a synergic effect of cations and anions. There have been studies of fructose conversion to HMF in ILs 3-(2-chloroethyl)-1-methylimidazolium ([ClEMIM]) and 3-(2-hydroxylethyl)-1-methylimidazolium ([HOEMIM]) coupling with anions of the small hydrogen-bond donor (Cl$^-$) or large non-coordinating [BF$_4^-$] or [NO$_3^-$]. The results indicated the most efficient IL for the conversion was [ClEMIM][Cl] with a 75.6% HMF yield at 100 °C for 40 min. It was 20.8% in [ClEMIM][BF$_4$] under the same conditions. These results are consistent with Binder and Raines' observation in the IL [EMIM][Cl] with DMA–LiCl–CuCl or DMA–H$_2$SO$_4$ solvent systems that chloride ions play a critical role in the reaction. In this study, the conversion of fructose remained very high at about 90% while the HMF yield decreased from about 90–70% when the fructose/[ClEMIM][Cl] weight ratio increased from 5% to 25%.

The IL [ClEMIM][Cl] may also act as a water scavenger because there was no effect on HMF yield if the water contents were kept below 25 wt.%. On the other hand, Qi and Smith, Jr *et al.* found that water had to be limited to 5 wt.% to achieve high efficiency HMF production in [BMIM][Cl] with Amberlyst 15 as the catalyst.[53] An HMF yield of 83.3% was achieved in 10 min at 80 °C. It was 82.2% in just 1 min when the temperature was increased to 120 °C. The resulting HMF was extracted with ethyl acetate and the IL and resin can be recycled without loss of activity. Dowex® 50WX8 resin and CuCl$_2$ were also effective in catalysing the reaction with about 80% yield. Chung *et al.* applied phosphorus pentoxide (P$_2$O$_5$) as a catalyst and water scavenger to mediate dehydration of fructose to HMF at low temperature in imidazolium IL, DMSO and DMA–KI reaction media. Maximum yield of 81.2% was achieved at 50 °C for 60 min in [BMIM][Cl]. Increase of temperature and reaction time did not enhance the yield due to the formation of humins.[54]

Efficient dehydration of fructose to yield 99.1% HMF was achieved using IL 1-(3-sulfonicacid)propyl-3-methyl imidazolium phosphotungstate ([MIMPS]$_3$PW$_{12}$O$_{40}$) as a catalyst in *sec*-butanol at 120 °C for 2 h.[48] The yield quickly reached 90% in 30 min when the temperature was increased to 160 °C. On the other hand, the yield was much lower in the case of *n*-butanol (74.6%) and iso-butanol (56.8%). The catalyst is a solid at room temperature and thus precipitates out at the end of reaction. A comparable yield of 96.7% was obtained for the homogeneous system containing $H_3PW_{12}O_{40}$ and DMSO. Dicationic ILs with oligo ethylene glycol linkages were found to be very effective in the conversion of fructose to HMF with a yield as high as 92.3% at 120 °C for 40 min.[39] The catalytic activities of dicationic ILs appeared to increase with the length of oligo ethylene glycol and tetra ethylene glycol-bis(3-methylimidazolium) dimesylate [TetraEG(MIM)$_2$][CH$_3$SO$_3$] provided the highest yield among the solvents studied. HMF yields were 69.8% and 77.2% for [DiEG(MIM)$_2$][CH$_3$SO$_3$] and [TriEG(MIM)$_2$][CH$_3$SO$_3$], respectively. In all three dicationic IL systems, conversion of fructose reached 100%. A series of co-catalysts, FeCl$_3$, CuCl$_2\cdot$6H$_2$O, NiCl$_2\cdot$6H$_2$O and CoCl$_2\cdot$6H$_2$O, were evaluated to enhance HMF yield at lower temperatures. The most active co-catalyst was NiCl$_2\cdot$6H$_2$O with an HMF yield of 81.2% at 100 °C. It was suggested that the ethylene glycol chain length was suitable for the formation of a complex with NiCl$_2\cdot$6H$_2$O and led to the improved performance. Dehydration of fructose of 10 wt.% loading was carried out using propylsulfonic acid functionalized mesoporous SBA-15 (SBA-15–SO$_3$H) as a catalyst in [BMIM][Cl] at 120 °C for 1 h with 99.2% conversion and an 81% HMF yield. The yield remained quite high at 69.5% at 50 wt.% fructose loading. The resulting HMF was separated by introducing THF to the reaction system. The conversion of fructose to HMF can be performed in a semi-continuous fashion using a THF/[BMIM][Cl] biphasic system. Fructose and THF were added to the reaction system at beginning of each consecutive operation. HMF isolated during the whole procedure was 93%.

Deep eutectic solvents (DESs), such as choline chloride (ChoCl)-based ILs, are considered promising reaction media for the synthesis of HMF due to their low cost, thermal stability and relatively low toxicity. König and co-workers started with their earlier development of the low melting carbohydrate/urea mixtures for the conversion of fructose to HMF.[55] The reactions were carried out with 40 wt.% of fructose and at 110 °C for 2 h. A very limited yield, 8%, was obtained for a D-fructose/*N,N'*-dimethyl urea (DMU) melt in the presence of FeCl$_3$ (10 mol%) as a catalyst. The yield was sharply increased to 89% when DMU was replaced with *N,N'*-tetramethyl urea (TMU). However, the authors abandoned the system and searched for alternatives due to an energy intensive separation process and the inherently toxic nature of the D-fructose/TMU melt. H-bond donors and acceptors evaluated for the formulation of eutectic mixtures include urea, DMU, imidazole, 4-methyl imidazole, pyrazole, choline chloride, guanidinium HCl and malonic acid. The fructose-based low-melting mixtures show pH values from acidic to basic. The low melting mixtures were evaluated for dehydration of fructose using Amberlyst 15 and FeCl$_3$ as catalysts at 100 °C for 1 h. For mixtures with low pH values, such as malonic acid (3.0) and maleic acid (3.2), rehydration of HMF occurred and levulinic acid was identified. The melting point for a eutectic mixture of choline chloride containing 40–60 wt.% fructose is in the range 79–82 °C. After optimization of the reaction conditions, the highest yield for fructose/choline chloride melt using FeCl$_3$ (10 wt.% based on sugar) as a catalyst is 59% at 100 °C for 0.5 h. However, the best HMF yield achieved for the D-fructose/choline chloride melt (4:6) is 67% by using *para*-toluene sulfonic acid (*p*TsOH) as a catalyst. The yield is about 60% in the presence of FeCl₃ or CrCl₃. Heterogeneous catalysts Amberlyst 15 and Montmorillonite, a weak acid, were evaluated for the dehydration of fructose, with HMF yields of 40% and 49%, respectively.

Tong and Li studied the dehydration of fructose to HMF in DMSO using a catalytic amount of (0.075–0.1 IL/fructose molar ratio) Brønsted-acidic ILs and achieved a 72.3% yield and 87.2% selectivity of HMF at 90 °C for 2 h.[38] The studies suggested the catalytic activities of *N*-methyl-2-pyrrolidonium methyl sulfonate [NMP][CH₃SO₃] and *N*-methyl-2-pyrrolidonium hydrogen-sulfate [NMP][HSO₄] were much higher than that of the imidazolium coun-terparts. [NMP][CH₃SO₃] provided the best yield. The [NMP]-based ILs have higher acidity than that of [MIM]-based ILs and thus exhibit higher dehy-dration activity. The solvent was recovered by distillation under reduced pressure after the dehydration. HMF was then extracted with ethyl acetate from the remaining mixture. The catalysts and solvent can be recycled with very little diminished activity. Study of the conversion of fructose to HMF in DMF–LiBr with a catalytic amount of *N*-methylmorpholium methylsulfonate [NMM][CH₃SO₃] provided a 74.8% yield under the same conditions.[56] The HMF yields were 25.4%, 70.5%, 69.2% and 67.5% using the same IL in DMF, DMF–LiCl, DMF–NaBr and DMF–KBr, respectively.

Cost, purification and toxicity are the major concerns of using ILs in large-scale production applications. Supported ionic liquid catalyst (SILCA) was developed for the ease of catalyst separation and reuse.[42] Immobilization of ILs was achieved by means of sol–gel chemistry, grafting, impregnation and polymerization. Catalytically active species, such as metal nanoparticles, metal complexes and enzymes, can be embedded in a thin layer of IL. With suitable design, ILs can have positive influence on the chemical properties of the embedded catalyst and synergetic effects on the reaction. Sidhpuria and co-workers demonstrated fructose dehydration with a 63% HMF yield in the presence of IL catalyst immobilized on silica nanoparticles (d ~610 nm) and DMSO as a solvent.[30] 1-(tri-ethoxy-silyl propyl)-3-methyl imidazolium hydro-gensulfate was attached to the surface of a silica nanoparticle by a sol–gel process.

7.3.2 Dehydration of Glucose in Ionic Liquids

Glucose is more abundant than fructose and can be converted to HMF in ILs containing isomerization catalysts, such as AlCl₃, CrCl₃, CrCl₂, boric acid, *etc.* Zhang and co-workers studied a series of metal chlorides, CrCl₂, CrCl₃, FeCl₂, FeCl₃, CuCl, CuCl₂, VCl₃, MoCl₃, PdCl₂, PtCl₂, PtCl₄, RuCl₃ and RhCl₃, for the dehydration of sugars.[36] For most of the catalysts studied, glucose conversion was high but only CrCl₂ afforded up to a 68–70% HMF yield and other non-Cr catalysts systems provided yields of 10% or less (Table 7.2). The authors observed that certain metal halides (CrCl₂, CuCl₂, VCl₃) and

Table 7.2 Dehydration of glucose in ILs.

Cation	Anion	Co-solvent	Catalyst	Temp (°C)	Time (min)	HMF Selectivity (%)	HMF Yield (%)	Loading	Ref.
[Cho]$^+$	[Cl$^-$]		CrCl$_2$	110	30		45	Glucose: 400 mg ChCl: 600 mg Catalyst: 10 mol%	55
[EMIM]$^+$	[Cl$^-$]		H$_3$BO$_3$	120	180	43	41	Glucose: 100 mg Solvent: 1 g Boric acid: 27.5 mg	58
[BMIM]$^+$	[Cl$^-$]	DMSO	LCC	160	50		68	ILs/DMSO: 4/6 Glucose: 10 wt.% Catalyst: 5 wt.%	61
[BMIM]$^+$	[Cl$^-$]		H$_2$SO$_4$	100	180	66	61	ILs: 2 g Glucose: 0.18 g Catalyst: 0.01 mmol	34
[BMIM]$^+$	[Cl$^-$]		12-TPA	100	180	81	66	ILs: 2 g Glucose: 0.18 g Catalyst: 0.01 mmol	34
[EMIM]$^+$	[Cl$^-$]		CrCl$_2$	100	180		70	DMSO: 0.3 mL	36
[EMIM]$^+$	[Cl$^-$]		CrCl$_3$ (n-BuOH)$_3$	100	180		69	ILs: 5 g Glucose: 0.5 g Catalyst: 60 mmol	57
[EMIM]$^+$	[Cl$^-$]		CrCl$_3$ (THF)$_3$	100	180		71	ILs: 5 g Glucose: 0.5 g Catalyst: 60 mmol	57
[EMIM]$^+$	[Cl$^-$]		CrCl$_3$	100	180		72		71
[Et$_4$N]$^+$	[Cl$^-$]		CrCl$_3$·6H$_2$O	130	10		71.5	CrCl$_3$·6H$_2$O: 10 mol% dissolved in 1 g of TEAC	72

H_2SO_4 appear to play a role in stabilizing HMF at 100 °C. $SnCl_4$ was also identified as an efficient catalyst for the conversion of glucose in [EMIM][BF$_4$]. The IL is a strong solvent for glucose. At 23 wt.% glucose loading, a 61% yield was achieved. Bali *et al.* studied a series of Cr^{II}, Cr^{III} and Cr^{IV} as well as other metal catalysts for the conversion of glucose to HMF in [EMIM][Cl] to understand the effect of the oxidation state of Cr on HMF production yield.[57] They achieved a 71% yield using $CrCl_3(THF)_3$ in ambient conditions at 100 °C. It was suggested that Cr^{II} and Cr^{IV} were inevitably converted to the more stable Cr^{III} oxidation state under the reaction conditions and thus Cr^{III} was the active species. Chidambaram and Bell studied a series of liquid acids and heteropoly acids for the dehydration of glucose in ILs, [BMIM][Cl] and [EMIM][Cl].[34] Among the liquid acids studied, H_2SO_4 and CF_3SO_3H afforded high glucose conversion but modest HMF selectivity. In addition to humins, 5-chloromethylfurfural (CMF) and furylhydroxymethyl ketone (FHMK) were found as by-products in about 10–20%. It is very likely that the formation of CMF is due to acid-catalysed substitution of chloride anions with the hydroxyl group of HMF. However, CMF only presented in the ILs containing HCl, CF_3SO_3H and CH_3SO_3H. This by-product was not generated in the presence of heteropoly acids. Actually, humins were the major by-products in case of heteropoly acids. It was found that a selected number of heteropoly acids were effective for the dehydration of glucose with high selectivity for the formation of HMF. Glucose conversions of 99% and HMF selectivity of 98% were achieved by using 12-molybdophosphoric acid (12-MPA) catalyst in a solution of [EMIM][Cl] with acetonitrile (CH_3CN) as a co-solvent. The conversion and selectivity were 71% and 89% respectively without the co-solvent. The authors speculated that addition of acetonitrile lowers the viscosity of the IL and thus suppressed the formation of humins. It is interesting to note that they proposed 12-MPA-catalysed glucose dehydration *via* ring opening instead of isomerization of glucose to fructose. Hydrogenation of HMF to DMF in the [EMIM][Cl]–CH_3CN without separation can be achieved after removal of 12-MPA and replacement with Pd/C.

Dehydration of glucose to HMF was performed in ILs using boric acid as an isomerization promoter for environmental considerations.[58] Boric acid forms stable chelate complexes with carbohydrates and stabilizes intermediates in the isomerization of glucose to fructose. The results of deuterium-labelling studies suggested that the isomerization of glucose in the presence of boric acid proceeded *via* a ring-opening route to generate an ene-diol intermediate. The mechanism differs from the enzyme-catalysed isomerization of glucose to fructose. The HMF yield reached a maximum of 40% after a 3 h reaction at 120 °C in [EMIM][Cl] with 1 equiv. of boric acid as catalyst. The glucose loading is 10 wt.%. The yield, 20%, is lower in [BMIM][Cl] as observed in the case of chromium catalysts. It is important to note that boric acid forms a stronger binding with fructose than with glucose. There would be a competition for boric acid after generation of fructose. Formation of fructose–boric acid complexes would very likely have adverse effects on the formation of HMF by blocking the dehydration of fructose and inhibiting the isomerization of glucose to fructose. The results of the investigation suggested the optimal boric acid concentration is between 0.8 and 1 equiv. Studies have suggested that chloride anions play a critical role in conversion of hexoses to HMF. The product only presents in the ILs containing Cl^-, $CH_3SO_4^-$ and $C_2H_5SO_4^-$ anions. Although the HMF selectivity was high, the yield was only 6% in the case of [EMIM][CH_3SO_4], [EMIM][$C_2H_5SO_4$] and [ChoCl][CH_3SO_4]. On the other hand, a mixture of 4:6 glucose/choline chloride produced HMF in 45% yield at 110 °C for 0.5 h using $CrCl_2$ as the catalyst.[55] An HMF yield of 45% was achieved for a D-glucose/choline chloride (4:6) eutectic mixture using $CrCl_2$ as the catalyst at 110 °C for 0.5 h.[55] The other carbohydrates and sugar alcohols for the low melting mixtures include D-mannose, D-sorbitol, isomaltose and glucosamine. Shih *et al.* achieved 94% glucose conversion and 61% HMF yield in diethylene glycol/choline chloride using $CrCl_2$ as the catalyst.[59]

Separation of HMF from reaction mixtures is a challenge because it is a thermally sensitive compound. Simple distillation is not feasible. Solvent extraction methods are time consuming and require a large amount of solvent. Wei *et al.* developed an entrainer-intensified vacuum reactive distillation (EIVRD) process for the separation of HMF from IL reaction media. The product, HMF, was removed continuously during the dehydration of fructose or glucose from IL by a vacuum and an entrainer bubbled through the mixture. Although dehydration reactivity increases with the decrease of the chain length of the alkyl group on the alkylmethylimidazolium ion, 1-methyl-3-octyl imidazolium chloride [OMIM][Cl] was used as a solvent in an EIVRD process to avoid escape of IL during distillation. Metal chlorides, $IrCl_3$ for fructose and $CrCl_3$ for glucose, were applied to catalyse dehydration and to stabilize HMF at high temperatures. The average recovery efficiencies of HMF from the dehydration of fructose and glucose were about 93% and 88%, respectively. The HMF yield from the dehydration of glucose increased more than 20% (from 42.1% to 64.8%) when the EIVRD process was applied for the dehydration of glucose at 150 °C for 30 min. This is due to removal of HMF and to minimize side reactions. Both recovery and yield improved when the temperature increased to 180 °C with a reduced reaction time of 10 min. The experimental results indicated that the type of entrainers, N_2, hexane and MIBK, have very little effect on HMF recovery and yield. This suggests that the efficiency is governed by the surface area of the liquid–gas interface rather than the entrainer polarity. There were very little changes in recovery and yield after five repeated uses of IrC_3-[OMIM][Cl] system for the dehydration of either fructose or glucose. However, small amounts of insoluble solid were present at the bottom of the reactor and 4.9–7.1% of [OMIM][Cl] was stripped out with HMF.

Alternatively, organic solvents, such as DMSO, were used as co-solvents with an IL to lower the viscosity of the reaction medium and limit the use of IL.[60] The viscosity of DMSO–[BMIM][Cl] at 4/6 weight ratio is 10.99 mPa which is lower than the 29.21 mPa of [BMIM][Cl] at 25 °C. Fang and co-workers investigated a series of biomass-derived solid carbonaceous acid catalysts for dehydration of fructose and conversion of glucose to HMF in DMSO–[BMIM]

[Cl] mixtures under 100 W microwave irradiation.[61] Biomass for the synthesis of carbonaceous catalysts includes glucose, fructose, cellulose, lignin, bamboo and *Jatropha* hulls. Experimental results suggested both acid density and morphology govern catalyst activity. The lignin-derived solid acid catalyst was found to be the most active for the dehydration of fructose. HMF yield increases from 61% to 82% in the presence of DMSO. Conversion of 99% of fructose was achieved at 130 °C while the maximum HMF yield was obtained at 110 °C. This is due to the rapid decomposition of HMF to levulinic acid, formic acid and furfural at temperatures above 120 °C in the presence of water generated from the dehydration process. The suggested optimal reaction time is 10 min. On the other hand, a 68% HMF yield at 99% glucose conversion was achieved at 160 °C for 50 min. The intermediate fructose was not detected due to a much higher conversion temperature than that of its dehydration. After the reaction, HMF was recovered by extracting with ethyl acetate. The catalyst was filtered, washed with water and air dried for reuse. After five uses, there was a very limited decrease in acid density of the catalyst and an HMF yield of 76% was obtained.

7.3.3 Conversion of Disaccharide to HMF

Sucrose is a disaccharide consisting of a fructose and a glucose unit linked together with a glycosidic bond which can be easily hydrolysed to generate monomers. Dicationic ionic acid $[TetraEG(MIM)_2][CH_3SO_3]_2$ showed efficient Lewis acidity not only for the dehydration of fructose but also for sucrose in the absence of catalyst.[39] A very high HMF yield of 67.2% was achieved using 2 equiv. of dicationic IL $[TetraEG(MIM)_2][CH_3SO_3]_2$ at 120 °C for 150 min. Again, the yield was reduced to 61.2% and 54.5% when the chain length of linkage ethylene glycol decreased for $[TriEG(MIM)_2][CH_3SO_3]_2$ and $[DiEG(MIM)_2][CH_3SO_3]_2$, respectively. The HMF yield was sharply decreased to 41% at 100 °C. A number of metal chlorides were evaluated as a co-catalyst for the conversion of sucrose to HMF. The yield was 62.2% when using $[TetraEGMIM][CH_3SO_3]$ with $NiCl_2 \cdot 6H_2O$ at 100 °C. The authors suggested that $NiCl_2 \cdot 6H_2O$ has suitable molecular size to interact with the oligo ethylene glycol chains, which leads to high catalytic activity. Microwave-assisted catalytic conversion of sucrose was carried out using scandium(III) chloride ($ScCl_3$) for the replacement of toxic heavy metal catalysts. Li *et al.* carried out dehydration of sucrose in DMF–LiBr using a catalytic amount (10.0 mol%) of $[NMM][CH_3SO_3]$ at 90 °C for 90 min and achieved 47.5% HMF yield.[56]

When isomerization promoter boric acid was used for the dehydration of sucrose, a maximum HMF yield of 66% was achieved after 8 h at 120 °C in [EMIM][Cl].[58] The boric acid concentration for the conversion of sucrose to HMF was selected to be 0.5 equiv. according to the results of glucose dehydration. In the case of maltose, a glucose dimer with α-1,4-linkage, the HMF yield is only about 33% after 8 h reaction at 120 °C in [EMIM][Cl] with 0.5 equiv. of boric acid as the catalyst.[58] A eutectic mixture of 1:1 sucrose/choline chloride was effective in producing HMF at a high yield of 63% at 100 °C in the presence of 10 mol% $CrCl_2$.[55] Isomaltose, a glucose dimer with α-1,6-linkage, also forms low melting mixtures with *N,N'*-dimethyl urea, imidazole, 4-methyl-imidazole, pyrazole and choline chloride. The HMF yield was 69% when a reaction mixture containing 10 wt.% sucrose and 6 mol% $CrCl_2$ in diethylene glycose/choline chloride was heated at 100 °C for 1 h.[59] ILs for the conversion of sucrose to HMF are summarized in Table 7.3.

7.3.4 Synthesis of HMF from Polysaccharide

Conversion of polysaccharides to HMF in one pot involves hydrolysis of feedstocks, isomerization of glucose and dehydration of fructose. A cascade process without intermediate product recovery reduces operating time, and consumption of energy and raw material. However, there will be more stringent operation conditions and combinations of solvents and catalysts to maximize the yield. The effects of ILs, catalysts and experimental conditions are illustrated in Table 7.4. Boric acid was also used as an isomerization promoter for the conversion of cellulose and starch to HMF with 33% yield at 120 °C in [EMIM][Cl] with 0.5 equiv. of boric acid as catalyst.[58] It took 8 h for cellulose to reach the maximum HMF yield while it was 24 h for starch. A eutectic mixture of diethylene glycol/choline chloride was applied for the conversion of starch to HMF with 39% yield at 120 °C for 2 h.[59]

Binder and Raines suggested and experimentally proved that non-aqueous solvents containing a high concentration of chloride ions could be as effective as [EMIM][Cl] for HMF synthesis.[32] DMA–LiCl can dissolve up to 15 wt.% cellulose. In the presence of $CrCl_3$ (25 mol%) and HCl (10 mol%) as catalysts, production of HMF from cellulose of up to 33% yield was obtained in DMA–LiCl (10 wt.%) solvent within 2 h at 140 °C. However, cellulose does not dissolve in either DMA–LiI or DMA–LiBr systems. A negligible quantity of HMF was produced in these solvents at the same conditions. There are 2.0 M of weakly ion-paired chloride ions in saturated anhydrous DMA–LiCl compared to the 6.8 M in [EMIM][Cl]. The loosely ion-paired chloride ions in the DMA–LiCl solvent can form hydrogen bonds with the hydroxyl groups of cellulose and facilitate its dissolution. A very high HMF yield of 54% was achieved for cellulose in DMA–LiCl using $CrCl_2$ and HCl as catalysts in combination with 60 wt.% [EMIM]Cl as an additive. The authors observed the overall reactions proceeded better with ILs than with lithium chloride. The process minimizes the use of IL and thus reduces the cost of HMF production. Under the same conditions without DMA–LiCl, the yield is 53% in [EMIM][Cl].

Biodegradable DES compositions were also developed. The solvents can be easily prepared by mixing choline chloride with citric acid monohydrate (ChoCl/citric acid) or oxalic acid dihydrate (ChoCl/oxalic acid). The above raw materials for DES are economical and renewable. Inulin is a non-digestible oligosaccharide consisting of glucose-(fructose)$_n$ with a high fructose/glucose ratio. It is available in large amounts and is ready to hydrolyse to produce fructose. Inulin is soluble in ChoCl/citric acid (28 mg g^{-1}) and ChoCl/oxalic acid (150 mg g^{-1}).[24] Both hydrolysis of inulin and dehydration of fructose are

Table 7.3 Conversion[58] of disaccharide to HMF.

Cation	Anion	Co-solvent	Catalyst	Temp (°C)	Time (min)	HMF Selectivity (%)	HMF Yield (%)	Loading	Ref.
$[Cho]^+$	$[Cl^-]$		$CrCl_2$	100	60		62	Sucrose: 500 mg ChoCl: 500 mg Catalyst: 10 mol%	55
$[EMIM]^+$	$[Cl^-]$		H_3BO_3	120	600		66	IL: 1 g Sucrose: 100 mg Boric acid: 0.5 equiv.	58
$[Tetra\text{-}EG(MIN)_2]^+$	$[OMs]_2^-$			120	40	67.2	67.2	1:1:1 Equimolar of [Tetra-EG(MIM)$_2$][OMs]$_2$, metal halide, fructose	39
$[BMIM]^+$	$[Cl]^-$	IBMK	$CrCl_3$	100			100		40
$[NMM]^+$	$[CH_3SO_3]^-$	DMF–LiBr		90	120		47.5	Sucrose: 1.0 g ILs: 10.0 mol% Solvent: 10 mL (DMF–LiBr: 70:1)	56
$[EMIM]^+$	$[Cl^-]$		$CuCl_2/CrCl_2$	120	480		57.5	CuCl$_2$/CrCl$_2$ catalyst ($\chi_{CuCl2} = 0:17$)	73
$[BMIM]^+$	$[Cl]^-$		$Cr([PSMIM]HSO_4)_3$	120	300		53	Cr([PSMIM]HSO$_4$)$_3$: 0.05 g [BMIM]Cl: 2 g	74

Table 7.4 Conversion of polysaccharide to HMF.

Cation	Anion	Co-solvent	Catalyst	Temp (°C)	Time (h)	HMF Selectivity (%)	HMF Yield (%)	Loading	Ref.
[Cho]$^+$	[Cl$^-$]	pTsOH	CrCl$_2$	90	1		57	Inulin: 500 mg ChoCl: 500 mg Catalyst: 10 mol%	55
[EMIM]$^+$	[Cl$^-$]		H$_3$BO$_3$	120	8		32	IL: 1 g Starch: 100 mg Boric acid: 0.5 equiv.	58
[EMIM]$^+$	[Cl$^-$]		H$_3$BO$_3$	120	24		32	IL: 1 g Starch: 100 mg Boric acid: 0.5 equiv.	58
Oxalic acid	ChoCl			80	2	64	87	Inulin: 81 mg ChoCl: 500 mg Catalyst: 10 mol%	24
[NMM]$^+$	[CH$_3$SO$_3$]$^-$	DMF–LiBr		90	2		47.5	1.0 g sucrose in 10.0 mol% protic ILs	56

catalysed by acids. The two DESs, therefore, can serve as catalysts and solvents to convert inulin to HMF in a one-pot reaction. HMF yields of 56% and 51% were achieved for ChoCl/oxalic acid and ChoCl/citric acid respectively. The HMF yield can be further increased to 57% in ChoCl/citric acid by applying a two-step heating process. Inulin was first converted to fructose at 50 °C and then hydrolysed to HMF at 80 °C. A biphasic system of ethyl acetate and ChoCl/oxalic acid was investigated for improving HMF productivity and 64% yield was achieved. Inulin can also form a eutectic mixture with choline chloride. HMF yields of greater than 50% were achieved with inulin/choline chloride eutectic mixtures at 90 °C using *para*-toluene sulfonic acid *p*TsOH, Amberlyst 15 or $FeCl_3$.[55] A comparable yield of 54% was achieved for this melt using a combination of Montmorillonite and Amberlyst 15 (1:1) as the catalyst.

7.3.5 Synthesis of HMF from Lignocellulosic Biomass

Agriculture residues, such as corn stover, corn stalk, wheat, rice straw, rice husk, bagasse and wood dust, are composed of cellulose, hemicellulose and lignin. These biomass resources can be converted to fuels and chemicals. Singh *et al.* demonstrated a 'closed loop' biorefinery by using tertiary amine-based ILs derived from lignin and hemicellulose for biomass pre-treatment. Aromatic aldehydes from the depolymerization of lignin and hemicellulose, as well as solid residues in ionic pre-treated lignocellulosic biomasses, were converted to IL *via* reductive amination. There are successful attempts to synthesize HMF directly from lignocellulosic biomass (Table 7.5). The processes include dissolution and hydrolysis of cellulose or lignocellulose followed by dehydration of the resulting sugars to generate HMF.

Li and Zhao demonstrated an efficient method for the hydrolysis of cellulose without pre-treatment using a low content of H_2SO_4 or other mineral acids as catalyst in IL [BMIM][Cl].[62] Cellulose hydrolysis followed a random fashion to produce total reducing sugars (TRS) in short reaction times while longer reaction times favoured formation of glucose. An acidic ILs-grafted nanoporous polymer, PDVB–SO_3H–[C_3vim][SO_3CF_3], showed improved catalytic activities in comparison with homogeneous and acidic resin catalysts for depolymerization of crystalline cellulose.[63] The grafted IL moiety improves the compatibility with IL [BMIM][Cl] and strong acidic functionality assists in destroying the intermolecular hydrogen bonds of crystalline cellulose. A one-pot synthesis to carry out these consecutive reaction steps, therefore, becomes possible by using ILs as solvents for the conversion of woody biomass to HMF and derivatives. On the other hand, conventional procedures to derive chemicals from crude lignocellulose require pre-treatment processes. One-pot synthesis minimizes use of chemicals, solvents, apparatus and equipment. The most important benefit is avoiding separation of intermediates. It may turn out to be an energy- and cost-saving processes. The concept of performing one-pot direct synthesis of HMF from lignocellulose is to carry out consecutive dissolution, hydrolysis and dehydration

Table 7.5 Synthesis of HMF from lignocellulosic biomass.

Cation	Anion	Co-solvent	Catalyst	Temp (°C)	Time (min)	HMF Selectivity (%)	HMF Yield (%)	Loading	Ref.
$[EMIM]^+$	$[Cl^-]$	DMA–LiCl	$CrCl_2$ 25 mol% HCl 6 mol%	140	120		60	Cellulose: 10 wt.% $CrCl_3$: 6 mol% ILs: 60 wt.%	32
$[BMIM]^+$	$[Cl^-]$		$MgCl_2$ H_3BO_3	105	120		40	Cotton: 5 wt.%	43
$[EMIM]^+$	$[Cl^-]$		HCl	105	120		31	Paper towel: 5 wt.%	43
$[EMIM]^+$	$[Cl^-]$		$CuCl_2/CrCl_2$	120	480		57.5	$CuCl_2/CrCl_2$ catalyst ($\chi_{CuCl2} = 0:17$)	73
$[BMIM]^+$	$[Cl]^-$	DMA–LiCl	$Zr(O)Cl_2/CrCl_3$	120	5		57	Fibre: 0.5 mg Microwave reactor	67
$[C_4MIM]^+$	$[Cl]^-$		$CrCl_3 \cdot 6H_2O$	140	2.5		62	Cellulose: 100 mg $[C_4MIM][Cl]$: 2 g $CrCl_3$: 10 mg	65
$[BMIM]^+$	$[Cl]^-$		$CrCl_3/LiCl$	140	40		61.9	Cellulose: 50 mg ILs: 2.0 g H_2O: 5 mg $CrCl_3/LiCl$: 50/50 mol	68
$[BMIM]^+$	$[Cl]^-$		$Cr([PSMIM]HSO_4)_3$	120	300		53	$Cr([PSMIM]HSO_4)_3$ 0.05 g ILs: 2 g	74
$[EMIM]^+$	$[Cl^-]$		$CrCl_3$	140	10	42	35.6		75

steps in a single reactor to avoid separation of intermediates.[1,42] Thus, it is a demanding task but also a resource- and cost-saving process. IL-assisted methods for the one-pot synthesis of HMF and derivatives from lignocellulosic biomass have been under intensive investigation in recent years.[42]

Long-chain alkyl glycosides were successfully synthesized in a one-pot reaction by coupling the hydrolysis of cellulose with the Fischer glycosidation with C_4–C_8 alcohols in [BMIM][Cl].[64] Both hydrolysis and glycosidation reactions are catalysed by acid. Solid acids Amberlyst-15Dry and $H_3PW_{12}O_{40}$ provided the best catalytic performance in the one-pot transformation reaction. A small amount of water was introduced to the system to facilitate hydrolysis of cellulose and to limit the formation of HMF. The presence of water has a negative influence on the following glycosidation. Therefore, the pressure of the reaction system was lowed from atmospheric pressure to 40 mbar when alcohol was introduced. The time to initiate glycosidation was set long before the time required to obtain a maximum yield of total reduced saccharides (TRS) to avoid accumulation and degradation of glucose. At optimal conditions, an 82% yield of alkyl-α,β-glycosides was achieved. The resulting non-ionic biodegradable surfactants have wide applications in cosmetics, foods, pharmaceuticals and detergents.

Corn stover, for example, contains about 34 wt.% cellulose. An HMF yield of 48% was achieved for the biomass feedstock by using DMA–LiCl as a solvent and $CrCl_3$ and HCl as catalysts in combination with 60 wt.% [EMIM]Cl as an additive at 140 °C.[32] Furfural was also obtained from this process at a yield of 34%. Direct production of HMF with yield of 45–52% from corn stalk, rice straw and pine wood was achieved by using IL [BMIM]Cl in the presence of $CrCl_3$ under microwave heating for 3 min.[65] A series of ILs, [BMIM][HSO_4], [HMIM][HSO_4], morpholium bisulfate [Morph][HSO_4], tetrabutylammonium bisulfate [Bu_4N][HSO_4], tetrabutylphosphonium bisulfate [Bu_4P][HSO_4] and choline bisulfate [ChoCl][HSO_4], were evaluated as catalysts in aqueous solutions for the conversion of agar from red seaweed to sugars.[66] [ChoCl][HSO_4] was found to be most effective for the production of galactose and 3,6-anhydrogalactose with limited HMF generated.

Abu-Omar, Saha and co-workers also applied the microwave heating method in DMA–LiCl solvent systems for conversion of cellulose, fibre and sugar cane bagasse to HMF and 5-ethoxymethyl-2-furfural (EMF) using Zr^0Cl_2/$CrCl_3$ (3:1) as a combined catalyst.[67] An HMF yield of 43% was achieved in the presence of 9 wt.% of [BMIM][Cl] as an additive at 120 °C for 5 min. The yield increased as the concentration of IL increased. For sugar cane bagasse (4 wt.%), the yield is 31% under the same reaction conditions. The yield increased to 42% in [BMIM][Cl] solution without DMA–LiCl. On the other hand, Yu and Zhan observed an HMF yield of greater than 60% for cellulose and wheat straw by using a dual catalyst system, $CrCl_3$/LiCl.[68] The average yield is 55.4 ± 4.0% for 12 repeated experiments using a $CuCl_2$/$CrCl_2$ (χ_{CuCl_2} = 0.17) catalyst pair in [EMIM][Cl] at 120 °C for 8 h. The resulting HMF was recovered at 90% by three successive MIBK extractions and the solution including the catalyst pair was then dried to remove moisture for repeat synthesis. The yield remained about the same for three reuses of the IL containing the catalyst. Humin by-products in the reused solvent did not affect dissolution and conversion of cellulose. Fluorous ILs were synthesized to facilitate the separation of products and catalysts.[69] The IL 3-methyl-1-(2′,2′,3′,3′-pentafluoropropyl)-imidazolium was evaluated for the hydrolysis of a cellulose and separated from glucose to recover 95%. Chen *et al.* used phosphotungstic acid-derived IL-based polyoxometalate salt (IL-POM) as a solid acid catalyst for the conversion of inulin and sucrose to HMF with high yield.[29]

In searching for benign processing chemicals, a comparable yield of 41% was achieved for the conversion of cellulose to HMF in a 2-methoxycarbonylphenyl boronic acid/$MgCl_2 \cdot 6H_2O$ catalyst and [EMIM][Cl] solvent system at 105 °C for 1 h.[43] The method is applicable for the production of HMF from cotton, paper towel and newspaper.

7.4 Conclusions and Outlook

The use of ILs broadens the selection of feedstocks for the synthesis of HMF. Additionally, the high viscosity of ILs suppresses side reactions during the synthesis of HMF. ILs consisting of 1-ethyl-3-methylimidazolium chloride [EMIM][Cl] are the most common ILs selected for the synthesis of HMF based on a patent mapping analysis. Chloride ions, as hydrogen-bond donors, play a critical role in the HMF syntheses using IL as a solvent. Acid ILs are effective in converting carbohydrates to HMF. Several attempts were successfully made to functionalize cations or anions in order to create acidity in ILs, including functionalization of the cation portion of the IL with a sulfonic acid moiety or using hydrogensulfate as the anion. ILs can be used as the catalyst alone, such as $[MIMPS]_3PW_{12}O_{40}$ and $[NMM][CH_3SO_3]$, or on a support, such as silica nanoparticles, to improve the yield or selectivity of HMF in organic solvents. ILs with high hydrogen bond basicity facilitate the breaking of intra- and inter-molecular hydrogen bonds in crystalline cellulose and thus improve conversion efficiency. Glucose, disaccharides, polysaccharides and lignocellulosic biomass in addition to fructose were explored with some success for HMF production in ILs with suitable selection of catalysts for dehydration and isomerization. Studies suggested that glucose needs to be transformed to fructose prior to conversion to HMF. Boric acid was used as an isomerization promoter instead of $CrCl_3$ for environmental considerations. Boric acid stabilizes intermediates in the isomerization of glucose to fructose by forming stable chelate complexes with carbohydrates. However, it was suggested that the optimal boric acid concentration is between 0.8 and 1 equiv. because the acid forms a stronger binding with fructose than with glucose and has adverse effects on HMF formation. Eutectic mixtures were evaluated for the conversion of sugars, native cellulose and lignocellulosic waste to HMF and furfural due to their low cost, thermal stability and relatively low toxicity. ILs are successfully applied in pre-treatment of lignocellulose prior to the enzymatic hydrolysis process for production of sugars. One-pot synthesis of HMF from polysaccharides and lignocellulose was achieved using a selected number of ILs. A cascade process of converting polysaccharides and lignocellulose to HMF reduces operating time, energy and raw material consumption. One-pot synthesis of HMF, however, requires stringent operation conditions to maximize the yield. The major concerns of using ILs in large-scale production applications are cost, purification and toxicity. Commercial applications of HMF rely on the development of isolation techniques to purify the thermally unstable chemicals. Extraction with a low boiling point organic solvent is an alternative but low concentration of HMF in the extract poses another challenge.

References

1. M. J. Climent, A. Corma and S. Iborra, *Green Chem.*, 2011, **13**, 520–540.
2. S. Dutta, S. De and B. Saha, *ChemPlusChem*, 2012, **77**, 259–272.
3. R.-J. van Putten, J. C. van der Waal, E. de Jong, C. B. Rasrendra, H. J. Heeres and J. G. de Vries, *Chem. Rev.*, 2013, **113**, 1499–1597.
4. S. P. Verevkin, V. N. Emel'yanenko, E. N. Stepurko, R. V. Ralys, D. H. Zaitsau and A. Stark, *Ind. Eng. Chem. Res.*, 2009, **48**, 10087–10093.
5. C. Triebl, V. Nikolakis and M. Ierapetritou, *Comput. Chem. Eng.*, 2013, **52**, 26–34.
6. B. Saha, D. Gupta, M. M. Abu-Omar, A. Modak and A. Bhaumik, *J. Catal.*, 2013, **299**, 316–320.
7. N. K. Gupta, S. Nishimura, A. Takagaki and K. Ebitani, *Green Chem.*, 2011, **13**, 824–827.
8. K. R. Vuyyuru and P. Strasser, *Catal. Today*, 2012, **195**, 144–154.
9. S. E. Davis, L. R. Houk, E. C. Tamargo, A. K. Datye and R. J. Davis, *Catal. Today*, 2011, **160**, 55–60.
10. S. E. Davis, B. N. Zope and R. J. Davis, *Green Chem.*, 2012, **14**, 143–147.
11. S. Albonetti, T. Pasini, A. Lolli, M. Blosi, M. Piccinini, N. Dimitratos, J. A. Lopez-Sanchez, D. J. Morgan, A. F. Carley, G. J. Hutchings and F. Cavani, *Catal. Today*, 2012, **195**, 120–126.
12. J. Ohyama, A. Esaki, Y. Yamamoto, S. Arai and A. Satsuma, *RSC Adv.*, 2013, **3**, 1033–1036.
13. A. Cukalovic and C. V. Stevens, *Green Chem.*, 2010, **12**, 1201–1206.
14. S. P. Teong, G. Yi and Y. Zhang, *Green Chem.*, 2014, **16**, 2015–2026.
15. B. Saha and M. M. Abu-Omar, *Green Chem.*, 2014, **16**, 24–38.
16. A. A. Rosatella, S. P. Simeonov, R. F. Frade and C. A. Afonso, *Green Chem.*, 2011, **13**, 754–793.
17. M. E. Zakrzewska, E. Bogel-Lukasik and R. Bogel-Lukasik, *Chem. Rev.*, 2011, **111**, 397–417.
18. R. Sanderson, D. Schneider and I. Schreuder, *J. Appl. Polym. Sci.*, 1994, **53**, 1785–1793.
19. E. Taarning, C. M. Osmundsen, X. Yang, B. Voss, S. I. Andersen and C. H. Christensen, *Energy Environ. Sci.*, 2011, **4**, 793–804.
20. C. A. Antonyraj, J. Jeong, B. Kim, S. Shin, S. Kim, K.-Y. Lee and J. K. Cho, *J. Ind. Eng. Chem.*, 2013, **19**, 1056–1059.
21. M. Bicker, J. Hirth and H. Vogel, *Green Chem.*, 2003, **5**, 280–284.

22. A. Eerhart, A. Faaij and M. K. Patel, *Energy Environ. Sci.*, 2012, **5**, 6407–6422.
23. J. Ma, Y. Pang, M. Wang, J. Xu, H. Ma and X. Nie, *J. Mater. Chem.*, 2012, **22**, 3457–3461.
24. S. Q. Hu, Z. F. Zhang, Y. X. Zhou, J. L. Song, H. L. Fan and B. X. Han, *Green Chem.*, 2009, **11**, 873–877.
25. T. Tuercke, S. Panic and S. Loebbecke, *Chem. Eng. Technol.*, 2009, **32**, 1815–1822.
26. R. L. de Souza, H. Yu, F. Rataboul and N. Essayem, *Challenges*, 2012, **3**, 212–232.
27. J. Jeong, C. A. Antonyraj, S. Shin, S. Kim, B. Kim, K.-Y. Lee and J. K. Cho, *J. Ind. Eng. Chem.*, 2013, **19**, 1106–1111.
28. B. R. Caes and R. T. Raines, *ChemSusChem*, 2011, **4**, 353–356.
29. Y. Qu, C. Huang, J. Zhang and B. Chen, *Bioresour. Technol.*, 2012, **106**, 170–172.
30. K. B. Sidhpuria, A. L. Daniel-da-Silva, T. Trindade and J. A. Coutinho, *Green Chem.*, 2011, **13**, 340–349.
31. F. Yang, Q. Liu, M. Yue, X. Bai and Y. Du, *Chem. Commun.*, 2011, **47**, 4469–4471.
32. J. B. Binder and R. T. Raines, *J. Am. Chem. Soc.*, 2009, **131**, 1979–1985.
33. E. Nikolla, Y. Román-Leshkov, M. Moliner and M. E. Davis, *ACS Catal.*, 2011, **1**, 408–410.
34. M. Chidambaram and A. T. Bell, *Green Chem.*, 2010, **12**, 1253–1262.
35. M. Yasuda, Y. Nakamura, J. Matsumoto, H. Yokoi and T. Shiragami, *Bull. Chem. Soc. Jpn.*, 2011, **84**, 416–418.
36. H. Zhao, J. E. Holladay, H. Brown and Z. C. Zhang, *Science*, 2007, **316**, 1597–1600.
37. C. Wang, L. Fu, X. Tong, Q. Yang and W. Zhang, *Carbohydr. Res.*, 2012, **347**, 182–185.
38. X. Tong and Y. Li, *ChemSusChem*, 2010, **3**, 350–355.
39. A. H. Jadhav, H. Kim and I. T. Hwang, *Catal. Commun.*, 2012, **21**, 96–103.
40. S. Lima, P. Neves, M. M. Antunes, M. Pillinger, N. Ignatyev and A. A. Valente, *Appl. Catal., A*, 2009, **363**, 93–99.
41. F. Yang, Q. Liu, X. Bai and Y. Du, *Bioresour. Technol.*, 2011, **102**, 3424–3429.
42. P. Mäki-Arvela, E. Salminen, T. Riittonen, P. Virtanen, N. Kumar and J.-P. Mikkola, *Int. J. Chem. Eng.*, 2012, **2012**, 1–10.
43. B. R. Caes, M. J. Palte and R. T. Raines, *Chem. Sci.*, 2013, **4**, 196–199.
44. M. Ohara, A. Takagaki, S. Nishimura and K. Ebitani, *Appl. Catal., A*, 2010, **383**, 149–155.
45. A. Yepez, A. Pineda, A. Garcia, A. A. Romero and R. Luque, *Phys. Chem. Chem. Phys.*, 2013, **15**, 12165–12172.
46. X. H. Qi, M. Watanabe, T. M. Aida and R. L. Smith, *Catal. Commun.*, 2009, **10**, 1771–1775.
47. T. Ståhlberg, W. Fu, J. M. Woodley and A. Riisager, *ChemSusChem*, 2011, **4**, 451–458.

48. J. N. Chheda, Y. Román-Leshkov and J. A. Dumesic, *Green Chem.*, 2007, **9**, 342–350.

49. Y. Román-Leshkov, J. N. Chheda and J. A. Dumesic, *Science*, 2006, **312**, 1933–1937.

50. K.-i. Shimizu, R. Uozumi and A. Satsuma, *Catal. Commun.*, 2009, **10**, 1849–1853.

51. C. Li, Z. K. Zhao, A. Wang, M. Zheng and T. Zhang, *Carbohydr. Res.*, 2010, **345**, 1846–1850.

52. H. Ma, B. Zhou, Y. Li and D. S. Argyropoulos, *BioResources*, 2012, 7, 0533–0544.

53. X. Qi, M. Watanabe, T. M. Aida and J. R. L. Smith, *Green Chem.*, 2009, **11**, 1327–1331.

54. D. Ray, N. Mittal and W.-J. Chung, *Carbohydr. Res.*, 2011, **346**, 2145–2148.

55. F. Ilgen, D. Ott, D. Kralisch, C. Reil, A. Palmberger and B. König, *Green Chem.*, 2009, **11**, 1948–1954.

56. X. Tong, Y. Ma and Y. Li, *Carbohydr. Res.*, 2010, **345**, 1698–1701.

57. S. Bali, M. A. Tofanelli, R. D. Ernst and E. M. Eyring, *Biomass Bioenergy*, 2012, **42**, 224–227.

58. T. Ståhlberg, S. Rodriguez-Rodriguez, P. Fristrup and A. Riisager, *Chem.-Eur. J.*, 2011, **17**, 1456–1464.

59. R. F. Shih, H. Y. Hsu and J. J. Wong, US Pat. 20130023679, 2013.

60. X. Zhou, Z. Zhang, B. Liu, Z. Xu and K. Deng, *Carbohydr. Res.*, 2013, **375**, 68–72.

61. F. Guo, Z. Fang and T.-J. Zhou, *Bioresour. Technol.*, 2012, **112**, 313–318.

62. C. Li and Z. K. Zhao, *Adv. Synth. Catal.*, 2007, **349**, 1847–1850.

63. F. Liu, R. K. Kamat, I. Noshadi, D. Peck, R. S. Parnas, A. Zheng, C. Qi and Y. Lin, *Chem. Commun.*, 2013, **49**, 8456–8458.

64. N. Villandier and A. Corma, *Chem. Commun.*, 2010, **46**, 4408–4410.

65. Z. Zhang and Z. K. Zhao, *Bioresour. Technol.*, 2010, **101**, 1111–1114.

66. C. Kim, H. J. Ryu, S. H. Kim, J.-J. Yoon, H. S. Kim and Y. J. Kim, *Bull. Korean Chem. Soc.*, 2010, **31**, 511–514.

67. S. Dutta, S. De, M. I. Alam, M. M. Abu-Omar and B. Saha, *J. Catal.*, 2012, **288**, 8–15.

68. P. Wang, H. Yu, S. Zhan and S. Wang, *Bioresour. Technol.*, 2011, **102**, 4179–4183.

69. B. R. Caes, J. B. Binder, J. J. Blank and R. T. Raines, *Green Chem.*, 2011, **13**, 2719–2722.

70. X. Guo, Q. Cao, Y. Jiang, J. Guan, X. Wang and X. Mu, *Carbohydr. Res.*, 2012, **351**, 35–41.

71. Y. Zhang, E. A. Pidko and E. J. Hensen, *Chem.-Eur. J.*, 2011, **17**, 5281–5288.

72. L. Hu, Y. Sun and L. Lin, *Ind. Eng. Chem. Res.*, 2012, **51**, 1099–1104.

73. C. Lansalot-Matras and C. Moreau, *Catal. Commun.*, 2003, **4**, 517–520.

74. L. Zhou, R. Liang, Z. Ma, T. Wu and Y. Wu, *Bioresour. Technol.*, 2013, **129**, 450–455.

75. H. Abou-Yousef and P. Steele, *J. Fuel Chem. Technol.*, 2013, **41**, 214–222.

Ionic Liquids as Efficient Tools for the Purification of Biomolecules and Bioproducts from Natural Sources

MATHEUS M. PEREIRA[a], JOÃO A. P. COUTINHO[a], AND MARA G. FREIRE*[a]

[a]CICECO-Aveiro Institute of Materials, Chemistry Department, University of Aveiro, 3810-193 Aveiro, Portugal
*E-mail: maragfreire@ua.pt

8.1   Introduction

Refineries, including petroleum refineries, are production facilities composed of a group of unit operations with the goal of refining raw materials or converting them into added-value products (fuels and chemicals). Crude oil has thousands of organic compounds in its composition, which are separated during the process, in accordance with market specifications, and that can be further converted into many valuable products.[1] In addition to being a finite source of raw materials, environmental problems associated with the production and consumption of fossil fuels, coupled with the need to seek more economically viable and environmentally sustainable alternatives, has prompted research on the use of biomass for the production of energy, chemicals and materials.[2]

RSC Green Chemistry No. 36
Ionic Liquids in the Biorefinery Concept: Challenges and Perspectives
Edited by Rafal Bogel-Lukasik

Published by the Royal Society of Chemistry, www.rsc.org

New opportunities for a more sustainable future based on renewable raw materials (*i.e.* biomass) for the production of bioproducts have been recognized.[3] Over the past few years, there has been a huge increase in research towards new materials, chemicals and fuels from renewable resources to reduce the dependence on fossil fuels (oil, natural gas and coal), along with the environmental impact associated with greenhouse gas emissions. Under this scenario, the concept of the biorefinery has attracted particular attention from the scientific community.[4] Biorefineries are industrial plants that integrate processes and equipment for a sustainable conversion of biomass into energy, fuels and chemicals.[5] Although the term 'biorefinery' is a recent hot topic of research, this concept is not new. The novelty appears in the search for a more complete use of biomass (without waste generation) and the production of a wider range of products (fuels, materials and chemicals) associated with a combination of technologies and processes (physical, chemical, biochemical and thermal) in an integrated approach similar to that used in oil refineries.[6] The facilities in biorefineries seek thereby to manage the use of biomass in order to maximize the production of energy and fuels coupled with the extraction/production of other platform chemicals while minimizing wastes.[5]

Nowadays, biomass is mostly used for production of pulp and paper and for combustion with the purpose of generating energy (heat and electricity). The thermal conversion of biomass into energy is responsible for over 97% of bioenergy production.[7] In general, biomass is composed of cellulose, hemicelluloses and lignin; yet, a large variety of extractable compounds can be found in its composition, such as proteins, lipids, vitamins, antioxidants and others natural compounds.[8–13] These can be divided into high value/low volume and low value/high volume compounds.[5] Currently, it is known that biomass is a potential source of biologically active natural products which play a significant role in the prevention and treatment of various diseases. More than 50 000 plant species are used for medicinal purposes and over 25% of approved drugs come from natural bioactive compounds. There are many examples of plant-derived drugs with remarkable antioxidant (*e.g.* liquiritin), analgesic (*e.g.* morphine), antiviral, anti-inflammatory, antitumoral (*e.g.* vincristine and taxol) and antimalarial (*e.g.* artemisinin) properties. In addition, in many cases, the isolation of these compounds from natural sources remains the only viable option because of the great complexity of these structures and the difficulty in synthesizing them.[14] However, there is a huge need for the exploration of new sources of bioactive compounds, since their large-scale supply is declining, either due to the limited availability of plants or their competition with other applications,[15] as well as on the development of cost-effective and more environmentally benign extraction processes.

In order to completely take advantage of the biomass-rich composition, the strategy for the recovery of these high-value compounds is mostly based on solid–liquid extraction processes.[16] For the extraction of compounds from biomass, volatile organic compounds (VOCs) are commonly used as solvents.[17–19] Nevertheless, the use of hazardous, toxic and sometimes carcinogenic VOCs in biorefineries goes against the principle of sustainable processes.[20] Furthermore, conventional extraction processes employing VOCs still present several drawbacks, such as a low yields, poor selectivity, are time-consuming or require a high energetic input, and may lead to the degradation of the targeted compounds since high temperatures are often employed. In this context, novel approaches have been proposed to address the use of cost-effective and safer solvents. Some examples include the use of solvents produced from renewable resources, water, supercritical fluids and, recently, ionic liquids (ILs).[21]

Ionic liquids belong to the molten salts group, with melting temperatures below 100 °C, usually composed of an organic cation and an organic or inorganic anion. The large dimensions of their ions do not allow their organization in a crystal structure and, therefore, these salts are liquid at lower temperatures than conventional salts. Due to their ionic character, ILs display a set of remarkable features, such as negligible volatility at atmospheric conditions, high thermal and chemical stabilities, and a strong solvation capability for a large variety of compounds, which allowed them to be proposed as suitable replacements for VOCs in diverse applications.[22–24] In addition to their 'green solvents' character, ILs are usually seen as 'designer solvents' since there is a large number of cation/anion combinations. This gives us the capacity of tuning their properties, such as their biodegradability or toxicological features, as well as their selective extractive potential when their use in extractions is envisaged.[25–27]

With the appearance of air- and water-stable ILs in the beginning of the 21st century, the research on the synthesis and application of novel ILs increased significantly. A new era on the study of ILs as candidates for the extraction of added-value compounds from natural sources has emerged. Their unique properties, which can be tailored, and their ability to solvate a large array of compounds, coupled with the need of finding more cost-effective and 'greener' solvents are the major reasons behind this tendency. Nowadays, amongst the vast variety of ILs that can be synthesized, the most commonly studied cations are imidazolium, phosphonium, ammonium, piperidinium, pyridinium and pyrrolidinium based, combined with anions such as chloride ([Cl]), bromide ([Br]), acetate ([CH$_3$CO$_2$]), bis(trifluoromethylsulfonyl)imide ([NTf$_2$]), hexafluorophosphate ([PF$_6$]) and tetrafluoroborate ([BF$_4$]). Nonetheless, the development of novel ILs is moving away from these fluorinated anions, some presenting poor water-stability,[28] towards renewable, non-toxic and biodegradable alternatives, namely based on carboxylic acids, amino acids and mandelic acid-derived anions,[29,30] often combined with the cholinium cation.[31,32]

In the past decade, much interest has been devoted to ILs and IL/molecular solvent mixtures as alternative solvents for extracting added-value compounds from biomass.[33] In general, ILs demonstrated superior performance leading to higher extraction yields and at moderate conditions (time of extraction, temperature and pressure) reducing therefore the overall costs of the IL-mediated process.[33] The pioneering work on the use of ILs as alternative solvents for solid–liquid extractions from biomass was published in 2007, by Du *et al.*,[34] which demonstrated the successful application of IL aqueous solutions in the microwave-assisted extraction (MAE) of *trans*-resveratrol from a Chinese traditional medicine herb. After this proof of principle, the number of works regarding the use of ILs for the extraction of high-value compounds increased significantly.[33] In general, either pure ILs or water–IL and alcohol–IL mixtures have been employed. Although a large number of different cation cores are available nowadays, researchers still focus on imidazolium-based fluids for this purpose. Within this group, halogen- and $[BF_4]$-based ILs were the most commonly employed.[33] Although it is well-established that $[BF_4]$-based ILs are not water stable,[28] even at room temperature, this anion remains the preferred choice of many authors – a tendency that urgently needs to be stopped and reversed. Nowadays, we are at an era where biodegradable and biocompatible ILs can be synthesized at low cost, meaning that further studies on this field should be focused on these more environmentally friendly alternatives. Actually, up to present, few works based on more biodegradable and non-toxic ILs for the extraction of added-value compounds from biomass can be found.[35–42]

The compounds extracted were mostly alkaloids, flavonoids and terpenoids (these representing 50% of all fine chemicals investigated).[33] However, after the extraction, the purification and isolation of the target compounds from complex extracts, as well as the recovery of the solvent, still need to be accomplished. Here, we present and discuss two potential strategies for purification of the products, namely using IL-supported materials or by liquid–liquid extractions employing aqueous biphasic systems. After isolation of the products, the viability of the solvents recovery and reusability should be also attempted, aiming at decreasing both the cost and environmental footprint of the whole process. In this context, recent advances on the ILs recovery are also outlined.

8.2 Purification of Added-Value Compounds Using SPE- or LLE-Based Techniques

8.2.1 Solid-Phase Extraction (SPE)

In addition to the well-established enhanced ability of ILs for the extraction of added-value compounds from biomass, the isolation/purification of the target compounds from the IL-based solvents still remains a challenge. Indeed, although a large amount of work reports the optimization of the extraction procedures, less than half of the authors studied the possibility of isolating the valuable compounds from the final IL solution.[33] Nevertheless, this step is of vital relevance when foreseeing the scale-up of an integrated extraction–recovery process. Albeit IL-mediated extractions display a high performance, the non-volatile nature of the aprotic ILs is a major drawback because a simple evaporation step cannot be applied to eliminate the solvent, aiming at recovering the extracts. Instead, protic ILs of high volatility can be applied, as successfully demonstrated by Chowdhury *et al.*[41] Nevertheless, this method is only applicable if the IL presents a high selectivity for the target compound, otherwise additional purification steps will be required. The isolation methods most applied to IL-based solvents consist of back-extractions using organic solvents, evaporation of the solvents or compounds (when applicable, for instance with essential oils) and precipitation with anti-solvents.[33] Other techniques, such as the use of macroporous materials, anion-exchange resins and supported ILs materials, were also proposed.[33]

Solid-phase extraction (SPE) is a pre-concentration and purification technique where solid materials are used as adsorbents. In order to increase the yields and purity afforded by solid-phase extractions, ILs were introduced to modify the chemical and physical characteristics of the adsorbent materials.[43] Supported IL phases (SILPs) combined the chemical functionalities provided by ILs with the advantages of solid supports, and led to significant improvements in the separation of bioactive compounds extracted from natural sources. SILPs in SPE can be classified as silica or polymer based, depending on the adsorbent material to which the ILs are covalently bonded. It should be highlighted that the liquid state of ILs is lost when immobilized onto a solid support. Nevertheless, under these conditions, a multitude of interactions provided by the ILs can be still exploited. A schematic representation of the surface and preparation route of these supported materials is depicted in Figure 8.1.

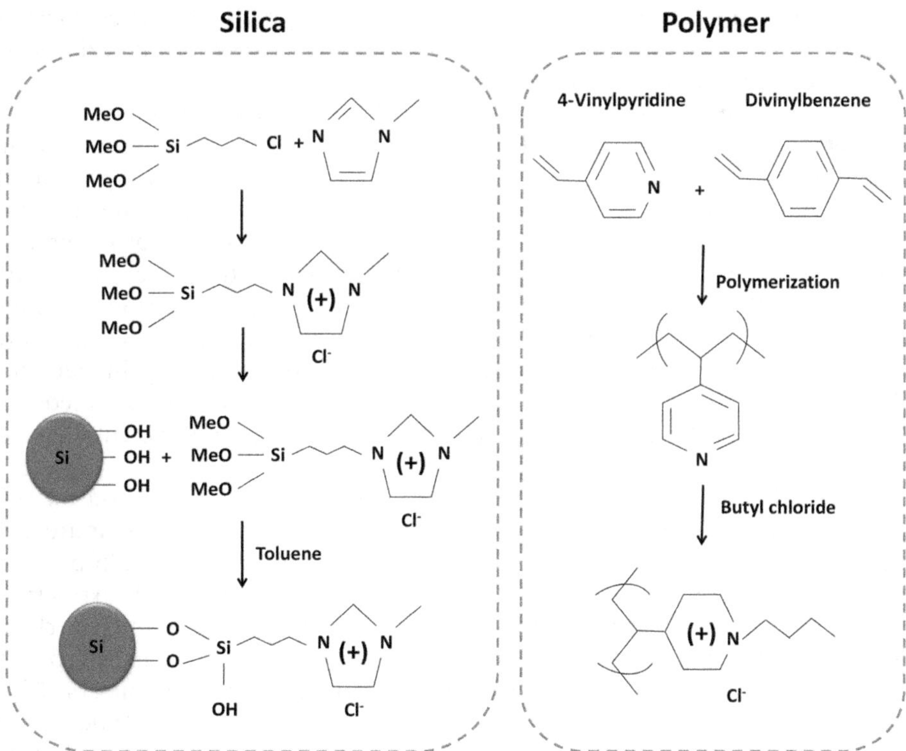

Figure 8.1   Preparation routes of IL-modified silica and IL-modified polymers.[44,45]

The first study comprising IL-modified silica for SPE was published in 2009 by Tian *et al.*[44] A methylimidazolium-modified silica, with chloride as counter ion, was used to isolate three tanshinone compounds from *Salvia miltiorrhiza* Bunge.[44] This traditional medicinal herb is currently used for the treatment of various diseases, particularly cardiovascular diseases.[45] Tanshinones, including tanshinone I, tanshinone IIA and cryptotanshinone, whose structures are depicted in Figure 8.2, display anti-cancer effects and are the major active constituents in *Salvia miltiorrhiza* Bunge.[44] In addition to the IL-supported material, the authors also used a commercial silica cartridge for comparison purposes, while demonstrating the highest recovery values attained with the SILP – due to more favourable interactions of the IL-modified silica with tanshinones. Extracted amounts of 0.0079, 0.18 and 0.12 mg g^{-1} for cryptotanshinone, tanshinone I and tanshinone IIA, respectively, have been reported. The same researchers[46] proposed the synthesis of a new IL-based silica sorbent to isolate liquiritin and glycyrrhizic acid from licorice (the root of the *Glycyrrhiza* plant species). Liquiritin is a flavonoid with antiviral and antioxidant properties, whereas glycyrrhizic acid has anti-inflammatory, anti-ulcer, anti-hepatotoxic and antiviral activities.[47] The molecular structures of liquiritin and glycyrrhizic acid are shown in Figure 8.2. Different washing and elution solvents, such as water, methanol and methanol/water mixtures were also investigated.[46] Furthermore, the IL-based silica sorbent was compared with traditional C_{18} sorbents and demonstrated to provide a higher selectivity. Extracted amounts of 0.18 mg g^{-1} and 1.0 mg g^{-1} for liquiritin and glycyrrhizic acid, respectively, were reported.[46]

The same group of authors[48] applied different IL-based silica materials for the SPE of three phenolic acids (protocatechuic, ferulic and caffeic acids) from *Salicornia herbacea* L., whose chemical structures are presented in Figure 8.2. Phenolic compounds display chemopreventive properties, *e.g.* antioxidant, anticarcinogenic, antimutagenic and anti-inflammatory effects, and also contribute to induce apoptosis by arresting the cell cycle and by inhibiting DNA binding and cell adhesion, migration, proliferation or differentiation.[49] An adequate sorbent for the phenolic acids extraction was firstly evaluated, and then the plant extracts were applied with the optimized sorbent and solvent to achieve multiphase dispersive extraction. The recovery yields obtained of purified protocatechuic acid, ferulic acid and caffeic acid were 94.69%, 79.09% and 87.32%, respectively.[48]

IL-based silica materials were also applied in a two-step method for the extraction and separation of the alkaloid oxymatrine, depicted in Figure 8.2, from a *S. flavescens* Ait. extract.[50] Alkaloids are amongst the most important chemical compounds present in plants, and the interest in their extraction arises from their pharmacological activities, such antifungal, antidiarrheal and anti-inflammatory characteristics.[51] Some of them, such as morphine and codeine, also display structural similarities with neurotransmitters of the human central nervous system and act as powerful analgesics. Indeed, the pioneering work where ILs were employed in the extraction of added-value compounds from natural sources was carried out with MAE, published in

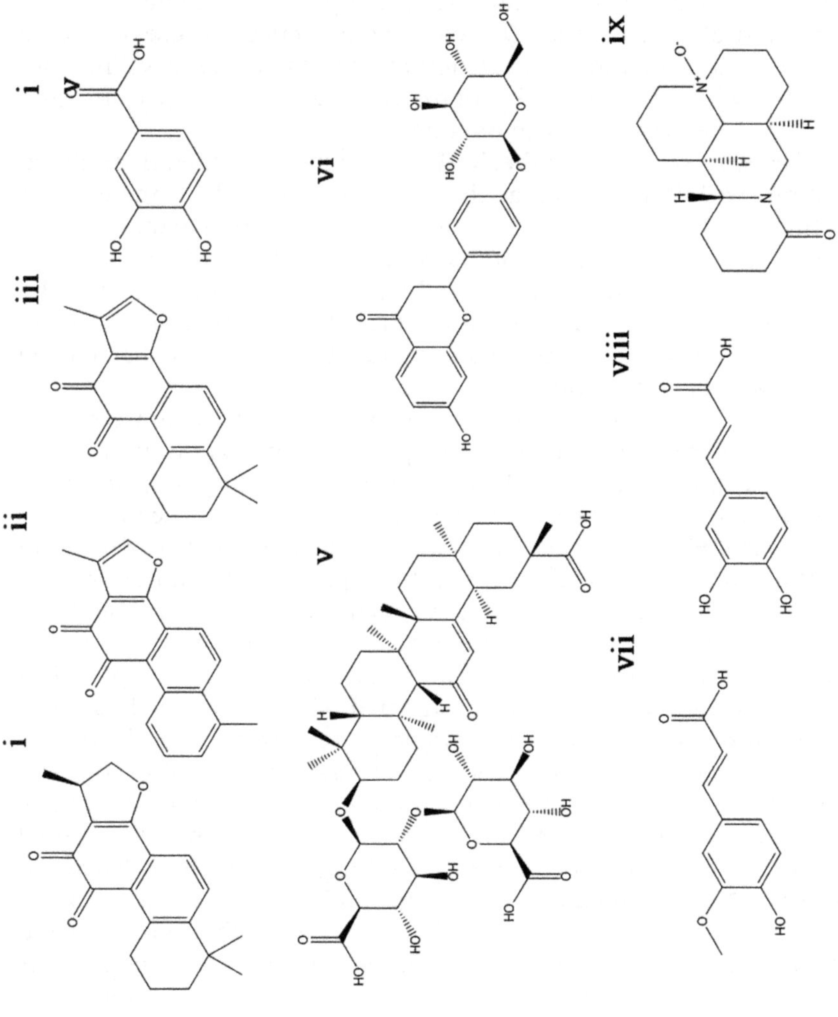

Figure 8.2   Chemical structures of the purified added-value compounds using IL-modified silica: (i) cryptotanshinone; (ii) tanshinone I; (iii) tanshinone II A; (iv) protocatechuic acid; (v) glycyrrhizic acid; (vi) liquiritin; (vii) ferulic acid; (viii) caffeic acid; and (ix) oxymatrine.

2007 by Du *et al.*[34] For purification and isolation purposes, the IL-based silica material was firstly mixed with the aqueous plant extract to adsorb oxymatrine.[50] The obtained suspension was then added to a cartridge containing the SILP for SPE.[50] Through these two-steps approach, the target alkaloid was separated from major interferences with a 93.4% recovery. Then, water was selected as the most adequate washing solvent for the removal of interferences, while most of the oxymatrine could be eluted from the SILP sorbent with methanol. The authors concluded that, compared with traditional SPE, the combined strategy accelerates the loading and reduces the use of organic solvents during washing. Finally, the authors evaluated the recycling ability of the SILP, where recovery yields of oxymatrine over four cycles ranging between 93.4 and 89.7% have been obtained.[50]

Row and co-workers[52] also synthesized a molecular imprinted IL-modified silica material with imidazolium as the cation and chloride as the counter ion with the aim of increasing the selectivity in the extraction of tanshinones from *Salvia miltiorrhiza* Bunge. A comparison with molecular non-imprinted IL-modified silica, commercial silica and C_{18} cartridges revealed that the molecular-imprinted-functionalized silica provides the highest selectivity towards tanshinones due to its organized functional groups that enable stronger and specific interactions.[52]

Some of the disadvantages of silica, such as the high price, low coverage of the functional groups and its restricted pH stability might limit its widespread use. These drawbacks can be circumvented by using polymers with high selectivity and a wide operating pH range. In addition, the properties of polymers as SPE materials can also be improved by their modification with ILs. Row and co-workers[53] studied the applicability of an aminopropylimidazolium polymer as a sorbent material for the extraction and isolation of the main active alkaloids, matrine and oxymatrine, from *Sophora flavescens* Ait. The chemical structures of both alkaloids are depicted in Figure 8.3. The raw samples were oven-dried, sliced and crushed, after which the powdered particles were subjected to extraction with water.[53] The aqueous solution containing the extract was then loaded onto a SPE cartridge and subsequently washed with ethanol, acetonitrile, methanol and methanol/triethylamine mixtures. Compared with conventional sorbents, the IL-modified polymer exhibited higher selectivity. Extracted amounts of 2.42 and 13.69 mg g^{-1} of matrine and oxymatrine, respectively, were achieved with the IL–polymer sorbent. The recyclability of the solid material was also ascertained, demonstrating that the extracted amounts of the target compounds do not significantly decrease.[53] The same group of researchers[54] studied the applicability of a methylimidazolium-polymer for the isolation of caffeine and theophylline, two alkaloids shown in Figure 8.3, from green tea extracts. A comparison of different SPE cartridges using the non-modified polymer and C_{18} showed that the highest recovery was obtained using the IL-modified polymer sorbent, maybe due to stronger interactions established between both alkaloids and the imidazolium-based material. The extracted amounts of caffeine and theophylline were 0.025 mg mL^{-1} and 0.58 mg mL^{-1}, respectively.[54]

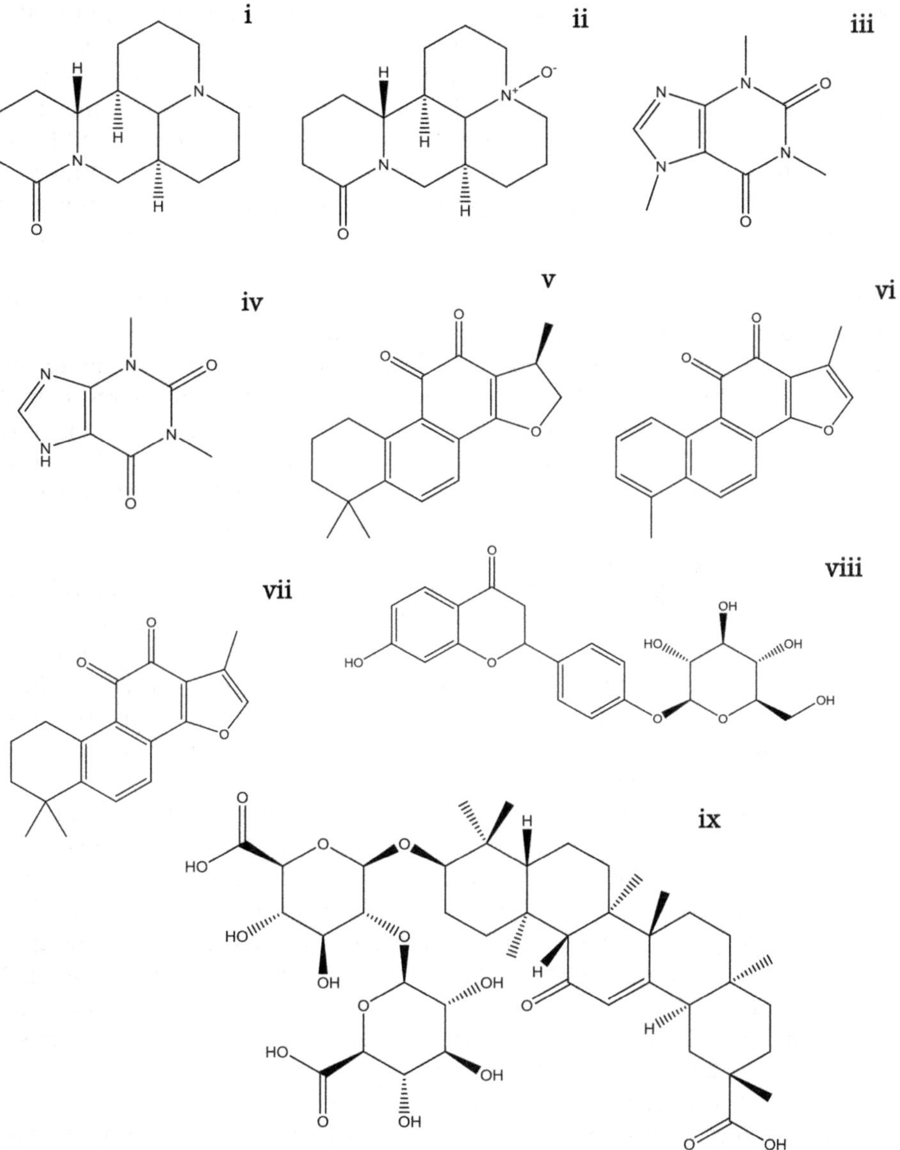

Figure 8.3   Chemical structures of the purified added-value compounds using IL-modified polymers: (i) matrine; (ii) oxymatrine; (iii) caffeine; (iv) theophylline; (v) cryptotanshinone; (vi) tanshinone I; (vii) tanshinone II A; (viii) liquiritin; and (ix) glycyrrhizin.

The molecular imprinting technique was also applied to IL-modified polymers using 9,10-phenanthrenequinone as the template[52] – similar to the work described before on the synthesis of IL-modified silica using tanshinones as the template. Five IL-modified porous polymers with different imidazolium-based functional groups were investigated by the authors.[52]

Adsorption isotherms were initially determined to work out the main interactions occurring between the polymers and the target compounds, revealing that the polymer with a carboxyl group displays the highest ability to selectively extract tanshinone I, tanshinone IIA and cryptotanshinone. Their chemical structures are depicted in Figure 8.3. Under the optimized SPE conditions (0.4 mL of loading volume, 4.0 mL of water/methanol (30:70, v/v) as washing solvent and 2.0 mL of methanol as elution solvent), 0.35 mg g^{-1} of cryptotanshinone, 0.33 mg g^{-1} of tanshinone I, and 0.27 mg g^{-1} of tanshinone IIA were obtained from the plant extract.[52]

Polymer-confined ILs were used for the separation of phenolic acids with antioxidant activity from natural plant extracts of *Salicornia herbacea* L. using an anion-exchange mechanism.[55] The phenolic acids investigated are shown in Figure 8.3. The functionalized polymers were synthesized using a molecular imprinting technique to reduce ion–ion interactions during anion exchange and other interactions with interference substances that could decrease the selectivity. The synthesized IL-based molecularly imprinted polymer led to high recoveries from the plant aqueous extract, namely 90.1% for protocatechuic acid, 95.5% for ferulic acid and 96.6% for caffeic acid.[55] These values are higher than those previously discussed and gathered by the same group of authors employing IL–silica particles (94.69%, 79.09% and 87.32% for protocatechuic acid, ferulic acid and caffeic acid, respectively[50]), demonstrating therefore the highest performance of polymer-modified material to recover added-value compounds from biomass extracts. Moreover, the authors confirmed the separation of the three phenolic acids by repeated solid-phase extraction cycles.[55]

An alkylpyridinium polymer material was also prepared for the isolation of liquiritin and glycyrrhizin from liquorice extract,[56] with their chemical structures provided in Figure 8.3. The porous alkylpyridinium polymer sorbent was compared with the C$_{18}$ conventional sorbent, showing that the pyridinium-based polymer displays a higher selectivity. In this study, 2.75 mg g^{-1} of liquiritin and 4.5 mg g^{-1} of glycyrrhizin were attained.[56] These values are substantially higher than the amounts described before, 0.18 mg g^{-1} and 1.0 mg g^{-1} for liquiritin and glycyrrhizic acid, respectively, using an IL-based silica sorbent reported by the same group of authors.[46]

In general, imidazolium-based ILs have been the most used ILs for the modification of silica or polymers in SPE. Only one report was found with the modification of the chemical properties of the polymer by introducing pyridinium-based ILs.[56] Since most of the added-value compounds extracted from biomass are aromatic, it seems that the aromatic imidazolium ring allows specific interactions with the target compounds resulting in high extraction yields and selectivity. Nevertheless, other cations, such as tetraalkylphosphonium and tetraalkylammonium, should be investigated to further confirm this trend. On the other hand, the counter anions of IL-modified materials mostly comprise chloride, bromide, hexafluorophosphate and tetrafluoroborate. A large number of different anions are nowadays available that deserve to be explored, aiming at working out the cation and anion influence through the isolation of target compounds from natural extracts.

8.2.2   Liquid–Liquid Extraction (LLE)

Aqueous solutions of several imidazolium-, quaternary ammonium- and pyrrolidinium-based ILs were used in the extraction of alkaloids, antioxidants and terpenoids.[33] The main reasons for this choice are related to the primary advantages afforded by IL–water mixtures due to the reduction of the viscosity of the extraction solvent, *i.e.* the problems derived from the high viscosity of most ILs are overcome by their use as mixtures with a low viscous fluid, such as water, thus enhancing the mass transfer and reducing energy consumption. Furthermore, the most green and low-cost solvent – water – is introduced while increasing the extraction selectivity for target high-value compounds, such as alkaloids that are moderately polar molecules. In addition, imidazolium-based fluids were the most used ILs, combined with chloride, bromide, acetate and tetrafluoroborate anions.[33] All of the ILs investigated are completely miscible with water and so can be used in the formation of aqueous biphasic systems (ABS) attempting at the purification of the extracted added-value compounds. These liquid–liquid systems are mostly composed of water and are formed by polymer–polymer, polymer–salt or salt–salt combinations dissolved in aqueous media, that above given concentrations separate into two aqueous-rich phases.[57] In 2003 Rogers and co-workers[58] demonstrated that ABS can be also formed by the combination of ILs and inorganic salts. Since this proof of principle, several works appeared in the literature revealing that IL-based ABS can be created with a wide plethora of salts, carbohydrates, amino acids and polymers.[59] The macroscopic appearance of a given ABS composed of an IL and an inorganic salt is provided in Figure 8.4.

The major advantage of IL-based ABS stands on their possibility of tailoring the phases' polarities so that selective and effective extractions, *i.e.* purification of the most diverse added-value compounds, can be straightforwardly

Figure 8.4   Macroscopic appearance of a given ABS composed of an IL and an inorganic salt. A synthetic dye was added to color the IL-rich phase.

achieved.[60] Furthermore, IL-based ABS can be applied as an integrated strategy by employing the aqueous solutions of ILs already used in the extraction of the most diverse compounds from biomass. On the other hand, IL-based ABS are of lower viscosity than pure ILs, and have already been studied in an attempt to purify crude extracts from biomass.[61–63] Although out of the scope of this literature overview, IL-based ABS have already been demonstrated to be a powerful route for the purification of exopolysaccharide antibiotics[64] and dyes[65] produced by fermentation, confirming therefore their potential as purification strategies when applied to complex matrices. Even though it has been with model systems, the performance of IL-based ABS has been widely investigated for the extraction of alkaloids,[66–68] antioxidants[69–72] and terpenoids.[73]

IL-based ABS have been applied to the extraction of a vast number of alkaloids, including codeine,[66] papaverine,[66] caffeine,[67,68] nicotine,[67,68] theophylline[68] and theobromine.[68] Their chemical structures are illustrated in Figure 8.5. The two most common methods for the purification of alkaloids comprise LLE and SPE, although using conventional approaches and materials.[74] While SPE presents good purification and concentration effects, it requires a relatively time consuming solvent desorption step (using traditional volatile organic solvents) and pre-treatment processes. LLE inherently involves the use of toxic and volatile organic compounds and the sample recovery is not always adequate. Accordingly, the development of cost-effective and environmentally friendly isolation methods for alkaloids is of great interest.

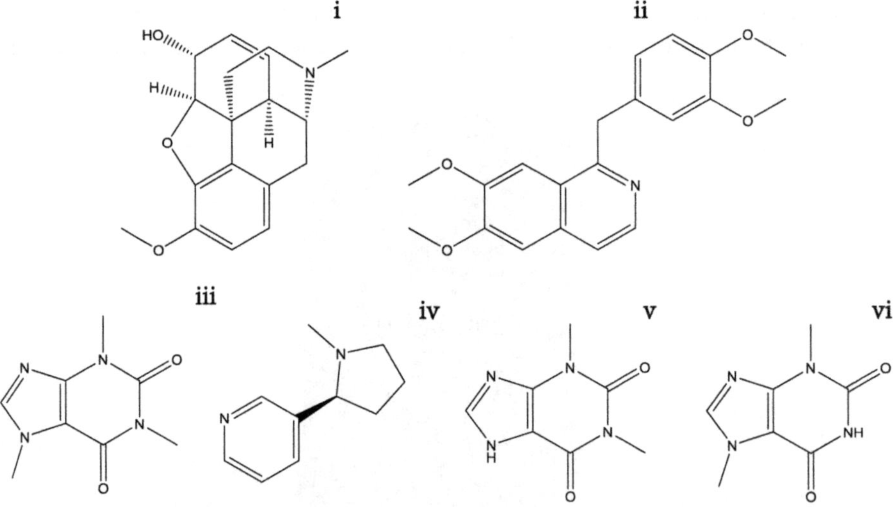

Figure 8.5   Chemical structures of the alkaloids extracted with IL-based ABS: (i) codeine; (ii) papaverine; (iii) caffeine; (iv) nicotine; (v) theophylline; and (vi) theobromine.

Li *et al.*,[66] in 2005, reported the pioneering application of IL-based ABS as a pre-treatment/extraction strategy in the analysis of opium alkaloids (codeine and papaverine). Under optimized conditions, the researchers reported extraction efficiencies of 93% for papaverine and 65% for codeine using an ABS formed by [C$_4$mim][Cl] and K$_2$HPO$_4$. The alkaloids in the coexisting phases were quantified by HPLC (high performance liquid chromatography) with no major interferences of the IL or salt employed. Later on, and although with a different aim, Freire *et al.*[67] demonstrated the complete extraction of archetypal alkaloids, such as caffeine and nicotine, achieved in a single step, by a proper tailoring of the IL employed in the ABS formulation and respective compositions. The higher efficiencies reported by Freire *et al.*[67] are a result of the investigation of a large number of ILs, while taking advantage of their tailoring ability, contrary to the work addressed by Li *et al.*[66] where only one IL and one salt were studied. In fact, the wide plethora of ILs available nowadays allows the tailoring of the IL chemical structure which better fits a target extraction, and this could result in a significantly lower consumption of ILs and salts. The study of Freire *et al.*[67] comprised 17 imidazolium-based ILs which formed ABS in the presence of K$_3$PO$_4$ aqueous solutions. In general, nicotine presented higher partition coefficients (when compared to caffeine) due to its higher hydrophobicity (with a methyl pyrrolidine ring) and its preferential affinity for the most hydrophobic (IL-rich) phase. Nevertheless, after a proper tailoring of the phase-forming compositions, the complete extraction of caffeine was also attained in a single step. The extraction efficiency of both alkaloids was shown to be strongly dependent on the IL employed in the formation of ABS, as well as on the composition of the phase-forming components.[67] Subsequent works demonstrated the ability of less-conventional ABS constituted by phosphonium-based ILs and inorganic salts,[73] imidazolium-based ILs and carbohydrates,[75] and imidazolium-based ILs and amino acids[76] to extract caffeine. Generally, the highest extraction efficiencies for the IL-rich phase were obtained with ABS formed by ILs and salts due to their strong salting-out effect which leads to the migration of the alkaloids to the opposite phase.[67] Due to the overall improved results obtained for alkaloids, IL-based ABS are thus viable options to carry out the isolation of caffeine, for instance as a following step in the work reported by Coutinho and co-workers[77] on the extraction of caffeine from guaraná seeds using aqueous solutions of ILs.

Based on the common trend typically observed in the partition coefficients of several added-value compounds,[67,71] where a maximum is attained in IL-based ABS composed of ILs with imidazolium cations of variable alkyl side chain length, Passos *et al.*[68] carried out a systematic study regarding the effect of the IL cation alkyl side chain length through the partitioning of a series of alkaloids of variable hydrophobicity, namely nicotine, caffeine, theophylline and theobromine. They studied ABS constituted by [C$_n$mim][Cl], with n = 4–10, and potassium citrate (at controlled pH) as the salt. The results obtained revealed that, for all the investigated alkaloids, the partition coefficients increase with the cation alkyl chain length up to [C$_6$mim][Cl]. On the other hand, cations with longer aliphatic chains are not favourable for the extraction of alkaloids for the IL-rich phase, inducing a change in the trend on the partition coefficients that start to decrease according to the increase of the aliphatic tail size. The authors[68] also concluded that the pH of the media, and consequently the charged/non-charged state of alkaloids, as well as the hydrophobicity of the molecules, do not alter the pattern observed. The decrease on the partition coefficients was justified by the self-aggregation of ILs with alkyl chains longer than hexyl in aqueous media, which was proved by the authors[68] using transmission electron microscopy (TEM). Up to a given alkyl side chain length the dispersive-type interactions between the alkaloids and the aliphatic moieties of the imidazolium-based cation seem to favour the extraction. With the self-aggregation of ILs, the alkyl chains become less available. This therefore decreases the dispersive-type interactions between these and the target alkaloids while inducing the observed shift in the trend.[68]

The extraction ability of IL-based ABS for antioxidants present in natural matrices, including phenolic aldehydes, *e.g.* vanillin,[70] phenolic acids,[69,71] such as gallic, vanillic and syringic acids, eugenol[72] and propyl gallate,[72] have been also investigated. Their chemical structures are provided in Figure 8.6.

Vanillin is one of the most valued fragrant substances, aiming at creating artificial flavors in a wide range of commercial products, from food to pharmaceutical goods.[78] To avoid the use of organic solvents in purification steps of vanillin-containing extracts, Cláudio *et al.*[70] investigated the application of IL-based ABS formed by a wide variety of imidazolium-based ILs and K_3PO_4. They also studied the effects of the IL chemical structure, the temperature of equilibrium and the initial concentration of vanillin through the antioxidant partitioning. In all situations, it was demonstrated that vanillin preferentially migrates to the IL-rich phase.[70] The partition coefficients

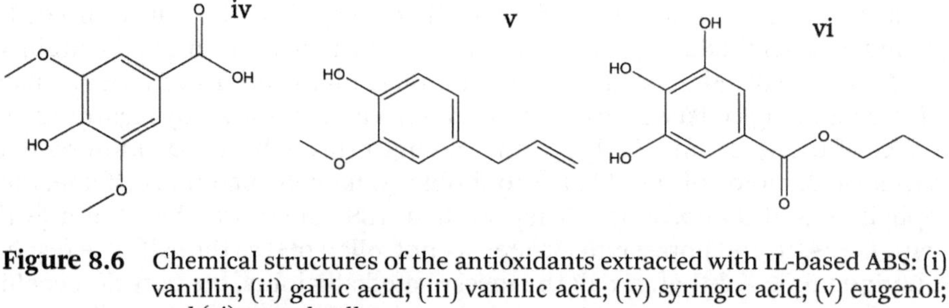

Figure 8.6 Chemical structures of the antioxidants extracted with IL-based ABS: (i) vanillin; (ii) gallic acid; (iii) vanillic acid; (iv) syringic acid; (v) eugenol; and (vi) propyl gallate.

of vanillin at 298 K ranged between 2.72 and 49.59. For ILs with a fixed anion and an imidazolium cation of variable alkyl side chain it was demonstrated that an increase in the size of the aliphatic moiety leads to an increase in the preferential migration of vanillin for the IL-rich phase. However, a maximum was observed at [C_6mim][Cl], followed by a decrease in the partition coefficients up to [C_{10}mim][Cl]. These results are in full agreement with the data provided by Passos *et al.*[68] on the investigation of the partition behavior of a series of alkaloids discussed before. Based on the molar thermodynamic functions of transfer of vanillin, it was concluded that the partition of vanillin results from an interplay between enthalpic and entropic contributions where both the IL anion and more complex cations play a crucial role. The authors[70] also determined and reported the viscosities and densities of the coexisting phases at the compositions for which the partitioning of vanillin was evaluated. On the whole, it was found that the viscosities of the IL-rich phase in all investigated ABS are substantially lower than those observed in typical polymer-based ABS, representing thus an improved advantage when foreseeing the scale-up of these systems.[70]

Gallic acid is present in relatively high concentrations in a large number of biomass sources, from where it could be extracted. It displays antioxidant, anti-inflammatory, antifungal and antitumoral properties.[79] Aiming at developing more benign and efficient extraction/purification processes for gallic acid, ABS formed by a wide variety of ILs and three salts (Na_2SO_4, K_3PO_4 and K_2HPO_4/KH_2PO_4) were investigated by Cláudio *et al.*[70] Several combinations of ILs and salts were used to work out the influence of the IL chemical structure and of the pH of the aqueous medium on the gallic acid partitioning.[71] With few exceptions, the partition coefficients of gallic acid for the IL-rich phase decrease in the following order: Na_2SO_4 (pH *ca.* 3-8) $\gg$ K_2HPO_4/KH_2PO_4 (pH *ca.* 7) > K_3PO_4 (pH *ca.* 13). Although K_3PO_4 is the strongest salting-out species amongst the salts investigated, the authors demonstrated that the pH of the aqueous media plays a major role in the partition behavior observed, particularly for compounds that exhibit acidic dissociation constants. In summary, it was found that at low pH values, the non-charged form of gallic acid (or other phenolic compounds) preferentially migrates for the IL-rich phase whereas its conjugate base preferentially partitions for the salt-rich phase.[71]

Based on the tailoring ability and high extraction efficiencies offered by ILs, as discussed above, yet using low IL amounts to reduce the costs of the overall ABS extraction/purification process, Almeida *et al.*[69] suggested the use of ABS composed of polyethylene glycol (PEG) of different molecular weights (200, 300, 400 and 600 g mol^{-1}) and Na_2SO_4, using ILs as additives (at 5 or 10 wt.%) for the extraction of gallic, vanillic and syringic acids from aqueous media. The results obtained by the authors disclosed that all the antioxidants investigated preferentially migrate to the PEG-rich phase and depend on the PEG molecular weight and IL employed. These results are in agreement with those presented by Cláudio *et al.*[70] who demonstrated the preferential migration of gallic acid for the most hydrophobic phase in ABS composed of ILs and salts. Nevertheless, the authors[69] revealed that the addition of only 5 wt.% of IL leads to extraction efficiencies of all phenolic acids ranging between 80% and 99%, validating thus the aptitude of the IL to tune the polarity of the PEG-rich phase. It should be highlighted that the authors also determined the partitioning of the ILs investigated and confirmed that they are enriched in the PEG-rich phase. The overall results confirm a promising substitution of high amounts of ILs in ABS (where similar extraction efficiencies are obtained[70,71]) by the less expensive and benign PEG-based systems, with the addition of only 5–10 wt.% of IL, without losing their extractive performance.

Santos *et al.*[72] studied ABS composed of diverse ILs and a potassium citrate $(C_6H_5K_3O_7/C_6H_8O_7)$ buffer at pH 7 for the extraction of two antioxidants, namely eugenol and propyl gallate. They also compared the extraction ability of IL-based ABS with more conventional systems formed by polyethylene glycol (PEG) and a potassium phosphate (K_2HPO_4/KH_2PO_4) buffer at pH 7 using imidazolium-based ILs as adjuvants, *i.e.* at lower concentrations in the overall system. In all situations, the authors[72] revealed the possibility of optimizing the extraction efficiency of both antioxidants up to 100% using either IL-based or PEG-based (with the IL as adjuvant) ABS. These results are in agreement with the findings of Almeida *et al.*[69] supporting that polymer/salt/IL-based systems, in which only 5 wt.% of IL was added, are able to lead to high and tailored extraction efficiencies, while decreasing the cost and the environmental impact of the proposed extraction/purification technology.[72]

Terpenoids are similar to terpenes, derived from isoprene units, and which may include some oxygen functionality or different rearrangements of the carbon skeleton. A large number of terpenoids present biological activities against cancer, inflammation and malaria, justifying the large interest devoted to their extraction from natural sources.[80] Amongst this group of high-value compounds, only two works employing IL-based ABS for the extraction of β-carotene were found, namely using ABS formed by phosphonium-based ILs and K_3PO_4[73] and imidazolium-based ILs and carbohydrates.[75] The chemical structure of β-carotene is shown in Figure 8.7. Partition coefficients of β-carotene in the order of 61 and between 5.5 and 24, in a single step at 298 K, were obtained applying ABS formed by $[P_{i(444)1}][Tos]$ and K_3PO_4[73] and $[C_4mim][CF_3SO_3]$ and several carbohydrates (sucrose, glucose, mannose, xylose, maltitol, xylitol and sorbitol),[75] respectively. Albeit the extraction of

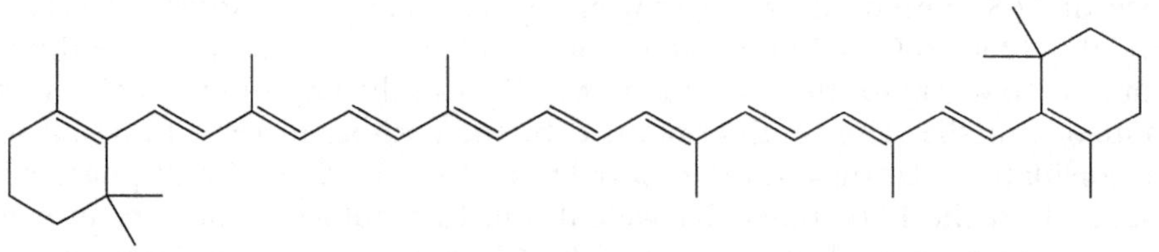

Figure 8.7 Chemical structure of β-carotene (terpenoid) extracted using IL-based ABS.

alkaloids was also investigated with both types of ABS,[73,75] both revealed a higher performance for extracting more hydrophobic substances, such as β-carotene.

Though the previous works were carried out with less complex matrices, they demonstrated that IL-based ABS are real and enhanced alternatives to the traditional extraction methods, offering simpler, greener, quicker and more efficient procedures. On the other hand, the extraction of added-value compounds from biomass, followed by the formation of ABS for purification, although scarce, are also available in the literature. Three examples were found, namely concerning the purification/isolation of pharmaceutical ingredients, such as anthraquinones derivatives[81] and polysaccharides,[82] the extraction of puerarin from *Radix Puerariae lobatae* extracts,[83] and on the extraction of saponins and polyphenols from biomass using IL–water mixtures, followed by the use of IL-based ABS for purification.[40]

Tan *et al.*[81,82] published two works on the purification of added-value compounds from *Aloe vera* L. (Liliaceae) employing IL-based ABS. A prominent feature of *Aloe vera* fillet is its high water content, ranging from 98.5% to 99.5% of fresh matter, and where more than 60% of the remaining dried biomass is composed of polysaccharides.[84] Aloe polysaccharides are the major active ingredients in the aloe gel, and are responsible for their curative or healing qualities, being thus used in the area of dermatology, especially for treating radiation-caused skin conditions.[85] On the other hand, aloe anthraquinone derivatives display antibacterial and antifungal activities.[86] Tan and co-workers[81,82] started by obtaining the crude materials, where aloe peel powder was soaked with an aqueous solution of 60% of ethanol. The solution and residue were then isolated by centrifugation, and the ethanol and water were removed by evaporation. For the aloe polysaccharides and proteins study, the colloid was just dried and used.[82] For the anthraquinone-rich extract, further sulfuric acid and chloroform were added into the extract, refluxed, and chloroform was finally removed. After evaporation, a yellowish-brown colloid as the crude extract was used and dissolved in methanol as the stock solution.[81] The stock solutions containing both extracts were then used in the formation of IL-based ABS, and the fractionation and purification of the added-value compounds under study were investigated. In both works,[81,82] ABS formed by [C₄mim][BF₄] and sodium-based salts were employed. For anthraquinones, the best results were obtained with the systems composed of Na_2SO_4,[81] whereas for the polysaccharides, superior extractions were obtained with NaH_2PO_4 as the salt phase-forming component.[82] It should be highlighted that a large number of imidazolium-based ILs, of different alkyl side chain length combined with different anions ([BF₄], [N(CN)₂] and [Br]), as well as a large battery of salts, were investigated by the authors.[81,82] After selecting the best ABS, the authors additionally studied the impact of the pH, temperature, centrifugation and equilibrium time. The extraction efficiencies of aloe-emodin and chrysophanol under the optimized conditions were of 92.34% and 90.46%, respectively.[81] The results provided by the authors demonstrated that IL-based ABS allow a good recovery of aloe anthraquinones with few impurities present. Unfortunately, the authors did not comment on the purification levels achieved by the use of IL-based ABS compared to traditional approaches, and only focused on the extraction efficiencies.[81] Finally, a methodology for the IL recovery by reverse extraction experiments was proposed, carried out by taking the IL-rich solution rich in antraquinones and forming a new ABS with the addition of a salt with alkaline characteristics.[81] The alkaline medium leads to the speciation of the antraquinones and further migration to the salt-rich phase, allowing thus the IL-rich phase to be recovered and reused. Still, the authors[87] did not evaluate the extraction performance of the recovered IL-rich phase. In the same line, the work of Cláudio *et al.*[87] confirmed the back-extraction of antioxidants present in biomass and the viability of the IL-rich solutions to be recovered and reused without losses on their extraction efficiencies. Two types of IL-based ABS were evaluated regarding their extraction efficiencies for phenolic acids, namely gallic, syringic and vanillic acids.[87] These ABS were formed either with Na_2CO_3 or Na_2SO_4 to afford different pH values at the coexisting phases, allowing therefore the back-extraction of phenolic acids ruled by their speciation. The most promising IL-based ABS were then used in sequential two-step cycles (product extraction/IL recovery) aiming at addressing the efficacy on the IL recyclability and reusability. Extraction efficiencies ranging between 73% and 99% were obtained in four sequential partitioning experiments involving the phenolic acids, while allowing the regeneration of 94–95% of the IL and further reutilization.[87] A flow chart of the two-step approach for a greener IL-recyclable ABS extraction/purification

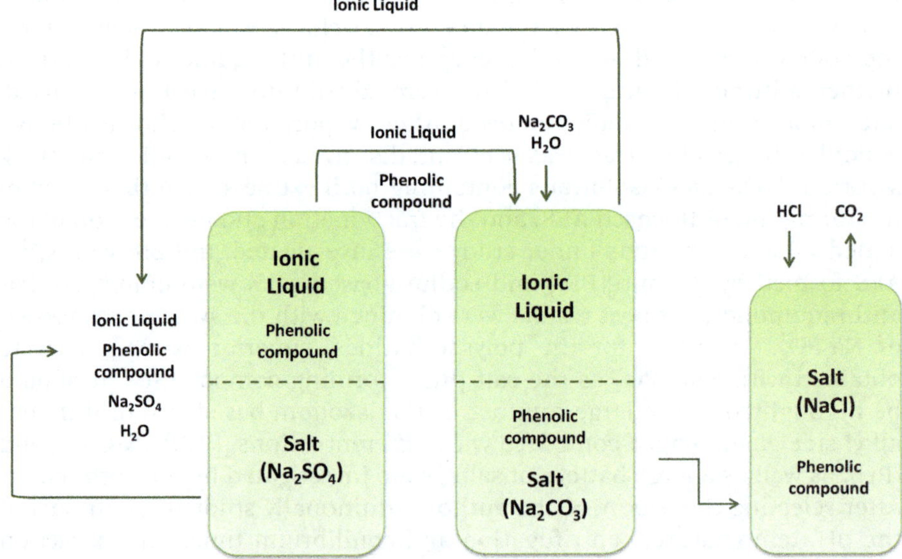

Figure 8.8 Flow chart of the two-step approach for a greener IL-recyclable ABS extraction of phenolic compounds.

of phenolic acids is depicted in Figure 8.8. These results support the establishment of IL-based ABS as greener cost-effective strategies with a substantial reduction in the environmental footprint and also dealing with the economic issues.

Tan *et al.*[82] also investigated the purification of aloe polysaccharides and proteins, starting by the optimization of parameters affecting the extraction efficiency, such as the type and concentration of the phase-forming salt, temperature, pH and addition of inorganic electrolytes. They also studied the performance of polymer-, surfactant- and alcohol-based ABS, revealing that IL-based ABS lead to higher extraction efficiencies. Then, the best IL-based ABS was applied to a crude polysaccharide extract obtained from a precipitation step carried out with ethanol from the aloe gel juice. The authors verified that, under the optimized conditions, aloe polysaccharides mostly partition into the salt-rich phase while the majority of aloe proteins and other impurities are enriched in the IL-rich phase. Aloe polysaccharides were further purified using a dialysis membrane to remove the salt and IL. The purity of the final product was demonstrated by thermogravimetric analysis (TGA). Finally, the authors proposed the recycling of the IL-rich phase by a simple solvent extraction method using dichloromethane. Nevertheless, the viability of the IL-rich phase for a subsequent extraction with fresh extracts was not tested.[82]

Although the works of Tan *et al.*[81,82] demonstrate the high potential of IL-based ABS for the purification of added-value compounds from biomass, the authors always focused on the use of $[C_4mim][BF_4]$ that is not water-stable; it hydrolyses to produce hydrofluoridric acid even at moderate conditions.[28] This is a major drawback when the recycling of the IL-rich phase is recommended.[81,82] Nowadays, we are facing an era of widespread availability of more benign ILs, such as those composed of cholinium-based cations combined with anions derived from carboxylic acids,[88] amino acids[89] and biological buffers,[90] and an urgent change towards the exploitation of this type of ILs for the extraction of added-value compounds from biomass is required. On the other hand, the authors only applied IL-based ABS to the purification of the target compounds whereas the extractions were carried out with traditional organic solvents. In this context, and based on the high performance of IL–water mixtures for the extraction of added-value compounds from biomass, an integrated strategy could be explored, *i.e.* an extraction from biomass carried out with IL aqueous solutions that could be further used (directly) in the formation of ABS for the purification step.

Puerarin (shown in Figure 8.9) is an important isoflavone with many beneficial effects on hypertension, arteriosclerosis and diabetes mellitus.[91] Due to these potential features, Fan *et al.*[83] investigated IL-based ABS to extract/isolate puerarin from *Radix Puerariae lobatae* extracts. The authors firstly evaluated model systems employing pure and commercial puerarin, revealing that the ILs' nature, the addition of short-chain alcohols, the salting-out ability of the salt, and the acidity and basicity of the aqueous media have an important role towards the extraction efficiency. Under the optimized conditions, an

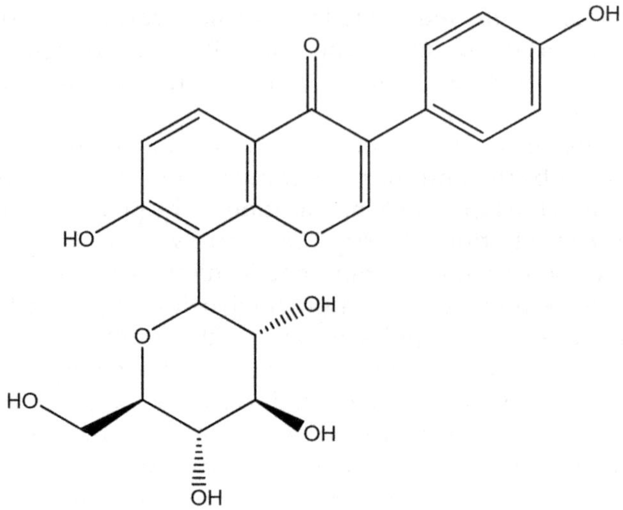

Figure 8.9   Chemical structure of puerarin (isoflavone) extracted/isolated using IL-based ABS.

extraction efficiency of the order of 99% for puerarin was achieved. Finally, crude *Radix Puerariae lobatae* extracts (puerarin purity of 40.3%) were added to an ABS formed by [C$_4$mim][Br] and K$_2$HPO$_4$ for isolation purposes. The results obtained indicated an extraction efficiency above 99% of puerarin for the IL-rich phase. Nevertheless, the authors[83] didn't report on the purity levels afforded by the IL-based ABS technique.

Ribeiro *et al.*[40] investigated the extraction of saponins and polyphenols from dried leaves and aerial parts of tea and mate using aqueous solutions of ILs. Saponins display interesting biological features (hemolytic, antimicrobial, insecticide, *etc.*) and are commercially explored in many applications by the food, cosmetic and pharmaceutical industries.[92] A large number of ILs was explored, ranging from imidazolium-based ILs with variable length in the side alkyl chains to more sustainable ILs, such as those based on the cholinium cation, while combined with several anions in order to ascertain on the IL anion effect through the extraction efficiencies.[40] The most efficient IL was then selected to set up a central composite experimental design in order to determine the IL/raw material ratio and IL concentration which lead to the maximum extraction efficiency. On the whole, most of the ILs investigated yielded higher extraction efficiencies when compared to the aqueous solution of 30 wt.% of ethanol commonly applied. K$_3$PO$_4$ and Na$_2$CO$_3$ were then added as salting-out salts to the aqueous solutions of cholinium chloride containing the extracts to create ABS. However, only K$_3$PO$_4$ was able to render phase separation. The ABS efficiency was further evaluated through the partition coefficients and the concentration factors achievable with these systems. The aqueous solution at 30 wt.% of ethanol was also tested for aqueous biphasic system formation, which was achieved using both inorganic salts. For the extracts from tea, partition coefficients two orders of magnitude higher were obtained for both saponins and phenolic compounds using [Ch][Cl] when compared to the ethanol-water mixtures. For the mate extracted compounds, similar results were obtained. On the other hand, the concentration factors of tea saponins are higher than those obtained for tea saponins, although they are below one for all ABS tested. Single compositions of the phase forming-components were applied at this stage,[40] and higher concentration factors could be explored by a manipulation of the overall mixture composition along the same tie-line.[93] After separation of the coexisting phases, a non-water miscible IL, [Ch][NTf$_2$], was added to the [Ch][Cl]-rich phase, resulting in the formation of a new phase mainly composed of [Ch][NTf$_2$] and [Ch][Cl], and an aqueous phase where saponins and phenols remain. However, the authors[40] did not discuss the recovery of the added-value compounds from the salt-rich phase neither the approach to separate and recover the two ILs for further use. Finally, they did not comment on the purity levels afforded by the use of IL-based ABS and whether they can be seen as a promising route.

In general, imidazolium-based ILs have been the most used ILs, combined with chloride, bromide, acetate, dicyanimide and tetrafluoroborate anions, in the formation of ABS. The work of Ribeiro *et al.*[40] highlights the potential on overcoming this trend by substituting the well-studied imidazolium-based fluids with more biocompatible and biodegradable cholinium-based ones. Still, a large number of different IL ions can be explored in the future towards the use of ABS in the isolation of target compounds from natural extracts. Moreover, an integrated strategy comprising the IL aqueous solutions used in the extraction of added-value compounds from biomass followed by their direct use in the formation of ABS for the purification step is still far from being accomplished. Although Ribeiro *et al.*[40] provided some pioneering results on this approach, the purification levels, number of cycles required and concentration factors still need to be investigated in more detail.

8.3 Recyclability Strategies for Ionic Liquid-Based Solvents

Even though ILs have an enhanced potential for the extraction, separation and purification of value-added compounds, the economic and environmental goals of the whole process can only be met by the recycling and reuse of the ILs employed as solvents. Furthermore, only by fulfilling these requirements, can the large-scale application of ILs be anticipated. In this context, several methods have been applied to recover ILs from aqueous media, namely distillation, adsorption, nanofiltration, ion-exchange and liquid–liquid extraction approaches. The most adequate technique to be used in the recovery of a particular IL is dependent on the characteristics of the medium and on the IL chemical structure and concentration.

Due to the negligible vapor pressure of ILs, their separation from an IL–solvent mixture can be easily attempted by evaporation of the molecular solvent. However, the concentration of the IL and the boiling point of the solvent are crucial factors in the economic feasibility of the process. On the other hand, most added-value compounds extracted from biomass are non-volatile or are thermo-sensitive, often leading to an IL–extract mixture that is difficult to separate. This problem can be overcome by the use of protic and distillable ILs, such as $[N_{111(2OH)}][CH_3CO_2]$, which was used in aqueous solution for the pre-treatment and fractionation of lignocellulosic biomass, and further recovered by distillation (IL recovery of 80% after 5 cycles).[94]

Hydrophobic ILs, which are water immiscible, can be easily separated from water-rich extracts by decantation. Nevertheless, most ILs investigated in the extraction of added-value compounds[33] are highly hydrophilic and completely miscible with water in the whole composition range and at temperatures close to room temperature. Even so, surfactant-like ILs, *e.g.* ILs possessing long aliphatic moieties, tend to form micelles in water, and these can be separated by membrane-based methods (*e.g.* filtration) or force field separation (*e.g.* centrifugation).[95] On the other hand, non-surfactant-based hydrophilic ILs can be recovered from aqueous media by a novel induced phase separation, *i.e.* by the addition of strong salting-out species or supercritical CO_2, or by back-extraction, adsorption or membrane-based techniques.

The extraction of non-volatile or thermally sensitive products from ILs can be carried out by liquid–liquid extraction, allowing therefore the recycling and reuse of ILs. Many molecular solvents are immiscible with ILs and they can be used to recover the added-value compounds from the IL-rich phase. For instance, hydrophilic products are easily extracted from hydrophobic ILs with water.[96] On the contrary, for hydrophilic ILs, organic solvents immiscible with water and with the IL–water mixtures should be employed. For instance, Dibble *et al.*[97] recovered $[C_2mim][CH_3CO_2]$ and lignin from the IL-pre-treated biomass by liquid–liquid extraction using acetone, 2-propanol and water. This mixture was shown to form a phase switchable solvent system providing the concentration and separation of hydrophobic solutes, short-chain carbohydrates and lignin. The extraction process led to a recovery of 89% of the IL. Coutinho and co-workers[77] also recovered the IL used in the extraction of caffeine from guaraná seeds by a back-extraction procedure. The caffeine was re-extracted with butanol allowing the IL aqueous solutions to be recovered and reused.[77] A flowchart of the process used in the re-extraction of caffeine and on the solvents recyclability and reusability is depicted in Figure 8.10. The authors demonstrated that aqueous solutions of ILs allow the selective extraction of caffeine from biomass and that these IL–water solvents can be recycled and reused without loss of their extraction performance.[77]

After the recovery of the added-value compounds, ILs can be also recovered from aqueous media by adsorption. Activated carbon (AC), which is widely used to remove organics from water, has been used to separate ILs from aqueous solutions. Anthony *et al.*[98] reported the pioneering work on the use of AC for the removal of $[C_4mim][PF_6]$ from waste waters. The adsorption of imidazolium-based ILs from aqueous solutions with commercial AC and modified AC adsorbents was extensively investigated in subsequent studies

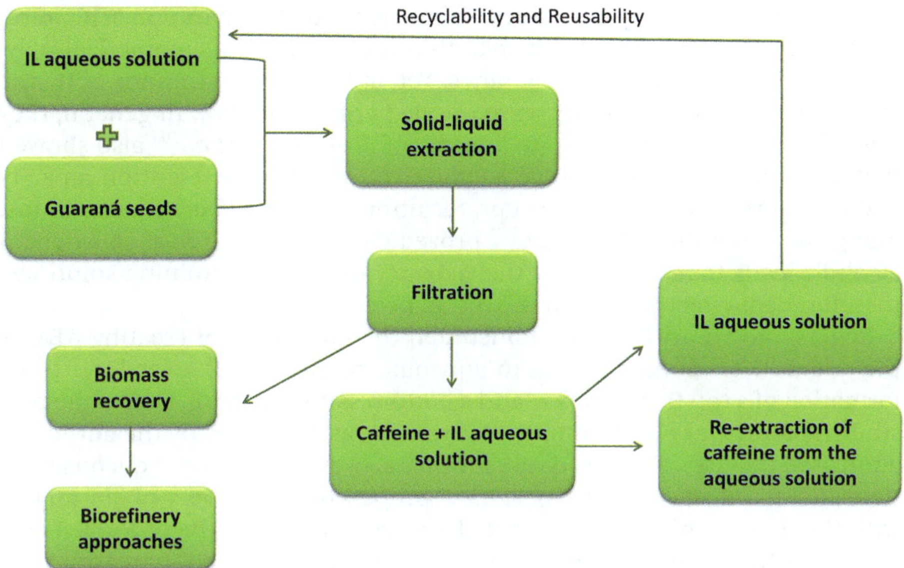

Figure 8.10   Process flowchart for the extraction of caffeine from guaraná seeds.

by Palomar and co-workers.[99–101] However, AC is more adequate to remove hydrophobic ILs from aqueous solutions and at low concentrations. Trying to overcome this drawback, the adsorption of hydrophilic ILs was shown to be enhanced by modifying the amount and nature of oxygen groups on the surface of AC and by the addition of inorganic (salting-out) salts to the aqueous solutions.[100,102] Nevertheless, although these works showed that AC could be used to remove the IL from the aqueous solutions, most of them did not address the recovery of the IL. Only Palomar and co-workers[100] suggested the use of acetone to regenerate AC and to recover the ILs. More recently, Qi *et al.*[103] used a novel functionalized carbonaceous material, loaded with carboxylic groups, to remove $[C_4mim][Cl]$ from aqueous media, which exhibited an adsorption efficiency comparable to commercial AC. Furthermore, the adsorbent was regenerated and recycled, for at least three times, without losses in the adsorption capacity.[103]

Binder and Raines[104] demonstrated the use of cation-exchange resins for the recovery of ILs from a biomass hydrolysate mixture by liquid chromatography. The authors investigated the exchange of H^+ of a commercial resin by the $[C_2mim]$ cation, while recovering up to 92% of $[C_2mim][Cl]$. In the same line, Mai *et al.*[105] employed the same ion-exchange approach and recovered 98.92% of $[C_2mim][CH_3CO_2]$ from an aqueous biomass hydrolysate. However, and although the adsorption/desorption process is robust and relatively easy to operate, it requires the use of additional desorption solvents and more complex chromatographic equipment.

Nanofiltration membranes are able to separate charged and neutral compounds or mono and divalent ions, and are thus feasible to separate ILs.

Based on this possibility, Abels *et al.*[106] investigated the separation of [C$_1$mim] [(CH$_3$)$_2$PO$_4$] from saccharides by nanofiltration using two commercial polyamide and one polyimide membranes. Yet, at high concentrations of IL, the authors verified a marked decrease on the permeation flow. In general, they were able to recover the IL up to a purity of 80%. Abels *et al.*[106] also showed that the addition of inorganic salts reduces the membrane retention capacity and the aggregation of ILs in water, resulting in more effective separations. Along the same line, Gan *et al.*[107] proved that the addition of short-chain alcohols leads to a decrease in the viscosity of the IL-containing solutions, therefore reducing the membrane resistance.

Rogers and co-workers[58] demonstrated the possibility of creating ABS by the addition of inorganic salts to aqueous solutions of ILs, leading to the formation of a top IL-rich phase and a bottom salt-rich phase. One of the possibilities for the application of these systems, as suggested by the authors,[58] addressed the use of IL-based ABS for the recovery of ILs from aqueous solutions. Based on this idea, Deng *et al.*[108] proposed the recovery of [amim][Cl] with three inorganic salts, with the IL recovery following the order: K$_3$PO$_4$ > K$_2$HPO$_4$ > K$_2$CO$_3$. Moreover, the authors concluded that an increase of the salt concentration leads to an increase in the recovery efficiency of the ILs, where a maximum recovery efficiency of 96.80% was attained using 46.48 wt.% of K$_2$HPO$_4$. Li *et al.*[109] also studied the effect of sodium-based salts (Na$_3$PO$_4$, Na$_2$CO$_3$, Na$_2$SO$_4$, NaH$_2$PO$_4$ and NaCl) to recover [C$_4$mim][BF$_4$] from aqueous solutions with the highest extraction efficiency (98.77%) achieved with 16.94 wt.% of Na$_2$CO$_3$. In addition to inorganic salts, Wu *et al.*[110,111] reported the recovery of different ILs from aqueous solutions by forming ABS through the addition of carbohydrates. A recovery efficiency of 65% of [amim][Br], 63% of [amim][Cl] and 74% of [C$_4$mim][BF$_4$] was achieved by the addition of sucrose.[110,111] Albeit carbohydrates have been used with the aim of replacing the high-charge density salts, it should be remarked that the addition of carbohydrates to remove ILs from aqueous media leads, on the other hand, to an increase in the amount of organic matter in the aqueous streams if applied in a large scale.

More recently, Neves *et al.*[112] demonstrated the use of aluminium-based salts to concentrate and remove ILs from aqueous solutions. Al$_2$(SO$_4$)$_3$ and AlK(SO$_4$)$_2$ were used to concentrate and recover several ILs from aqueous media, including imidazolium-, pyridinium- and phosphonium-based fluids. These salts were chosen due to their remarkable salting-out aptitude and because they are actually used in water treatment processes.[113] With concentrations of salts ranging from 2 to 16 wt.%, recovery efficiencies of IL between 96% and 100% were accomplished. The authors also demonstrated that the aluminium-based salts are present in negligible concentrations (insignificant cross-contamination) at the IL-rich phase. Finally, and based on the enhanced recovery efficiencies obtained, the authors proposed the scale-up of the process, as depicted in Figure 8.11. The proposed process was tested with one IL up to 4 recovery cycles, while proving the recovery of 100% and the recyclability of the inorganic salt.[113]

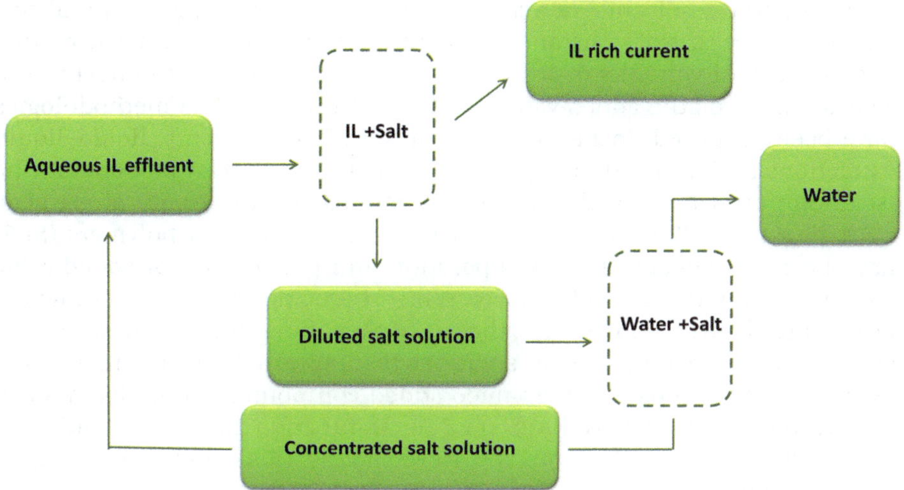

Figure 8.11   Configuration of the global process for the recovery of ionic liquids from aqueous effluents.

In a more complex method, Shill *et al.*[114] employed aqueous solutions of K_3PO_4 and K_2HPO_4 to precipitate the IL pre-treated cellulosic material while inducing the formation of a three-phase system (solid phase containing the biomass, and liquid–liquid IL-rich and salt-rich phases). This process reduces the amount of water to be evaporated from the recycled IL. Moreover, the addition of a salting-out salt with alkaline characteristics during the precipitation results in the partial delignification of the biomass, turning the substrate more accessible and enhancing the enzymatic hydrolysis.[114]

Recently, magnetic ILs have been disclosed and used as catalysts, reaction media and solvents, among other uses.[115] Magnetic ILs, comprising $FeCl_4^-$, display a magnetic response and can be easily recovered from mixtures by applying a magnetic field. For instance, Lee *et al.*[116] applied a magnetic field to recover $[C_4mim][FeCl_4]$ from aqueous solution. Although the two-phase mixture ($[C_4mim][FeCl_4]$-rich phase and water-rich phase) was easily separated by the magnetic field, the separation of the homogeneous IL–water phase was more difficult to accomplish and still requires the application of other methods.

8.4   Conclusions

ILs have attracted considerable attention as potential solvents in the extraction and separation of bioactive compounds from plants owing to their outstanding properties. In general, ILs can be tailored for the extraction of target chemicals and high extraction yields of alkaloids, flavonoids, terpenoids and lipids, among others, from the most diverse natural sources, have been reported. After the extraction, the crucial steps are is the product recovery and the solvents' recyclability, aiming at developing an overall sustainable process. Notwithstanding some attempts that have been carried out, the lack of volatility of ILs makes it nearly impossible to directly concentrate non-volatile added-value products. Although some methodologies have been proposed, including the addition of anti-solvents, liquid–liquid extractions and adsorption approaches, further work must be done on the isolation/purification of the target compounds from the natural extracts. Both IL-based ABS (LLE) and IL-functionalized silica and polymers (SPE) have been investigated for the separation and purification of added-value chemicals from the natural extracts. While these approaches have shown to be enhanced purification routes, the use of a combined/integrated strategy is still missing. In fact, IL aqueous solutions that revealed high potential for the extraction of the most diverse valued-added compounds from biomass can be directly used in IL-based ABS (LLE) or IL-functionalized silica and polymers (SPE) for the purification step. Nevertheless, this combined strategy is far from being completely addressed.

Both in SPE and ABS purification stages, imidazolium-based ILs have been the preferred choice – justified by their favorable and specific interactions with the target compounds. Nevertheless, at this stage, no major conclusions can be drawn due to the lack of a significant number of studies employing tetraalkylammonium- and tetraalkylphosphonium-based ILs. In this line, more targeted IL-based ABS and materials, particularly using more benign ILs, need to be investigated, aiming at guaranteeing the development of sustainable processes. It needs to be kept in mind that the selectivity and ability of SPE and ABS processes can be fine tuned by selecting adequate ILs.

Although several works on the purification stage have been published, few addressed the IL recovery or evaluated their use in a new cycle or application. Some works have however attempted this crucial task, and distillation, adsorption, nanofiltration, ion-exchange and liquid–liquid extraction approaches have been proposed to recover ILs. Nevertheless, the exploitation of the IL efficiency after the recovery step is still seldom investigated. Only after the development of suitable methodologies can ILs be reused several times, thereby decreasing the cost and the environmental footprint of the whole process while foreseeing their use on a large scale under the biorefinery framework.

Acknowledgements

This work was developed in the scope of the project CICECO-Aveiro Institute of Materials (Ref. FCT UID/CTM/50011/2013) financed by national funds through the FCT/MEC. M. M. Pereira acknowledges the Ph.D grant (2740-13-3) and financial support from Coordenação de Aperfeiçoamento de Pessoal de Nível Superior - Capes. M. G. Freire acknowledges the European Research Council (ERC) for the Starting Grant ERC-2013-StG-337753.

References

1. A. Stark, *Energy Environ. Sci.*, 2011, **4**, 19–32.
2. L. R. Lynd, C. E. Wyman and T. U. Gerngross, *Biotechnol. Prog.*, 1999, **15**, 777–793.
3. M. FitzPatrick, P. Champagne, M. F. Cunningham and R. A. Whitney, *Bioresour. Technol.*, 2010, **101**, 8915–8922.
4. F. Cherubini, *Energy Convers. Manage.*, 2010, **51**, 1412–1421.
5. S. Fernando, S. Adhikari, C. Chandrapal and N. Murali, *Energy Fuels*, 2006, **20**, 1727–1737.
6. J. H. Clark, R. Luque and A. S. Matharu, *Annu. Rev. Chem. Biomol. Eng.*, 2012, **3**, 183–207.
7. A. Demirbas, *Energy Convers. Manage.*, 2001, **42**, 279–294.
8. S. Moncheva, S. Gorinstein, G. Shtereva, F. Toledo, P. Arancibia-Avila, I. Goshev and S. Trakhtenberg, *Hydrobiologia*, 2003, **501**, 23–28.
9. L. Laurens, N. Nagle, R. Davis, N. Sweeney, S. Van Wychen, A. Lowell and P. Pienkos, *Green Chem.*, 2015, **17**, 1145–1158.
10. L. Han, H. Y. Pei, W. R. Hu, L. Q. Jiang, G. X. Ma, S. Zhang and F. Han, *Bioresour. Technol.*, 2015, **175**, 262–268.
11. R. G. Berger, *Curr. Opin. Food Sci.*, 2015, **1**, 38–43.
12. M. J. Barbosa, J. W. Zijffers, A. Nisworo, W. Vaes, J. van Schoonhoven and R. H. Wijffels, *Biotechnol. Bioeng.*, 2005, **89**, 233–242.
13. N. Březinováá Belcredi, J. Ehrenbergerova, V. Fiedlerova, S. Belakova and K. Vaculova, *J. Agric. Food Chem.*, 2010, **58**, 11755–11761.
14. S. Gomez-Galera, A. M. Pelacho, A. Gene, T. Capell and P. Christou, *Plant Cell Rep.*, 2007, **26**, 1689–1715.
15. R. M. A. Domingues, A. R. Guerra, M. Duarte, C. S. R. Freire, C. P. Neto, C. M. S. Silva and A. J. D. Silvestre, *Mini-Rev. Org. Chem.*, 2014, **11**, 382–399.
16. R. W. H. Sargent and G. R. Sullivan, *Ind. Eng. Chem. Process Des. Dev.*, 1979, **18**, 113–124.
17. S. Mukherjee, P. Das and R. Sen, *Trends Biotechnol.*, 2006, **24**, 509–515.
18. L. S. White, *J. Membr. Sci.*, 2006, **286**, 26–35.
19. S. Le Borgne and R. Quintero, *Fuel Process. Technol.*, 2003, **81**, 155–169.
20. J. G. Watson, J. C. Chow and E. M. Fujita, *Atmos. Environ.*, 2001, **35**, 1567–1584.
21. C. Capello, U. Fischer and K. Hungerbühler, *Green Chem.*, 2007, **9**, 927–934.
22. L. J. A. Conceiçao, E. Bogel-Lukasik and R. Bogel-Lukasik, *RSC Adv.*, 2012, **2**, 1846–1855.
23. J. MSS Esperança, J. N. Canongia Lopes, M. Tariq, L. s. M. Santos, J. W. Magee and L. s. P. N. Rebelo, *J. Chem. Eng. Data*, 2009, **55**, 3–12.
24. C. G. Cassity, A. Mirjafari, N. Mobarrez, K. J. Strickland, R. A. O'Brien and J. H. Davis, *Chem. Commun.*, 2013, **49**, 7590–7592.
25. L. S. Wang, X. X. Wang, Y. Li, K. Jiang, X. Z. Shao and C. J. Du, *AIChE J.*, 2013, **59**, 3034–3041.

26. J. Ranke, S. Stolte, R. Stormann, J. Arning and B. Jastorff, *Chem. Rev.*, 2007, **107**, 2183–2206.

27. M. Petkovic, J. L. Ferguson, H. N. Gunaratne, R. Ferreira, M. C. Leitao, K. R. Seddon, L. P. N. Rebelo and C. S. Pereira, *Green Chem.*, 2010, **12**, 643–649.

28. M. G. Freire, C. M. Neves, I. M. Marrucho, J. A. Coutinho and A. M. Fernandes, *J. Phys. Chem. A*, 2010, **114**, 3744–3749.

29. K. Fukumoto, M. Yoshizawa and H. Ohno, *J. Am. Chem. Soc.*, 2005, **127**, 2398–2399.

30. C. R. Allen, P. L. Richard, A. J. Ward, L. G. van de Water, A. F. Masters and T. Maschmeyer, *Tetrahedron Lett.*, 2006, **47**, 7367–7370.

31. S. P. M. Ventura, M. Gurbisz, M. Ghavre, F. M. M. Ferreira, F. Gonçalves, I. Beadham, B. Quilty, J. A. P. Coutinho and N. Gathergood, *ACS Sustainable Chem. Eng.*, 2013, **1**, 393–402.

32. D.-J. Tao, Z. Cheng, F.-F. Chen, Z.-M. Li, N. Hu and X.-S. Chen, *J. Chem. Eng. Data*, 2013, **58**, 1542–1548.

33. H. Passos, M. G. Freire and J. A. Coutinho, *Green Chem.*, 2014, **16**, 4786–4815.

34. F.-Y. Du, X.-H. Xiao and G.-K. Li, *J. Chromatogr. A*, 2007, **1140**, 56–62.

35. M. G. Bogdanov and I. Svinyarov, *Sep. Purif. Technol.*, 2013, **103**, 279–288.

36. K. Wu, Q. Zhang, Q. Liu, F. Tang, Y. Long and S. Yao, *J. Sep. Sci.*, 2009, **32**, 4220–4226.

37. H. Garcia, R. Ferreira, M. Petkovic, J. L. Ferguson, M. C. Leitao, H. Q. N. Gunaratne, K. R. Seddon, L. P. N. Rebelo and C. S. Pereira, *Green Chem.*, 2010, **12**, 367–369.

38. A. K. Ressmann, R. Zirbs, M. Pressler, P. Gaertner and K. Bica, *Z. Naturforsch., B: J. Chem. Sci.*, 2013, **68**, 1129–1137.

39. R. Ferreira, H. Garcia, A. F. Sousa, M. Petkovic, P. Lamosa, C. S. R. Freire, A. J. D. Silvestre, L. P. N. Rebelo and C. S. Pereira, *New J. Chem.*, 2012, **36**, 2014–2024.

40. B. D. Ribeiro, M. A. Z. Coelho, L. P. N. Rebelo and I. M. Marrucho, *Ind. Eng. Chem. Res.*, 2013, **52**, 12146–12153.

41. S. A. Chowdhury, R. Vijayaraghavan and D. R. MacFarlane, *Green Chem.*, 2010, **12**, 1023–1028.

42. C. Yansheng, Z. Zhida, L. Changping, L. Qingshan, Y. Peifang and U. Welz-Biermann, *Green Chem.*, 2011, **13**, 666–670.

43. N. Fontanals, F. Borrull and R. M. Marcé, *Trends Anal. Chem.*, 2012, **41**, 15–26.

44. M. Tian, H. Yan and K. H. Row, *J. Chromatogr. B*, 2009, **877**, 738–742.

45. H.-W. Luo, B.-J. Wu, M.-Y. Wu, Z.-G. Yong, M. Niwa and Y. Hirata, *Phytochemistry*, 1985, **24**, 815–817.

46. M. Tian, W. Bi and K. H. Row, *J. Sep. Sci.*, 2009, **32**, 4033–4039.

47. T. Hatano, T. Yasuhara and K. Miyamotob, *Chem. Pharm. Bull.*, 1988, **36**, 2286–2288.

48. W. Bi, J. Zhou and K. H. Row, *J. Liq. Chromatogr. Relat. Technol.*, 2012, **35**, 723–736.

49. W.-Y. Huang, Y.-Z. Cai and Y. Zhang, *Nutr. Cancer*, 2009, **62**, 1–20.

50. W. Bi, M. Tian and K. H. Row, *J. Chromatogr. B*, 2012, **880**, 108–113.

51. P. Williams, *J. Nat. Prod.*, 2012, **75**, 1261–1261.

52. M. Tian and K. H. Row, *Chromatographia*, 2011, **73**, 25–31.

53. W. Bi, M. Tian and K. H. Row, *J. Sep. Sci.*, 2010, **33**, 1739–1745.

54. M. Tian, H. Yan and K. H. Row, *Anal. Lett.*, 2009, **43**, 110–118.

55. W. Bi, M. Tian and K. H. Row, *J. Chromatogr. A*, 2012, **1232**, 37–42.

56. W. Bi, M. Tian and K. H. Row, *Phytochem. Anal.*, 2010, **21**, 496–501.

57. P.-A. Albertsson, *Partition of cell particles and macromolecules: Separation and purification of biomolecules, cell organelles, membranes and cells in aqueous polymer two phase systems and their use in biochemical analysis and biotechnology*, John Wiley and Sons, Chichester, 1986.

58. K. E. Gutowski, G. A. Broker, H. D. Willauer, J. G. Huddleston, R. P. Swatloski, J. D. Holbrey and R. D. Rogers, *J. Am. Chem. Soc.*, 2003, **125**, 6632–6633.

59. M. G. Freire, A. F. M. Claudio, J. M. M. Araujo, J. A. P. Coutinho, I. M. Marrucho, J. N. C. Lopes and L. P. N. Rebelo, *Chem. Soc. Rev.*, 2012, **41**, 4966–4995.

60. J. F. Pereira, L. P. N. Rebelo, R. D. Rogers, J. A. Coutinho and M. G. Freire, *Phys. Chem. Chem. Phys.*, 2013, **15**, 19580–19583.

61. Y. Fan, M. Chen, C. Shentu, F. El-Sepai, K. Wang, Y. Zhu and M. Ye, *Anal. Chim. Acta*, 2009, **650**, 65–69.

62. A. Arce, A. Pobudkowska, O. Rodríguez and A. Soto, *Chem. Eng. J.*, 2007, **133**, 213–218.

63. M. G. Freire, A. R. R. Teles, J. N. Canongia Lopes, L. P. N. Rebelo, I. M. Marrucho and J. A. Coutinho, *Sep. Sci. Technol.*, 2012, **47**, 284–291.

64. J. F. B. Pereira, F. Vicente, V. C. Santos-Ebinuma, J. M. Araujo, A. Pessoa, M. G. Freire and J. A. P. Coutinho, *Process Biochem.*, 2013, **48**, 716–722.

65. S. P. Ventura, V. C. Santos-Ebinuma, J. F. Pereira, M. F. Teixeira, A. Pessoa and J. A. Coutinho, *J. Ind. Microbiol. Biotechnol.*, 2013, **40**, 507–516.

66. S. Li, C. He, H. Liu, K. Li and F. Liu, *J. Chromatogr. B*, 2005, **826**, 58–62.

67. M. G. Freire, C. M. Neves, I. M. Marrucho, J. N. C. Lopes, L. P. N. Rebelo and J. A. Coutinho, *Green Chem.*, 2010, **12**, 1715–1718.

68. H. Passos, M. P. Trindade, T. S. Vaz, L. P. da Costa, M. G. Freire and J. A. Coutinho, *Sep. Purif. Technol.*, 2013, **108**, 174–180.

69. M. R. Almeida, H. Passos, M. M. Pereira, Á. S. Lima, J. A. Coutinho and M. G. Freire, *Sep. Purif. Technol.*, 2014, **128**, 1–10.

70. A. F. M. Cláudio, M. G. Freire, C. S. R. Freire, A. J. D. Silvestre and J. A. P. Coutinho, *Sep. Purif. Technol.*, 2010, **75**, 39–47.

71. A. F. M. Cláudio, A. M. Ferreira, C. S. Freire, A. J. Silvestre, M. G. Freire and J. A. Coutinho, *Sep. Purif. Technol.*, 2012, **97**, 142–149.

72. J. H. Santos, S. P. Ventura, J. A. Coutinho, R. L. Souza, C. M. Soares and Á. S. Lima, *Biotechnol. Prog.*, 2015, **31**, 70–77.

73. C. L. S. Louros, A. F. M. Claudio, C. M. S. S. Neves, M. G. Freire, I. M. Marrucho, J. Pauly and J. A. P. Coutinho, *Int. J. Mol. Sci.*, 2010, **11**, 1777–1791.

74. B. Tang, W. Bi, M. Tian and K. H. Row, *J. Chromatogr. B*, 2012, **904**, 1–21.
75. M. G. Freire, C. L. S. Louros, L. P. N. Rebelo and J. A. P. Coutinho, *Green Chem.*, 2011, **13**, 1536–1545.
76. M. Domínguez-Pérez, L. I. Tomé, M. G. Freire, I. M. Marrucho, O. Cabeza and J. A. Coutinho, *Sep. Purif. Technol.*, 2010, **72**, 85–91.
77. A. F. M. Claudio, A. M. Ferreira, M. G. Freire and J. A. P. Coutinho, *Green Chem.*, 2013, **15**, 2002–2010.
78. J.-P. Vidal, in *Kirk-Othmer Encyclopedia of Chemical Technology*, John Wiley & Sons, 2006.
79. Y. Zuo, H. Chen and Y. Deng, *Talanta*, 2002, **57**, 307–316.
80. P. K. Ajikumar, K. Tyo, S. Carlsen, O. Mucha, T. H. Phon and G. Stephanopoulos, *Mol. Pharm.*, 2008, **5**, 167–190.
81. Z. Tan, F. Li and X. Xu, *Sep. Purif. Technol.*, 2012, **98**, 150–157.
82. Z. J. Tan, F. F. Li, X. L. Xu and J. M. Xing, *Desalination*, 2012, **286**, 389–393.
83. J.-P. Fan, J. Cao, X.-H. Zhang, J.-Z. Huang, T. Kong, S. Tong, Z.-Y. Tian, J.-H. Zhu and X.-K. Ouyang, *Sep. Sci. Technol.*, 2012, **47**, 1740–1747.
84. B. H. Mcanalley, EP0356484, 1993.
85. R. H. Davis, M. Leitner, J. Russo and M. Byrne, *J. Am. Podiatric Med. Assoc.*, 1989, **79**, 559–562.
86. M. H. Radha and N. P. Laxmipriya, *J. Tradit., Complementary Med.*, 2015, **5**, 21–26.
87. A. F. M. Cláudio, C. F. Marques, I. Boal-Palheiros, M. G. Freire and J. A. Coutinho, *Green Chem.*, 2014, **16**, 259–268.
88. S. P. Ventura, F. A. e Silva, A. M. Gonçalves, J. L. Pereira, F. Gonçalves and J. A. Coutinho, *Ecotoxicol. Environ. Saf.*, 2014, **102**, 48–54.
89. X. D. Hou, Q. P. Liu, T. J. Smith, N. Li and M. H. Zong, *PLoS One*, 2013, **8**, e59145-1–e59145-7.
90. M. Taha, F. A. e Silva, M. V. Quental, S. P. Ventura, M. G. Freire and J. A. Coutinho, *Green Chem.*, 2014, **16**, 3149–3159.
91. K. H. Wong, G. Q. Li, K. M. Li, V. Razmovski-Naumovski and K. Chan, *J. Ethnopharmacol.*, 2011, **134**, 584–607.
92. S. Sparg, M. Light and J. Van Staden, *J. Ethnopharmacol.*, 2004, **94**, 219–243.
93. H. Passos, A. C. Sousa, M. R. Pastorinho, A. J. Nogueira, L. P. N. Rebelo, J. A. Coutinho and M. G. Freire, *Anal. Methods*, 2012, **4**, 2664–2667.
94. K. Ninomiya, K. Inoue, Y. Aomori, A. Ohnishi, C. Ogino, N. Shimizu and K. Takahashi, *Chem. Eng. J.*, 2015, **259**, 323–329.
95. N. L. Mai, K. Ahn and Y.-M. Koo, *Process Biochem.*, 2014, **49**, 872–881.
96. J. G. Huddleston and R. D. Rogers, *Chem. Commun.*, 1998, 1765–1766.
97. D. C. Dibble, C. L. Li, L. Sun, A. George, A. R. L. Cheng, O. P. Cetinkol, P. Benke, B. M. Holmes, S. Singh and B. A. Simmons, *Green Chem.*, 2011, **13**, 3255–3264.
98. J. L. Anthony, E. J. Maginn and J. F. Brennecke, *J. Phys. Chem. B*, 2001, **105**, 10942–10949.
99. J. Lemus, C. M. Neves, C. F. Marques, M. G. Freire, J. A. Coutinho and J. Palomar, *Environ. Sci.: Processes Impacts*, 2013, **15**, 1752–1759.

100. J. Lemus, J. Palomar, F. Heras, M. A. Gilarranz and J. J. Rodriguez, *Sep. Purif. Technol.*, 2012, **97**, 11–19.
101. J. Palomar, J. Lemus, M. Gilarranz and J. Rodriguez, *Carbon*, 2009, **47**, 1846–1856.
102. C. M. Neves, J. Lemus, M. G. Freire, J. Palomar and J. A. Coutinho, *Chem. Eng. J.*, 2014, **252**, 305–310.
103. X. Qi, L. Li, T. Tan, W. Chen and R. L. Smith Jr, *Environ. Sci. Technol.*, 2013, **47**, 2792–2798.
104. J. B. Binder and R. T. Raines, *Proc. Natl. Acad. Sci. U. S. A.*, 2010, **107**, 4516–4521.
105. N. L. Mai, N. T. Nguyen, J. I. Kim, H. M. Park, S. K. Lee and Y. M. Koo, *J. Chromatogr. A*, 2012, **1227**, 67–72.
106. C. Abels, C. Redepenning, A. Moll, T. Melin and M. Wessling, *J. Membr. Sci.*, 2012, **405**, 1–10.
107. Q. Gan, M. Xue and D. Rooney, *Sep. Purif. Technol.*, 2006, **51**, 185–192.
108. Y. Deng, T. Long, D. Zhang, J. Chen and S. Gan, *J. Chem. Eng. Data*, 2009, **54**, 2470–2473.
109. C. Li, J. Han, Y. Wang, Y. Yan, J. Pan, X. Xu and Z. Zhang, *J. Chem. Eng. Data*, 2009, **55**, 1087–1092.
110. B. Wu, Y. Zhang and H. Wang, *J. Phys. Chem. B*, 2008, **112**, 6426–6429.
111. B. Wu, Y. M. Zhang and H. P. Wang, *J. Chem. Eng. Data*, 2008, **53**, 983–985.
112. C. M. Neves, M. G. Freire and J. A. Coutinho, *RSC Adv.*, 2012, **2**, 10882–10890.
113. American Water Works Association, *Water Quality and Treatment: A Handbook of Community Water Supplied*, Mcgraw-Hill, Texas, USA, 1990.
114. K. Shill, S. Padmanabhan, Q. Xin, J. M. Prausnitz, D. S. Clark and H. W. Blanch, *Biotechnol. Bioeng.*, 2011, **108**, 511–520.
115. C. Biao, L. Quan and Z. Baozhong, *Prog. Chem.*, 2012, **24**, 225–234.
116. S. H. Lee, S. H. Ha, C.-Y. You and Y.-M. Koo, *Korean J. Chem. Eng.*, 2007, **24**, 436–437.

Ionic Liquids in the Biorefinery: How Green and Sustainable Are They?

ROGER A. SHELDON[*a]

[a]Department of Biotechnology, Delft University of Technology, Julianalaan 136, 2628BL Delft, Netherlands
*E-mail: r.a.sheldon@tudelft.nl

9.1 Introduction: Green and Sustainable Chemicals from Renewable Biomass

9.1.1 The Bio-Based Economy, Green Chemistry and Sustainability

The pressing need for climate change mitigation is a major driver in the current focus on the development of green and sustainable technologies for the conversion of waste biomass to biofuels, to commodity chemicals and to new bio-based materials such as bioplastics. *Green chemistry* is concerned with the design of environmentally benign products and processes.[1,2]

Green chemistry comprises three basic elements:

(i) minimization of waste through efficient utilization of raw materials
(ii) avoiding the use of toxic and/or hazardous substances, including solvents, and

RSC Green Chemistry No. 36
Ionic Liquids in the Biorefinery Concept: Challenges and Perspectives
Edited by Rafal Bogel-Lukasik
© The Royal Society of Chemistry 2016
Published by the Royal Society of Chemistry, www.rsc.org

(iii) using renewable biomass instead of non-renewable fossil feedstocks in the form of crude oil, coal or natural gas.

Green chemistry is primary pollution prevention as opposed to end-of-pipe, waste remediation.

Sustainable development comprises the so-called three pillars of sustainability: people, planet and profit. It acknowledges the need for sustainable industrial and societal development, defined as *meeting the needs of the present generation without compromising the needs of future generations to meet their own needs.*[3] In order to be sustainable, our natural resources should not be used at rates that result in their depletion and residues should not be generated at rates that exceed the rate of their assimilation by the natural environment.

Although attention continues to be focused on waste minimization and avoiding the use of toxic and/or hazardous reagents and solvents, the last decade has witnessed a growing emphasis on the third element of green chemistry, namely the substitution of non-renewable fossil resources – crude oil, coal and natural gas – by renewable biomass as a sustainable feedstock for the manufacture of commodity chemicals and liquid fuels.[4] A bio-based economy is envisaged in which biofuels, commodity chemicals and novel materials such as bioplastics will be produced in integrated biorefineries, thus affording an environmentally beneficial reduction in the carbon footprint of chemicals and liquid fuels.[5–8] An additional benefit could be derived from the substitution of existing products by inherently safer alternatives with reduced environmental footprints, as exemplified by biocompatible and biodegradable plastics.[9]

9.1.2 First-Generation *Versus* Second-Generation Biomass

The use of first-generation biomass feedstocks, such as sugar cane, maize and edible vegetable oils, is not perceived as a sustainable option in the longer term as it competes, directly or indirectly, with food production. In contrast, second-generation integrated biorefineries will utilize mainly lignocellulosic biomass as feedstock,[10,11] produced by deliberate cultivation of fast-growing, non-edible crops or, better still, by valorization of waste biomass generated in the production of edible crops or forestry products. Examples include sugar cane bagasse, corn stover, wheat straw, rice husks and orange peel.[12] Indeed, the drive to avoid waste and find new renewable resources for fuels and chemicals has focused attention on a new and promising feedstock for biorefineries: food supply chain waste (FSCW).[13,14] Enormous amounts of organic waste are generated in the harvesting, processing and use of agricultural products, including food and beverages. In developed countries, additional waste is generated when food is discarded.[15] Most of this waste goes to landfill, which has a negative value. The proposed second-generation bio-based economy is founded on the full utilization of agricultural biomass for the production of fuels and chemicals by employing green and sustainable chemistry. Two decades ago the accent was firmly on waste prevention at source and waste remediation was not considered green chemistry. We now recognize, however, that in some cases, such as agricultural production, waste can't be avoided and there is a growing need for effective utilization or, better still, valorization of this waste, in order to create value from unavoidable waste.

9.1.3 Conversion of Lignocellulosic Biomass

Lignocellulose is much more difficult to process than the first-generation renewable feedstocks – sugars, starches and vegetable oils. It consists of three major polymeric components: lignin (*ca.* 20%), cellulose (*ca.* 40%) and hemicellulose (*ca.* 25%). Lignin is a three-dimensional polyphenolic biopolymer having a non-uniform structure that imparts rigidity and recalcitrance to plant cell walls. It is in volume the second largest biopolymer after cellulose and the only one composed entirely of aromatic subunits. Irrespective of whether the goal is to produce liquid fuels or commodity chemicals, the lignocellulose first has to be depolymerized and (partially) deoxygenated and there are basically two ways of achieving this: thermochemical and hydrolytic (Scheme 9.1).[16]

Thermochemical processing involves pyrolysis to a mixture of charcoal and pyrolysis oil or gasification to afford syngas (a mixture of carbon monoxide and hydrogen), analogous to syngas from coal gasification.[17] The syngas can be converted to liquid fuels or platform chemicals *via* established technologies such as the Fischer–Tropsch process[18] or methanol synthesis, respectively.[17] An interesting alternative is to use the syngas or mixtures of carbon dioxide and hydrogen as a fermentation feedstock for the microbial

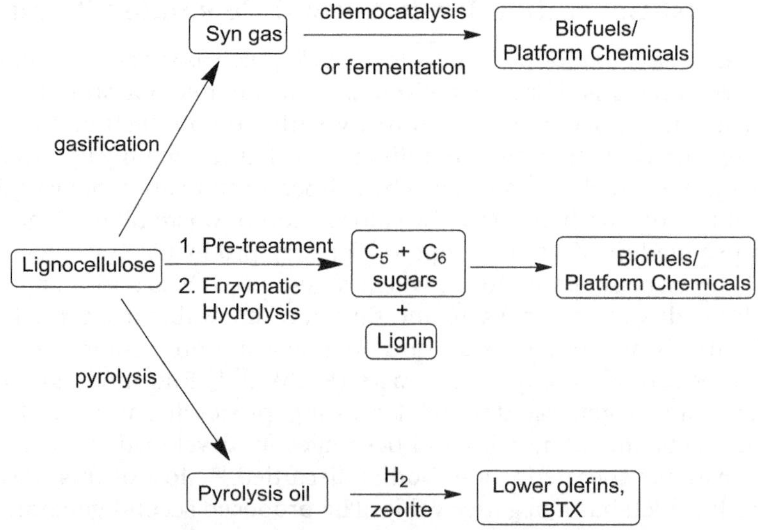

Scheme 9.1   Strategies for primary conversion of lignocellulose.

production[19,20] of biofuels and platform chemicals, a technology being developed and commercialized by, *inter alia*, Lanzatech[21] and Coskata.[22]

Hydrolytic conversion of lignocellulose, affording a mixture of lignin, hemicellulose and cellulose, is catalysed by mineral acids at elevated temperatures or enzyme cocktails under milder conditions. In the former case copious amounts of inorganic waste, *e.g.* chlorides or sulfates, are formed, as a result of neutralization of the dilute mineral acid used and attention is being focused on the design of solid acid catalysts for a cleaner conversion of biomass[23–25] by analogy with the processing of crude oil fractions in the petrochemical industry. In the case of enzymatic hydrolysis some form of pre-treatment, such as a steam or ammonia explosion, is necessary to open up the lignocellulose structure and render the targeted glycoside and ester bonds accessible to the enzyme cocktails.[26] In both cases, the lignin is generally used to generate electricity. Alternatively, more added value would be obtained by converting the lignin into chemically useful aromatic components.[27–29] The C_5 and C_6 sugars, derived from hydrolysis of cellulose and hemicellulose, can subsequently be used as raw materials for conversion to biofuels or platform commodity chemicals by either fermentation or chemocatalytic processes.[30–32]

9.2 Reaction Media for Lignocellulose Pre-Treatment and Conversion

The primary conversion of lignocelluloses is generally conducted in water, in which the cellulose, hemicellulose and lignin are present as suspended solids. The use of alternative reaction media that (partially) dissolve these polymeric substrates could have processing advantages. However, in order to be economically and environmentally viable, the solvent should be inexpensive, non-toxic, biodegradable, recyclable and preferably derived from renewable resources.

9.2.1 The Organosolv Process

In the Organosolv process (Scheme 9.2),[33] for example, 50% (w/w) aqueous ethanol is used as the solvent at elevated temperatures (190–210 °C) in the presence or absence of an acid catalyst. This results in hydrolysis of the hemicellulose and dissolution of the lignin. The remaining cellulose is separated and the dissolved lignin is precipitated by water addition or ethanol evaporation. Overall, the products comprise cellulose, solid lignin and an aqueous stream containing hemicellulose, C_5 sugars and derivatives thereof, such as furfural.

9.2.2 Ionic Liquids

In recent years attention has been focused on the use of ionic liquids as reaction media for conversion of renewable biomass, in particular lignocellulose.[34,35] Some ionic liquids are able to dissolve lignocellulose[36–38] and, in

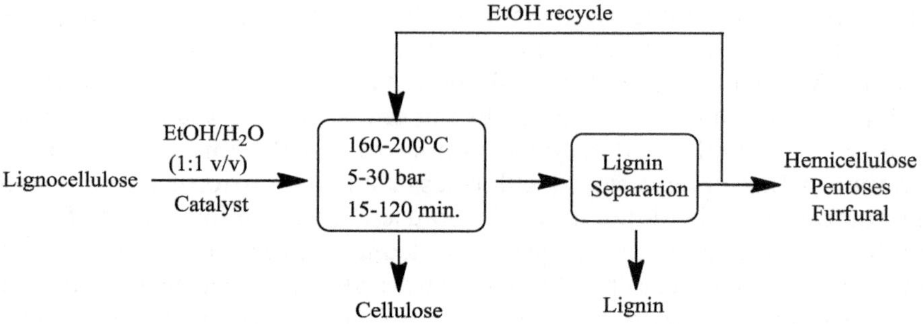

Scheme 9.2   The organosolv process.

combination with water, are potential reaction media for various steps in a second generation, integrated biorefinery: (i) pre-treatment of the lignocellulose, (ii) chemocatalytic[39,40] or biocatalytic[41–43] hydrolysis of the polysaccharides (saccharification), and (iii) further conversion of the resulting C_5 and C_6 sugars into biofuels and/or platform chemicals.

9.3   Catalysis in Ionic Liquids

Ionic liquids (ILs) are substances that are composed entirely of ions and are liquid at or close to room temperature. Interest in their use as reaction media,[44–47] in particular for catalytic processes,[48–50] has increased exponentially over the last two decades. This interest derives largely from their negligible vapour pressure, coupled with good thermal stability and widely tuneable properties, such as polarity, hydrophobicity and solvent-miscibility behaviour, through manipulation of the structures of both cation and anion. Consequently, they have been widely advocated as green alternatives to volatile organic solvents.

Interest in conducting catalytic processes in ILs dates back to the 1960s. Initial studies were conducted typically in ILs consisting of dialkylimidazolium and alkylpyridinium cations together with chloroaluminate anions. An important shortcoming of the latter was their extreme sensitivity towards hydrolysis and accompanying highly corrosive properties. In the 1990s these highly reactive anions were replaced by less reactive, weakly coordinating anions, in particular tetrafluoroborate (BF_4^-) and hexafluorophosphate (PF_6^-), which were much more stable towards air and water and, hence, more suitable as a medium for performing catalytic reactions.[48–50]

9.3.1   Biocatalysis in Ionic Liquids

Initial studies of biocatalysis in ionic liquids[51–53] were also conducted in water-miscible [bmim][BF₄] or water-immiscible [bmim][PF₆]. However, these anions are not completely stable towards hydrolysis, which can lead to the formation of traces of hydrofluoric acid. Hence, other weakly coordinating anions, such as trifluoroacetate, triflate, bistriflamide, dicyanamide and

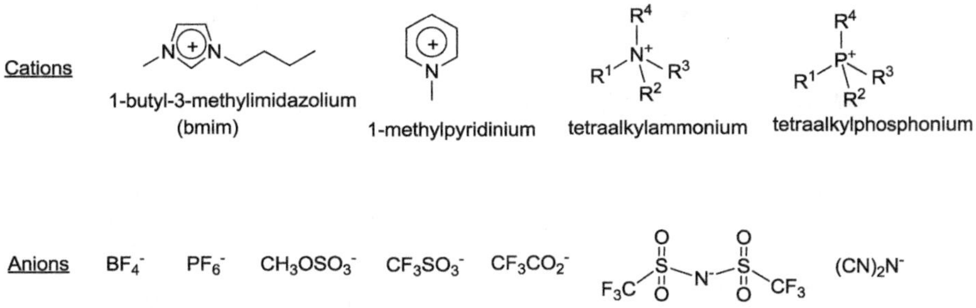

Cations

1-butyl-3-methylimidazolium (bmim)   1-methylpyridinium   tetraalkylammonium   tetraalkylphosphonium

Anions BF_4^- PF_6^- $CH_3OSO_3^-$ $CF_3SO_3^-$ $CF_3CO_2^-$ $(CN)_2N^-$

Scheme 9.3   Structures of first-generation ionic liquids.

methylsulfate, were introduced. In the context of biotransformations these ILs can be referred to as first-generation ILs (Scheme 9.3). From both a stability and cost viewpoint it would seem more logical to have used simple anions such as chloride and acetate but their use was prohibited by the fact that such coordinating anions can cause dissolution and accompanying deactivation of the enzymes (see later).

9.4 First- and Second-Generation Ionic Liquids

9.4.1 First-Generation

The use of IL media for catalytic processes was motivated by the possibility of replacing volatile organic solvents with non-volatile ILs with low flammability, whereby the risk of air pollution is largely circumvented. However, ILs have significant solubility in water and, hence, the environmental fate and possible ecotoxicity of ILs is a cause for concern.[54] Based on their high chemical and thermal stabilities one could expect problems with poor biodegradability, bioaccumulation and aquatic ecotoxicity. Indeed, the first-generation dialkylimidazolium ILs, such as [bmim][BF_4] and [bmim][PF_6], and tetraalkylammonium ILs exhibit acute toxicity towards a variety of aquatic organisms[55–57] and are poorly biodegradable.[58] A further shortcoming of these first-generation ILs, comprising quaternary ammonium cations in conjunction with relatively expensive anions, is that they are prohibitively expensive and many of the synthetic methods used to make them are considerably less green (high E factors) than is taken for granted.[59]

9.4.2 Second-Generation

Consequently, second-generation ILs containing more biocompatible cations and anions, often derived from more eco-friendly natural products,[60] such as carbohydrates[61,62] and amino acids,[63,64] preferably in a simple derivitization step, are emerging (see Scheme 9.4 for examples).

Choline-based ILs, for example, are prepared by reaction of inexpensive choline hydroxide with a carboxylic acid,[65,66] such as naturally occurring, biodegradable naphthenic acids,[66] affording the corresponding carboxylate salt

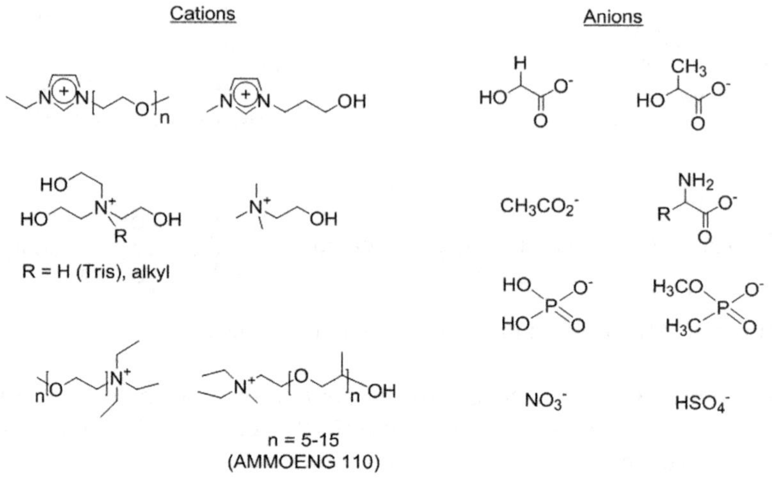

Cations   Anions

R = H (Tris), alkyl

$CH_3CO_2^-$

NO_3^- HSO_4^-

n = 5–15
(AMMOENG 110)

Scheme 9.4   Structures of second-generation ionic liquids.

with water as the sole co-product. An investigation of the toxicity and biodegradability of cholinium salts of various amino acids revealed that most of these ILs combined low toxicity with high biodegradability.[67] Moreover, the enzyme inhibitory potential of these ILs was an order of magnitude less than traditional [bmim][Cl]. Similarly, 2-hydroxyethylammonium lactate consists of a cation closely resembling that of the natural cation, choline, and a natural, readily biodegradable anion.[68]

An important driver for the use of ILs is the fact that they lend themselves to fine tuning of their properties by an appropriate selection of cation and anion. The current trend, therefore, is towards the deliberate design of ILs that can be used for particular biotransformations while maintaining a low environmental footprint. For example, AMMOENG 110™, containing a choline-like cation combined with acetate as the anion (see Scheme 9.4 for the structure) is compatible with enzymes and, at the same time, is able to dissolve carbohydrates (including cellulose), triglycerides and amino acids.[69]

9.4.3 Protic Ionic Liquids

The quest for inexpensive ILs with reduced ecotoxicity, improved biodegradability and compatibility with enzymes, recently led to the use of protic ionic liquids (PILs) containing a protonated N-atom in the cation, as a solvent for biotransformations, *e.g.* transesterifications catalysed by *Candida antarctica* lipase B (CaLB) (Scheme 9.5).[70] PILs are exquisitely simple to prepare by simply mixing a tertiary amine with an acid, such as a carboxylic acid, and are known[71,72] to exhibit better biodegradability and lower toxicity than the corresponding quaternary ammonium salts. Moreover, they have suitable H-bond donating properties for interaction with and stabilization of enzymes and, when combined with alkanoate anions, they are self-buffering.

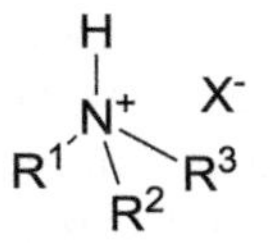

$$R^1, R^2, R^3 = CH_3, \text{n-}C_4H_9, \text{i-}C_4H_9,$$
$$\text{n-}C_8H_{17}, \text{n-}C_{12}H_{25}$$

$$X = CH_3CO_2, C_2H_5CO_2,$$
$$C_5H_{11}CO_2, C_7H_{15}CO_2$$

Scheme 9.5   Structures of protic ionic liquids (PILs).

9.5   Enzymes in Ionic Liquids

In the last decade a plethora of enzymatic transformations has been shown to be feasible in ILs.[73] This involved mainly hydrolytic enzymes, such as lipases, esterases, proteases and glycosidases but oxidoreductases – ketoreductases, peroxidises, Baeyer Villiger monooxygenases and laccases – and lyases, such as hydroxynitrile lyases, have also been extensively studied.[74] Most of the early examples involved reactions with suspensions of free enzymes, as powders, in ILs containing weakly coordinating anions, in particular BF_4^-, PF_6^- and $(CF_3SO_2)N^-$, which are not able to dissolve proteins. Enzymes exhibited remarkable storage and operational stabilities in such non-coordinating ILs.[75–77] In contrast, hydrophilic ILs containing coordinating anions, such as nitrate, sulfate and chloride, are able to dissolve biopolymers by breaking intermolecular hydrogen bonds. For example, [bmim][Cl] effectively dissolves cellulose because chloride ions interact as H-acceptors with the cellulose OH groups, thereby breaking the H-bonding network of cellulose.[78,79] Similarly, proteins dissolve in ILs containing coordinating anions. In many cases, this leads to deactivation of an enzyme by disruption of intramolecular hydrogen bonds that are essential for the tertiary structure and, hence, the activity of the enzyme.[80] Dissolution of *Trichoderma reesei* cellulase in [bmim][Cl], for example, resulted in its deactivation, presumably owing to interactions with the strongly coordinating chloride ion.[81] Interestingly, substantial recovery of hydrolytic activity was observed when the solution of an inactivated enzyme in an IL was diluted with a large excess of water,[80] suggesting that the enzyme unfolds and denatures on dissolving in the IL but on the addition of water it (partially) refolds into its active form.

9.5.1   Strategies for Maintaining Activity

There are basically two strategies for maintaining the activity of an enzyme after dissolution in an IL: (i) by designing enzyme compatible ILs or (ii) by modifying the enzyme to make it more resistant to denaturation by the IL.

The first example of the former strategy was, to our knowledge, reported by Bruce and co-workers[82,83] who designed an IL containing a hydroxyl functionality in both the cation and the anion, based on the assumption that the enzyme would be stable in an IL that more closely resembled an aqueous environment. Indeed, an IL consisting of a 3-hydroxypropyl-1-methylimidazolium cation (I) and a glyoxylate anion (II) was designed for the enzymatic dehydrogenation of codeine to codeinone.[82,83] It dissolved the substrate, product, enzyme and the cofactor and the dissolved enzyme was more active, even at a water content of 100 ppm, than as a suspension in other ILs. Similarly, Das and co-workers[84] showed that horseradish peroxidase was compatible with tetrakis (2-hydroxyethyl) ammonium triflate.

An example of the second approach is provided by the reported stabilization of enzymes towards denaturation in ILs, by immobilization, as cross-linked enzyme aggregates (CLEAs).[85–89]

9.6 Deep Eutectic Solvents

In addition to this second generation of IL media for biocatalysis, another class of interesting solvents has emerged in recent years: so-called deep eutectic solvents (DES).[90–99] The latter are formed by mixing certain solid salts, in varying proportions, with a hydrogen bond donor such as urea and glycerol. For example, combining choline chloride (melting point 302 °C) with urea (melting point 132 °C) affords a DES which is a liquid at room temperature (melting point 12 °C). Although DESs are, strictly speaking, not ILs (since they contain uncharged moieties, *e.g.* urea), they have properties resembling those of ILs and are eminently suitable as reaction media for biocatalysis.[100–105] In particular DESs derived from mixtures of cholinium salts with natural H-bond acceptors, such as glucose and glycerol (see Scheme 9.6

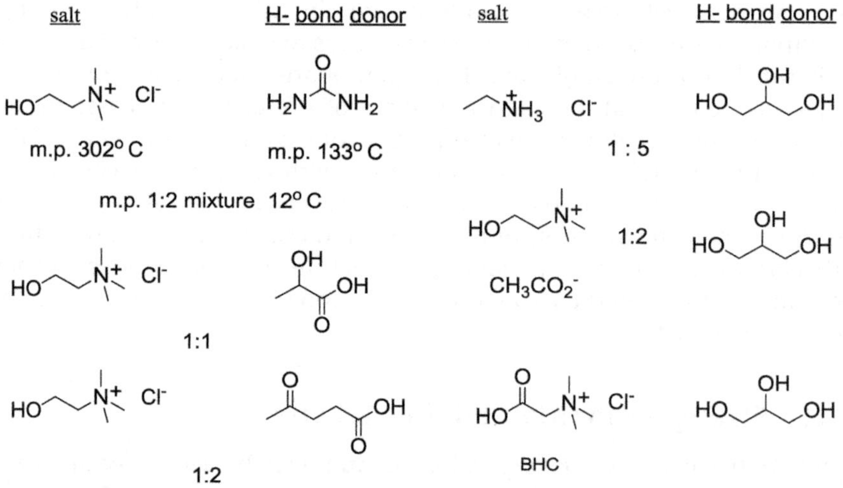

Scheme 9.6 Structures of deep eutectic solvents (DESs).

for examples) have an activating and stabilizing effect on enzymes.[106] For example, Kazlauskas and co-workers[107] reported CaLB catalysed transesterifications and amidations in a 1:2 mixture of choline chloride and glycerol. Similarly, Zhao and co-workers[108] reported a process for biodiesel production in the same DES using free CaLB as the catalyst.

9.7 Biotransformations of Highly Polar Substrates Such as Carbohydrates

Some reactions, such as (trans)esterifications cannot be performed in water owing to equilibrium limitations and/or product hydrolysis, which necessitates conducting such reactions in an organic solvent. This is a challenge with highly polar substrates such as carbohydrates and nucleosides which are very soluble in water but have negligible solubilities in common organic solvents. The few exceptions, such as dimethyl formamide (DMF), dimethyl sulfoxide (DMSO) and pyridine, have many undesirable features and/or are incompatible with enzymes. In the last decade increasing interest has been focused on the design of ILs that are able to dissolve substantial amounts of mono-, oligo- and poly-saccharides[109,110] and, hence, can be used as non-aqueous media for performing reactions of a wide variety of carbohydrates,[111] in the biocatalytic synthesis of sugar derivatives,[112] for example. The solubilities of carbohydrates in ILs are largely determined by the nature of the anion. Strong H-bond accepting anions, such as chloride and acetate, are able to break the intermolecular H-bonds between carbohydrate molecules and, hence, cause their dissolution in the IL. MacFarlane and co-workers[113] were the first to note that ILs containing the dicyanamide anion, $((CN)_2N^-$, dca) dissolved glucose in concentrations in excess of 100 g L^{-1}. We subsequently showed[114] that [bmim][dca] is an excellent solvent for sucrose, lactose and β-cyclodextrin and that it could be used as a reaction medium for enzymatic esterifications of sugars. Interestingly, Rogers and co-workers[115] showed that [bmim][Cl] can even dissolve cellulose to the extent of 100 g L^{-1} at 100 °C. More recently, Ohno and co-workers[116] reported that ILs composed of dialkylimidazolium cations in combination with dimethyl phosphate, $(MeO)_2PO_2^-$, and dimethyl phosphonate, $Me(MeO)PO_2^-$, are able to dissolve cellulose under mild conditions. Similarly, other polysaccharides, such as hemicellulose,[117] starch[118] and chitin,[119,120] as well as other biopolymers, such as lignin[121,122] and proteins, *e.g.* silk[123] and wool,[124] dissolve in ILs.

9.7.1 Enzyme Compatible Ionic Liquids

Zhao and co-workers[125] designed a series of ILs, consisting of acetate anions and an imidazolium cation containing an oligoethylene (or oligopropylene) glycol side chain (Scheme 9.7), that are both enzyme-compatible and dissolve more than 10 wt.% cellulose and up to 80 wt.% glucose.

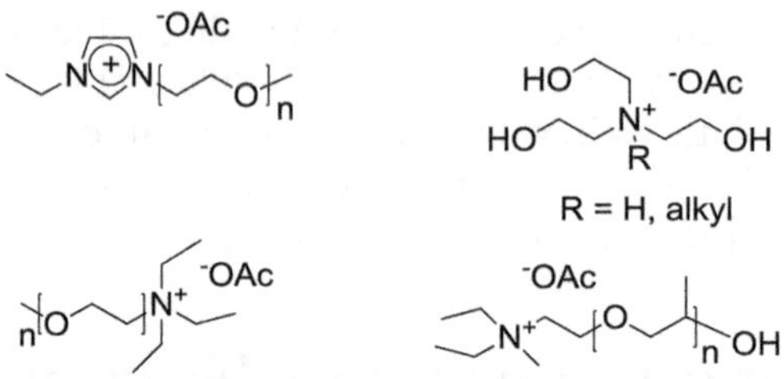

R = H, alkyl

Scheme 9.7　Enzyme compatible ILs.

The length and steric bulk of the glycol chain in the cation could be designed to dissolve sufficient amounts of carbohydrates, while, at the same time, being compatible with enzymes. Free CaLB dissolved in these ILs with retention of activity, thus providing the possibility of conducting homogeneous enzymatic reactions such as the acylation of glucose and the steroid betulinic acid.[125]

9.7.2　Enzymatic Acylation of Carbohydrates

Fatty acid esters of sugars are commercially important products with a wide variety of applications in food, cosmetics and pharmaceutical formulations because, in addition to being derived from renewable raw materials, they are tasteless, odourless, nontoxic, non-irritant, biodegradable and they have a hydrophilic–lipophilic balance (HLB) which is tuneable by a suitable choice of fatty acid and carbohydrate. They are generally manufactured using traditional chemical processes but there is increasing interest[126,127] in the use of enzymatic alternatives that can be conducted under milder conditions with higher selectivities and, consequently, higher product qualities.

In particular, sucrose fatty acid esters[128] find many applications in food and cosmetics, for example, as an emulsifier. They are currently manufactured by a chemical process at elevated temperatures, resulting in low selectivities and side reactions affording coloured impurities. However, owing to the very low solubility of sucrose in common organic solvents, it is a challenge to find a suitable reaction medium for the (trans)esterification of sucrose with a fatty acid (ester). We already showed[129] in 1996 that Novozyme 435 (immobilized CaLB) catalysed the selective acylation of a suspension of sucrose with ethyl dodecanoate in refluxing *tert*-butanol (82 °C), giving a 35% conversion to a 1:1 mixture of the 6- and 6′- regioisomers in 7 days. Obviously, such a slow reaction is not attractive for large-scale production. Therefore, attention turned to the design of ILs that can function as reaction media for the synthesis of sugar fatty acid esters in general and sucrose fatty acid esters in particular. However, most studies of sugar acylations concentrated mainly on monosaccharides, which are relatively easy substrates compared to disaccharides such as sucrose.

9.8 Lignocellulose Pre-Treatment and Saccharification in ILs and DES

Currently, much attention is focused on the use of neoteric solvents, such as ILs and DESs, as reaction media for the enzymatic hydrolysis of lignocellulose to fermentable sugars (saccharification) and for further conversion of the latter to biofuels and commodity chemicals.[130] A variety of lignocellulosic biomass,[38] such as wood chips,[36,131,132] corn stover,[133] wheat straw[134] and sugar cane bagasse,[135,136] has been shown to dissolve in ILs and the latter could be used as reaction media for subsequent hydrolysis of the cellulose and hemicellulose to fermentable sugars (saccharification) catalysed either by acids[137] or enzymes.

9.8.1 First-Generation Ionic Liquids

Much of this work was performed in first-generation dialkylimidazolium ILs. For example, [emim][OAc] was used as a solvent for the pre-treatment and fractionation of wheat straw into lignin and carbohydrate fractions.[138,139] The same IL was used for the saccharification of wood chips,[140] switch grass (*Miscanthus giganteus*)[141] and sugar cane bagasse[43] by combined pre-treatment and enzymatic hydrolysis. Similarly, Koo and co-workers[142] used a 1:1 (v/v) mixture of [emim][OAc] and DMSO to increase the solubility of the lignocellulose and decrease the viscosity of the medium in the pre-treatment and enzymatic hydrolysis of rice straw. Interestingly, Li and co-workers[37] employed a combination of two ILs, [bmim][Cl] to dissolve the lignocellulose and the strongly acidic IL, 1-(4-sulfobutyl)-3-methylimidazolium hydrogensulfate ($C_4H_8SO_3HmimHSO_4$), to catalyse the hydrolysis of the cellulose and hemicellulose.

9.8.2 Cellulase Enzymes in Ionic Liquids

In the enzymatic hydrolysis of mixtures of cellulose and hemicellulose, a cocktail of several enzymes having the collective name 'cellulase' is used. In order to be viable in IL media, it is necessary that the cellulase exhibits activity and stability in the presence of ILs, and this has been demonstrated with [emim][OAc].[143] Interestingly, cellulase CLEAs displaying good activity and stability have been described.[144–146] Hybrid cellulase CLEAs containing a silica core, prepared by physical adsorption of cellulose CLEAs on a highly porous silica support, exhibited twice as much activity as the regular CLEA, and had the added advantage that it settled better after the hydrolysis, thus facilitating its separation.[147] The immobilization of cellulase CLEAs[148,149] and xylanase CLEAs[150] (for hemicellulose hydrolysis) on magnetic nanoparticles (MNPs), consisting of an amine-functionalized magnetite (Fe_3O_4)-silica core, have also been described. They have the advantage that they can be magnetically separated from suspensions of other solids, as is the case with lignocellulosic biomass. However, to our knowledge, such magnetic cellulose CLEAs have not yet been used in IL-containing media.

9.8.3 Second-Generation Ionic Liquids

In the above examples, classical first-generation dialkylimidazolium ILs were used. However, as discussed earlier, from both an economic and an environmental viewpoint it would be more interesting to use ILs based on renewable raw materials, or PILs or DESs. Moreover, it has recently been shown[151] that dialkylimidazolium salts with carboxylate anions, such as the widely used [emim][OAc] are not completely stable under the operating conditions of lignocellulose deconstruction because they react with reducing sugars.

Zong and co-workers[152] used a 20:80 mixture (v/v) of the renewable cholinium lysinate IL, [Ch][Lys], and water for the pre-treatment and subsequent enzymatic hydrolysis of rice straw. Similarly, Singh and co-workers[153] compared four different ILs in the pre-treatment and enzymatic hydrolysis of switch grass: [emim][OAc], [emim][Lys], [Ch][OAc] and [Ch][Lys]. All four were effective but the ILs containing the lysinate anion afforded greater delignification and higher glucose yields (78–96% *versus* 56–90%).

In the final analysis, the sustainability and, hence, the success of ILs in lignocellulose pre-treatment and saccharification will depend on the cost of the IL, its environmental acceptability and its stability under operating conditions, the biomass loading and its recycling efficiency.[154] Welton and co-workers[155] used a mixture of the PIL, 1-butylimidazolium hydrogensulfate, [bhim][HSO$_4$], and water (80:20 v/v) for dissolution and subsequent cellulase-catalysed saccharification of switch grass. In a study[156] designed to identify low-cost PILs for lignocellulose pre-treatment, a range of PILs containing the hydrogensulfate anion were prepared by simply mixing the corresponding amines with sulfuric acid. The best results were obtained with triethylammonium hydrogensulfate, [Et$_3$NH][HSO$_4$]. The cost of such protic hydrogensulfate ILs is primarily determined by the choice of amine, owing to the very low cost of sulfuric acid and the simple synthesis which has an atom economy of 100% and a very low E factor.[157] Thus, the cost price of the [bhim][HSO$_4$] and [Et$_3$NH][HSO$_4$] were calculated to be \$2.96 per kg and \$1.24 per kg, respectively.[158]

9.8.4 Natural Deep Eutectic Solvents (NADES)

Deep eutectic solvents (DES) constitute an inexpensive alternative to ILs for lignocellulose pre-treatment (see Section 9.6) and a reengineered cellulase was recently shown[159] to be stable in a mixture of water and ChCl/glycerol (1:2 v/v). DESs such as ChCl/glycerol and ChCl/lactic acid are prepared by simply mixing raw materials costing *ca.* \$1 per kg or less and cannot, therefore, cost more than *ca.* \$1 per kg. Indeed, natural deep eutectic solvents (NADES), which are derived from primary metabolites, namely, amino acids, glycerol, certain carboxylic acids, sugars and cholinium and betaine salts (see Scheme 9.8), have been heralded as solvents for the 21st century[160] and new potential media for green technology.[161] Furthermore, determination of the concentrations of primary metabolites in cells, using NMR-based

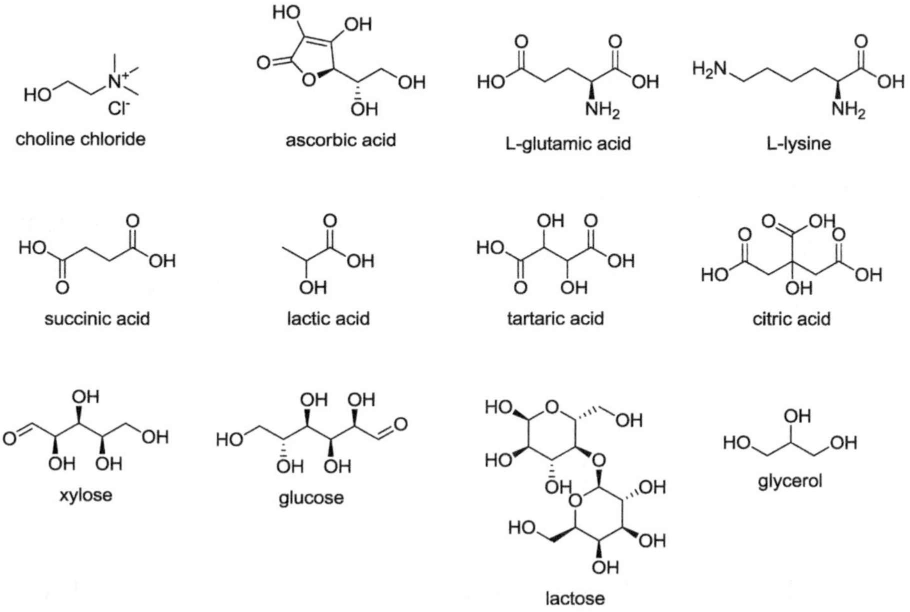

Scheme 9.8   Structures of natural deep eutectic solvents (NADES).

metabolomics, revealed that a few very simple molecules are always present in considerable amounts in all microbial, mammalian and plant cells, suggesting that they must serve some basic cellular function.[162] These include certain sugars, amino acids, organic acids (such as lactic, citric, malic and succinic acids) and choline. With the exception of sugars, which may serve as a source of energy, these compounds are present in such large amounts that it does not seem reasonable to consider them only as intermediates in metabolic pathways. Consequently, it was postulated[162] that they constitute a third type of liquid in the cell, in addition to water and lipids, and that they may play a role in various cellular processes such as the biosynthesis of water-insoluble small molecules, *e.g.* steroids and flavonoids, and of macromolecules. It would seem to be worthwhile, therefore, to investigate the use of inexpensive NADES–water mixtures as reaction media for lignocelluloses pre-treatment and saccharification.

9.9   Further Conversion of C_5 and C_6 Sugars in ILs and DESs

Interest in the use of ILs and DESs as reaction media for the conversion of lignocellulosic biomass to biofuels and chemicals is not limited to the pre-treatment and saccharification steps discussed in the preceding section. These neoteric solvents can also be used for subsequent conversion of the C_5 and C_6 sugars into platform chemicals and/or liquid fuels.

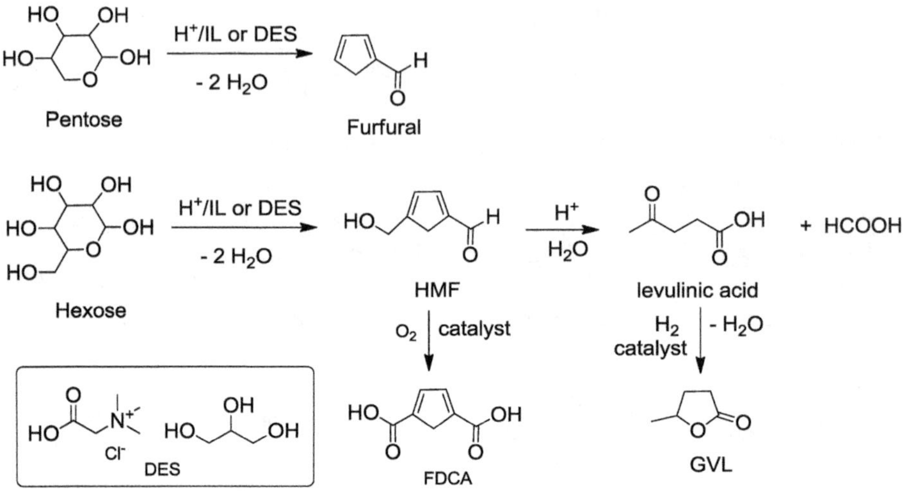

Pentose

H⁺/IL or DES
− 2 H₂O

Furfural

Hexose

H⁺/IL or DES
− 2 H₂O

HMF

H⁺
H₂O

levulinic acid

+ HCOOH

O₂ | catalyst

H₂ | − H₂O
catalyst

DES

FDCA

GVL

Scheme 9.9 Conversion of C₅ and C₆ sugars to furfural and HMF.

For example, acid-catalysed hydrolysis of glucose (from cellulose) and xylose (from hemicellulose) affords 5-hydroxymethyl furfural (HMF) and furfural, respectively (Scheme 9.9),[163] both of which have potential as platform chemicals.[31] Furfural is currently produced from oat hulls in an annual volume of more than 400 000 tons[164] and finds applications in, *inter alia*, solvents, polymers and fuel additives. HMF has enormous potential as a raw material for the production of chemicals, polymers and biofuels but its production has not yet been reduced to commercial practice.[165] For example, it can be further converted to levulinic acid (LA) and γ-valerolactone (GVL) or to furan-2,5-dicarboxylic acid (FDCA), a potential building block for polyesters.[166] Recently, catalytic processes for the direct conversion of cellulose to HMF, LA and lactic acid have been described.[167]

Acid-catalysed dehydration of hexoses to hydroxymethyl furfural (HMF), in industrially attractive selectivities, is challenging owing to the propensity of HMF towards further reaction under the acidic reaction conditions. The conversion of glucose to HMF proceeds *via* initial isomerization to fructose and higher yields are generally obtained by starting from fructose. Recent developments suggest that these dehydrations can also be performed more selectively in ionic liquids as reaction media,[168–170] using acids or metal chlorides as catalysts, but downstream processing is an issue that remains to be resolved.

DESs have also been used for the conversion of hexoses to HMF. For example, the DES comprising choline chloride and citric acid afforded HMF in 91% yield from fructose.[171] The zwitterionic glycine betaine, *N*,*N*,*N*-trimethylglycine, is a metabolite of choline and an important by-product of the sugar beet processing industry.[172] It is also available as the hydrochloride salt, betaine hydrochloride (BHC). Both products are biocompatible, biodegradable and inexpensive. BHC has a melting point of 242 °C and, hence, is not an IL but in combination with water or glycerol it forms DESs that can be used as a solvent for carbohydrate processing. Moreover, because of their acidic properties, BHC-based systems can act as both acid catalyst and solvent for carbohydrate conversions. For example, BHC, in combination with water or glycerol, was used as the reaction medium for the conversion of fructose and inulin to HMF in yields of up to 84%.[173] Similarly, by tuning the reaction temperature it was possible to control the conversion of the hemicellulose and cellulose in lignocellulosic biomass from wheat straw to furfural and LA, respectively.[174]

9.10 Conclusions and Prospects

Neoteric solvents – ILs and DESs – have the potential to play an important role as a (co)solvent in various stages of the conversion of lignocellulosic biomass in second-generation biorefineries: in pre-treatment and saccharification and in further conversion of the C_5 and C_6 sugars to platform chemicals and biofuels. However, this potential can only be realized if they are sustainable, *i.e.* they are biocompatible, biodegradable, economically attractive and preferably derived from renewable resources. Indeed, it appears to be possible to produce second-generation ILs, PILs and DESs from inexpensive raw materials for prices in the region of $1 per kg. However, even at this price level the IL or DES will have to be recovered and recycled multiple times in order to be economically and environmentally viable, *i.e.* to be sustainable, and this still needs to be convincingly and broadly demonstrated.

References

1. P. T. Anastas and J. C. Warner, *Green Chemistry: Theory and Practice*, Oxford University Press, New York, 1998.
2. R. A. Sheldon, I. Arends and U. Hanefeld, *Green Chemistry and Catalysis*, John Wiley & Sons, 2007.
3. *United Nations Report of the World Commission on Environment and Development*, Brundtland Commission, 1987.
4. P. Imhof and J. C. van der Waal, *Catalytic Process Development for Renewable Materials*, John Wiley & Sons, 2013.
5. S.-T. Yang, H. A. El-Enshasy and N. Thongchul, *Bioprocessing Technologies in Biorefinery for Sustainable Production of Fuels, Chemicals, and Polymers*, John Wiley & Sons, 2013.
6. P. Gallezot, *Chem. Soc. Rev.*, 2012, **41**, 1538–1558.
7. P. Gallezot, *Catal. Today*, 2007, **121**, 76–91.
8. A. Corma, S. Iborra and A. Velty, *Chem. Rev.*, 2007, **107**, 2411–2502.
9. E. Jong, A. Higson, P. Walsh and M. Wellisch, *Biofuels, Bioprod. Biorefin.*, 2012, **6**, 606–624.
10. J. H. Clark, *J. Chem. Technol. Biotechnol.*, 2007, **82**, 603–609.
11. C. H. Christensen, J. Rass-Hansen, C. C. Marsden, E. Taarning and K. Egeblad, *ChemSusChem*, 2008, **1**, 283–289.

12. C. O. Tuck, E. Perez, I. T. Horvath, R. A. Sheldon and M. Poliakoff, *Science*, 2012, **337**, 695–699.
13. C. S. K. Lin, L. A. Pfaltzgraff, L. Herrero-Davila, E. B. Mubofu, S. Abderrahim, J. H. Clark, A. A. Koutinas, N. Kopsahelis, K. Stamatelatou and F. Dickson, *Energy Environ. Sci.*, 2013, **6**, 426–464.
14. L. A. Pfaltzgraff, E. C. Cooper, V. Budarin and J. H. Clark, *Green Chem.*, 2013, **15**, 307–314.
15. T. Stuart, *Waste: Uncovering the Global Food Scandal*, Penguin, 2009.
16. J.-P. Lange, *Biofuels, Bioprod. Biorefin.*, 2007, **1**, 39–48.
17. R. A. Sheldon, *Chemicals from Synthesis Gas*, D. Reidel Publishing Company, Dordrecht, 1983.
18. E. Rytter, E. Ochoa-Fernández and A. Fahmi, in *Catalytic Process Development for Renewable Materials*, Wiley-VCH Verlag GmbH & Co. KGaA, 2013, pp. 265–308.
19. P. C. Munasinghe and S. K. Khanal, *Bioresour. Technol.*, 2010, **101**, 5013–5022.
20. A. M. Henstra, J. Sipma, A. Rinzema and A. J. Stams, *Curr. Opin. Biotechnol.*, 2007, **18**, 200–206.
21. J. Daniell, M. Köpke and S. D. Simpson, *Energies*, 2012, **5**, 5372–5417.
22. Coskata, Inc., http://www.coskata.com/.
23. R. Rinaldi and F. Schüth, *Energy Environ. Sci.*, 2009, **2**, 610–626.
24. J. M. Thomas, J. C. Hernandez-Garrido and R. G. Bell, *Top. Catal.*, 2009, **52**, 1630–1639.
25. P. L. Dhepe and A. Fukuoka, *ChemSusChem*, 2008, **1**, 969–975.
26. P. Kumar, D. M. Barrett, M. J. Delwiche and P. Stroeve, *Ind. Eng. Chem. Res.*, 2009, **48**, 3713–3729.
27. H. Leisch, S. Grosse, K. Morley, K. Abokitse, F. Perrin, J. Denault and P. C. K. Lau, *Green Process. Synth.*, 2013, **2**, 7–17.
28. J. Zakzeski, P. C. Bruijnincx, A. L. Jongerius and B. M. Weckhuysen, *Chem. Rev.*, 2010, **110**, 3552–3599.
29. M. Kosa and A. J. Ragauskas, *Green Chem.*, 2013, **15**, 2070–2074.
30. T. M. Carole, J. Pellegrino and M. D. Paster, *Appl. Biochem. Biotechnol.*, 2004, **115**, 871–885.
31. J. J. Bozell and G. R. Petersen, *Green Chem.*, 2010, **12**, 539–554.
32. P. Y. Dapsens, C. Mondelli and J. Pérez-Ramírez, *ACS Catal.*, 2012, **2**, 1487–1499.
33. J. Wildschut, A. T. Smit, J. H. Reith and W. J. Huijgen, *Bioresour. Technol.*, 2013, **135**, 58–66.
34. C.-Z. Liu, F. Wang, A. R. Stiles and C. Guo, *Appl. Energ.*, 2012, **92**, 406–414.
35. Z. Fang, *Production of Biofuels and Chemicals with Ionic Liquids*, Springer, 2014.
36. M. Zavrel, D. Bross, M. Funke, J. Buchs and A. C. Spiess, *Bioresour. Technol.*, 2009, **100**, 2580–2587.
37. J. X. Long, X. H. Li, B. Guo, F. R. Wang, Y. H. Yu and L. F. Wang, *Green Chem.*, 2012, **14**, 1935–1941.

38. T. Vancov, A.-S. Alston, T. Brown and S. McIntosh, *Renewable Energy*, 2012, **45**, 1–6.
39. R. Rinaldi, R. Palkovits and F. Schüth, *Angew. Chem., Int. Ed.*, 2008, **47**, 8047–8050.
40. C. Z. Li, Q. Wang and Z. K. Zhao, *Green Chem.*, 2008, **10**, 177–182.
41. K. Nakashima, K. Yamaguchi, N. Taniguchi, S. Arai, R. Yamada, S. Katahira, N. Ishida, H. Takahashi, C. Ogino and A. Kondo, *Green Chem.*, 2011, **13**, 2948–2953.
42. K. Ninomiya, K. Kamide, K. Takahashi and N. Shimizu, *Bioresour. Technol.*, 2012, **103**, 259–265.
43. Z. H. Qiu, G. M. Aita and M. S. Walker, *Bioresour. Technol.*, 2012, **117**, 251–256.
44. P. Wasserscheid and T. Welton, *Ionic Liquids in Synthesis*, Wiley-VCH, Weinheim, Germany, 2nd edn, 2008.
45. M. Freemantle, *An Introduction to Ionic Liquids*, Royal Society of Chemistry, 2009.
46. P. T. Anastas, P. Wasserscheid and A. Stark, *Handbook of Green Chemistry, Green Solvents, Ionic Liquids*, John Wiley & Sons, 2014.
47. J. P. Hallett and T. Welton, *Chem. Rev.*, 2011, **111**, 3508–3576.
48. R. Sheldon, *Chem. Commun.*, 2001, 2399–2407.
49. D. Betz, P. Altmann, M. Cokoja, W. A. Herrmann and F. E. Kühn, *Coord. Chem. Rev.*, 2011, **255**, 1518–1540.
50. V. I. Pârvulescu and C. Hardacre, *Chem. Rev.*, 2007, **107**, 2615–2665.
51. S. Cull, J. Holbrey, V. Vargas-Mora, K. Seddon and G. Lye, *Biotechnol. Bioeng.*, 2000, **69**, 227–233.
52. M. Erbeldinger, A. J. Mesiano and A. J. Russell, *Biotechnol. Prog.*, 2000, **16**, 1129–1131.
53. R. Madeira Lau, F. Van Rantwijk, K. Seddon and R. Sheldon, *Org. Lett.*, 2000, **2**, 4189–4191.
54. T. P. Pham, C. W. Cho and Y. S. Yun, *Water Res.*, 2010, **44**, 352–372.
55. S. Stolte, M. Matzke, J. Arning, A. Boschen, W. R. Pitner, U. Welz-Biermann, B. Jastorff and J. Ranke, *Green Chem.*, 2007, **9**, 1170–1179.
56. R. F. Frade and C. A. Afonso, *Hum. Exp. Toxicol.*, 2010, **29**(12), 1038–1054.
57. S. Bruzzone, C. Chiappe, S. Focardi, C. Pretti and M. Renzi, *Chem. Eng. J.*, 2011, **175**, 17–23.
58. D. Coleman and N. Gathergood, *Chem. Soc. Rev.*, 2010, **39**, 600–637.
59. M. Deetlefs and K. R. Seddon, *Green Chem.*, 2010, **12**, 17–30.
60. G. Imperato, B. Koenig and C. Chiappe, *Eur. J. Org. Chem.*, 2007, **2007**, 1049–1058.
61. S. T. Handy, *Chem.–Eur. J.*, 2003, **9**, 2938–2944.
62. C. Chiappe, A. Marra and A. Mele, in *Carbohydrates in Sustainable Development II*, Springer, 2010, pp. 177–195.
63. K. Fukumoto, M. Yoshizawa and H. Ohno, *J. Am. Chem. Soc.*, 2005, **127**, 2398–2399.
64. X. Chen, X. Li, A. Hu and F. Wang, *Tetrahedron: Asymmetry*, 2008, **19**, 1–14.

65. Y. Fukaya, Y. Iizuka, K. Sekikawa and H. Ohno, *Green Chem.*, 2007, **9**, 1155–1157.
66. Y. Yu, X. Lu, Q. Zhou, K. Dong, H. Yao and S. Zhang, *Chem.–Eur. J.*, 2008, **14**, 11174–11182.
67. X. D. Hou, Q. P. Liu, T. J. Smith, N. Li and M. H. Zong, *PLoS One*, 2013, **8**(3), e59145.
68. S. Pavlovica, A. Zicmanis, E. Gzibovska, M. Klavins and P. Mekss, *Green Sustainable Chem.*, 2011, **1**, 103–110.
69. H. Zhao, G. A. Baker, Z. Song, O. Olubajo, T. Crittle and D. Peters, *Green Chem.*, 2008, **10**, 696–705.
70. P. Antonia, F. van Rantwijk and R. A. Sheldon, *Green Chem.*, 2012, **14**, 1584–1588.
71. T. L. Greaves and C. J. Drummond, *Chem. Rev.*, 2008, **108**, 206–237.
72. C. Pretti, C. Chiappe, I. Baldetti, S. Brunini, G. Monni and L. Intorre, *Ecotoxicol. Environ. Saf.*, 2009, **72**, 1170–1176.
73. P. Domínguez de María, in *Ionic Liquids in Biotransformations and Organocatalysis*, John Wiley & Sons, Inc., 2012, pp. 1–14.
74. F. van Rantwijk and R. A. Sheldon, *Chem. Rev.*, 2007, **107**, 2757–2785.
75. P. Lozano, T. De Diego, D. Carrie, M. Vaultier and J. Iborra, *Biotechnol. Lett.*, 2001, **23**, 1529–1533.
76. P. Lozano, T. de Diego, J. P. Guegan, M. Vaultier and J. L. Iborra, *Biotechnol. Bioeng.*, 2001, **75**, 563–569.
77. T. Fráter, O. Ulbert, K. Bélafi-Bakó and L. Gubicza, *Commun. Agric. Appl. Biol. Sci.*, 2002, **68**, 293–296.
78. S. D. Zhu, Y. X. Wu, Q. M. Chen, Z. N. Yu, C. W. Wang, S. W. Jin, Y. G. Ding and G. Wu, *Green Chem.*, 2006, **8**, 325–327.
79. R. C. Remsing, R. P. Swatloski, R. D. Rogers and G. Moyna, *Chem. Commun.*, 2006, 1271–1273.
80. R. A. Sheldon, R. M. Lau, M. J. Sorgedrager, F. van Rantwijk and K. R. Seddon, *Green Chem.*, 2002, **4**, 147–151.
81. M. B. Turner, S. K. Spear, J. G. Huddleston, J. D. Holbrey and R. D. Rogers, *Green Chem.*, 2003, **5**, 443–447.
82. A. J. Walker and N. C. Bruce, *Chem. Commun.*, 2004, 2570–2571.
83. A. J. Walker and N. C. Bruce, *Tetrahedron*, 2004, **60**, 561–568.
84. D. Das, A. Dasgupta and P. K. Das, *Tetrahedron Lett.*, 2007, **48**, 5635–5639.
85. A. R. Toral, P. Antonia, F. J. Hernández, M. H. Janssen, R. Schoevaart, F. van Rantwijk and R. A. Sheldon, *Enzyme Microb. Technol.*, 2007, **40**, 1095–1099.
86. E. Topakas, C. Vafiadi and P. Christakopoulos, *Process Biochem.*, 2007, **42**, 497–509.
87. S. Shah and M. N. Gupta, *Bioorg. Med. Chem. Lett.*, 2007, **17**, 921–924.
88. P. Hara, U. Hanefeld and L. T. Kanerva, *Green Chem.*, 2009, **11**, 250–256.
89. J.-Q. Lai, Z.-L. Hu, R. A. Sheldon and Z. Yang, *Process Biochem.*, 2012, **47**, 2058–2063.

90. A. P. Abbott, D. Boothby, G. Capper, D. L. Davies and R. K. Rasheed, *J. Am. Chem. Soc.*, 2004, **126**, 9142–9147.

91. A. P. Abbott, G. Capper, D. L. Davies, R. K. Rasheed and V. Tambyrajah, *Chem. Commun.*, 2003, 70–71.

92. A. P. Abbott, R. C. Harris, K. S. Ryder, C. D'Agostino, L. F. Gladden and M. D. Mantle, *Green Chem.*, 2011, **13**, 82–90.

93. E. L. Smith, A. P. Abbott and K. S. Ryder, *Chem. Rev.*, 2014, **114**, 11060–11082.

94. C. Ruß and B. König, *Green Chem.*, 2012, **14**, 2969–2982.

95. A. Hayyan, M. A. Hashim, M. Hayyan, F. S. Mjalli and I. M. AlNashef, *Ind. Crops Prod.*, 2013, **46**, 392–398.

96. A. Hayyan, M. A. Hashim, F. S. Mjalli, M. Hayyan and I. M. AlNashef, *Chem. Eng. Sci.*, 2013, **92**, 81–88.

97. A. Hayyan, F. S. Mjalli, I. M. AlNashef, Y. M. Al-Wahaibi, T. Al-Wahaibi and M. A. Hashim, *J. Mol. Liq.*, 2013, **178**, 137–141.

98. M. Francisco, A. van den Bruinhorst and M. C. Kroon, *Angew. Chem., Int. Ed.*, 2013, **52**, 3074–3085.

99. Q. Zhang, M. Benoit, K. De Oliveira Vigier, J. Barrault and F. Jérôme, *Chem.–Eur. J.*, 2012, **18**, 1043–1046.

100. J. Gorke, F. Srienc and R. Kazlauskas, *Biotechnol. Bioprocess Eng.*, 2010, **15**, 40–53.

101. E. Durand, J. Lecomte, B. Baréa, G. Piombo, E. Dubreucq and P. Villeneuve, *Process Biochem.*, 2012, **47**, 2081–2089.

102. H. Zhao, G. A. Baker and S. Holmes, *J. Mol. Catal. B: Enzym.*, 2011, **72**, 163–167.

103. M. Krystof, M. Pérez-Sánchez and P. Domínguez de María, *ChemSusChem*, 2013, **6**, 630–634.

104. B.-P. Wu, Q. Wen, H. Xu and Z. Yang, *J. Mol. Catal. B: Enzym.*, 2014, **101**, 101–107.

105. M. Perez-Sanchez, M. Sandoval, M. J. Hernaiz and P. D. d. Maria, *Curr. Org. Chem.*, 2013, **17**, 1188–1199.

106. Z.-L. Huang, B.-P. Wu, Q. Wen, T.-X. Yang and Z. Yang, *J. Chem. Technol. Biotechnol.*, 2014, **89**, 1975–1981.

107. J. T. Gorke, F. Srienc and R. J. Kazlauskas, *Chem. Commun.*, 2008, 1235–1237.

108. H. Zhao, C. Zhang and T. D. Crittle, *J. Mol. Catal. B: Enzym.*, 2013, **85**, 243–247.

109. M. E. Zakrzewska, E. Bogel-Lukasik and R. Bogel-Lukasik, *Energy Fuels*, 2010, **24**, 737–745.

110. A. A. Rosatella, L. C. Branco and C. A. M. Afonso, *Green Chem.*, 2009, **11**, 1406–1413.

111. S. Murugesan and R. J. Linhardt, *Curr. Org. Synth.*, 2005, **2**, 437–451.

112. N. Galonde, K. Nott, A. Debuigne, M. Deleu, C. Jerôme, M. Paquot and J. P. Wathelet, *J. Chem. Technol. Biotechnol.*, 2012, **87**, 451–471.

113. D. R. MacFarlane, J. Golding, S. Forsyth, M. Forsyth and G. B. Deacon, *Chem. Commun.*, 2001, 1430–1431.

114. Q. Liu, M. H. Janssen, F. van Rantwijk and R. A. Sheldon, *Green Chem.*, 2005, **7**, 39–42.
115. R. P. Swatloski, S. K. Spear, J. D. Holbrey and R. D. Rogers, *J. Am. Chem. Soc.*, 2002, **124**, 4974–4975.
116. Y. Fukaya, K. Hayashi, M. Wada and H. Ohno, *Green Chem.*, 2008, **10**, 44–46.
117. N. Sun, H. Rodríguez, M. Rahman and R. D. Rogers, *Chem. Commun.*, 2011, **47**, 1405–1421.
118. Q. Xu, J. F. Kennedy and L. Liu, *Carbohydr. Polym.*, 2008, **72**, 113–121.
119. Y. Qin, X. M. Lu, N. Sun and R. D. Rogers, *Green Chem.*, 2010, **12**, 968–971.
120. Y. Wu, T. Sasaki, S. Irie and K. Sakurai, *Polymer*, 2008, **49**, 2321–2327.
121. Y. Pu, N. Jiang and A. J. Ragauskas, *J. Wood Chem. Technol.*, 2007, **27**, 23–33.
122. M. Rahman, Y. Qin, M. L. Maxim and R. D. Rogers, WO2010056790 A1, 2010.
123. D. M. Phillips, L. F. Drummy, D. G. Conrady, D. M. Fox, R. R. Naik, M. O. Stone, P. C. Trulove, H. C. De Long and R. A. Mantz, *J. Am. Chem. Soc.*, 2004, **126**, 14350–14351.
124. H. Xie, S. Li and S. Zhang, *Green Chem.*, 2005, **7**, 606–608.
125. H. Zhao, C. L. Jones and J. V. Cowins, *Green Chem.*, 2009, **11**, 1128–1138.
126. Z. Yang and Z.-L. Huang, *Catal. Sci. Technol.*, 2012, **2**, 1767–1775.
127. A. Gumel, M. Annuar, T. Heidelberg and Y. Chisti, *Process Biochem.*, 2011, **46**, 2079–2090.
128. Y.-G. Shi, J.-R. Li and Y.-H. Chu, *J. Chem. Technol. Biotechnol.*, 2011, **86**, 1457–1468.
129. M. Woudenberg-van Oosterom, F. van Rantwijk and R. A. Sheldon, *Biotechnol. Bioeng.*, 1996, **49**, 328–333.
130. P. Domínguez de María, *J. Chem. Technol. Biotechnol.*, 2014, **89**, 11–18.
131. D. A. Fort, R. C. Remsing, R. P. Swatloski, P. Moyna, G. Moyna and R. D. Rogers, *Green Chem.*, 2007, **9**, 63–69.
132. N. Sun, M. Rahman, Y. Qin, M. L. Maxim, H. Rodriguez and R. D. Rogers, *Green Chem.*, 2009, **11**, 646–655.
133. Y. Cao, H. Li, Y. Zhang, J. Zhang and J. He, *J. Appl. Polym. Sci.*, 2010, **116**, 547–554.
134. Q. Li, Y. C. He, M. Xian, G. Jun, X. Xu, J. M. Yang and L. Z. Li, *Bioresour. Technol.*, 2009, **100**, 3570–3575.
135. Z.-M. Wang, L. Li, K.-J. Xiao and J.-Y. Wu, *Bioresour. Technol.*, 2009, **100**, 1687–1690.
136. W. Li, N. Sun, B. Stoner, X. Jiang, X. Lu and R. D. Rogers, *Green Chem.*, 2011, **13**, 2038–2047.
137. S. Morales-delaRosa, J. M. Campos-Martin and J. L. Fierro, *Chem. Eng. J.*, 2012, **181**, 538–541.
138. A. M. da Costa Lopes, K. G. Joao, D. F. Rubik, E. Bogel-Lukasik, L. C. Duarte, J. Andreaus and R. Bogel-Lukasik, *Bioresour. Technol.*, 2013, **142**, 198–208.

139. S. P. Magalhães da Silva, A. M. da Costa Lopes, L. B. Roseiro and R. Bogel-Lukasik, *RSC Adv.*, 2013, **3**, 16040–16050.

140. J. Viell, H. Wulfhorst, T. Schmidt, U. Commandeur, R. Fischer, A. Spiess and W. Marquardt, *Bioresour. Technol.*, 2013, **146**, 144–151.

141. J. Shi, J. M. Gladden, N. Sathitsuksanoh, P. Kambam, L. Sandoval, D. Mitra, S. Zhang, A. George, S. W. Singer, B. A. Simmons and S. Singh, *Green Chem.*, 2013, **15**, 2579–2589.

142. N. L. Mai, K. Ahn and Y.-M. Koo, *Process Biochem.*, 2014, **49**, 872–881.

143. S. Datta, B. Holmes, J. I. Park, Z. W. Chen, D. C. Dibble, M. Hadi, H. W. Blanch, B. A. Simmons and R. Sapra, *Green Chem.*, 2010, **12**, 338–345.

144. P. O. Jones and P. T. Vasudevan, *Biotechnol. Lett.*, 2010, **32**, 103–106.

145. S. Dalal, A. Sharma and M. N. Gupta, *Chem. Cent. J.*, 2007, **1**, 16.

146. B. Li, S. L. Dong, X. L. Xie, Z. B. Xu and L. Li, *Adv. Mater. Res.*, 2012, **581**, 257–260.

147. L. Sutarlie and K.-L. Yang, *J. Colloid Interface Sci.*, 2013, **411**, 76–81.

148. K. J. Khorshidi, H. Lenjannezhadian, M. Jamalan and M. Zeinali, *J. Chem. Technol. Biotechnol.*, 2015, DOI: 10.1002/jctb.4615.

149. R. A. Sheldon, M. J. Sorgedrager and B. Kondor, WO2012023847 A3, 2013.

150. A. Bhattacharya and B. I. Pletschke, *Enzyme Microb. Technol.*, 2014, **61**, 17–27.

151. M. T. Clough, K. Geyer, P. A. Hunt, S. Son, U. Vagt and T. Welton, *Green Chem.*, 2015, **17**, 231–243.

152. X.-D. Hou, N. Li and M.-H. Zong, *Bioresour. Technol.*, 2013, **136**, 469–474.

153. N. Sun, R. Parthasarathi, A. M. Socha, J. Shi, S. Zhang, V. Stavila, K. L. Sale, B. A. Simmons and S. Singh, *Green Chem.*, 2014, **16**, 2546–2557.

154. A. Brandt, J. Grasvik, J. P. Hallett and T. Welton, *Green Chem.*, 2013, **15**, 550–583.

155. P. Verdia, A. Brandt, J. P. Hallett, M. J. Ray and T. Welton, *Green Chem.*, 2014, **16**, 1617–1627.

156. A. George, A. Brandt, K. Tran, S. N. S. M. S. Zahari, D. Klein-Marcuschamer, N. Sun, N. Sathitsuksanoh, J. Shi, V. Stavila, R. Parthasarathi, S. Singh, B. M. Holmes, T. Welton, B. A. Simmons and J. P. Hallett, *Green Chem.*, 2015, **17**, 1728–1734.

157. R. A. Sheldon, *Green Chem.*, 2007, **9**, 1273–1283 and references cited therein.

158. L. Chen, M. Sharifzadeh, N. Mac Dowell, T. Welton, N. Shah and J. P. Hallett, *Green Chem.*, 2014, **16**, 3098–3106.

159. C. Lehmann, F. Sibilla, Z. Maugeri, W. R. Streit, P. Dominguez de Maria, R. Martinez and U. Schwaneberg, *Green Chem.*, 2012, **14**, 2719–2726.

160. A. Paiva, R. Craveiro, I. Aroso, M. Martins, R. L. Reis and A. R. C. Duarte, *ACS Sustainable Chem. Eng.*, 2014, **2**, 1063–1071.

161. Y. Dai, J. van Spronsen, G.-J. Witkamp, R. Verpoorte and Y. H. Choi, *Anal. Chim. Acta*, 2013, **766**, 61–68.

162. Y. H. Choi, J. van Spronsen, Y. Dai, M. Verberne, F. Hollmann, I. W. Arends, G.-J. Witkamp and R. Verpoorte, *Plant Physiol.*, 2011, **156**, 1701–1705.

163. R.-J. van Putten, A. S. Dias and E. de Jong, in *Catalytic Process Development for Renewable Materials*, Wiley-VCH Verlag GmbH & Co. KGaA, 2013, pp. 81–117.

164. B. Kamm, P. R. Gruber and M. Kamm, *Biorefineries – Industrial Processes and Products. Status Quo and Future Directions*, Wiley-VCH Verlag GmbH & Co. KGaA, Weinheim, 2006.

165. R.-J. van Putten, J. C. van der Waal, E. de Jong, C. B. Rasrendra, H. J. Heeres and J. G. de Vries, *Chem. Rev.*, 2013, **113**, 1499–1597.

166. R. A. Sheldon, *Green Chem.*, 2014, **16**, 950–963 and references cited therein.

167. W. Deng, Q. Zhang and Y. Wang, *Sci. China Chem.*, 2015, **58**, 29–46.

168. Y.-B. Yi, J.-W. Lee and C.-H. Chung, *Curr. Org. Chem.*, 2014, **18**, 1149–1158.

169. M. Dashtban, A. Gilbert and P. Fatehi, *RSC Adv.*, 2014, **4**, 2037–2050.

170. T. Ståhlberg, W. Fu, J. M. Woodley and A. Riisager, *ChemSusChem*, 2011, **4**, 451–458.

171. S. Q. Hu, Z. F. Zhang, Y. X. Zhou, B. X. Han, H. L. Fan, W. J. Li, J. L. Song and Y. Xie, *Green Chem.*, 2008, **10**, 1280–1283.

172. F. Goursaud, M. Berchel, J. Guilbot, N. Legros, L. Lemiegre, J. Marcilloux, D. Plusquellec and T. Benvegnu, *Green Chem.*, 2008, **10**, 310–320.

173. K. D. O. Vigier, A. Benguerba, J. Barrault and F. Jerome, *Green Chem.*, 2012, **14**, 285–289.

174. F. Liu, F. Boissou, A. Vignault, L. Lemee, S. Marinkovic, B. Estrine, K. De Oliveira Vigier and F. Jerome, *RSC Adv.*, 2014, **4**, 28836–28841.

CHAPTER 10

Ionic Liquid-Based Processes in the Biorefinery: A SWOT Analysis

ANNEGRET STARK*[a]

[a]SMRI Sugarcane Biorefinery Research Chair, University of KwaZulu-Natal, College of Agriculture, Engineering and Science, School of Engineering, Howard College Campus, Durban, South Africa
*E-mail: starka@ukzn.ac.za

10.1 Introduction

In this chapter, a SWOT analysis[1] is presented which evaluates the strengths and weaknesses, and identifies the opportunities and threats of using ionic liquids (ILs) in an integrated biorefinery. Tables 10.1 and 10.2 provide summaries. Here an integrated biorefinery is defined as a plant converting various bio-based feeds into a number of products, which may include composite materials and formulations, platform chemicals, fine and specialty chemicals, but also fuels, steam and electricity. Optimally, it is 'integrated' also in terms of its processes being interlinked, *e.g.* the waste stream of one becomes the feed of another product.

In the preceding chapters, the state-of-the-art concerning ILs in the biorefinery has been summarized. The application areas of ILs in the biorefinery can be diverse, and include their use as solvents, catalysts, absorbents, *etc.* Chapters 1 to 4 focussed on potential large-scale usage of ILs in the

RSC Green Chemistry No. 36
Ionic Liquids in the Biorefinery Concept: Challenges and Perspectives
Edited by Rafal Bogel-Lukasik
© The Royal Society of Chemistry 2016
Published by the Royal Society of Chemistry, www.rsc.org

Table 10.1   SWOT analysis (strengths and weaknesses): application potential of ionic liquids in the biorefinery.

Strengths	Weaknesses
- Variability of structure allowing for fine-tuning of physicochemical properties for a specific application	- Little information on gate-to-gate life-cycle assessment
- Unique dissolution properties for bio-based feeds (prerequisite for regeneration; unique types and morphologies of regenerates; high selectivity homogeneous phase derivatization)	- High IL costs
	- Dissolution capacities for biomass moderate
	- Side reactions/strong adsorption on polar surfaces (IL loss)
- Reduction of waste generation	- Long-term stability/decomposition
- Accessibility of waste and complex biomass as feed	- Viscosity (affecting processing costs)/ melting points
- Reduction of number of processing steps	- Unusual physicochemical properties
- Higher extraction or reaction selectivity	- Applicability of conventional recovery and purification methods (in particular distillation)
- Applicability of binary (IL + organic solvent) mixtures to reduce IL inventory	- Affinity and accumulation of polar compounds (including water) in ILs
- Adjustability of dissolution/extraction selectivities and capacities by structural design of ILs (unique selling point)	- Ion leaching
	- Know-how transfer bottleneck from fundamental research to application
- Availability of substantial amount of knowledge on IL design and handling	- Availability of data only for specific feedstock, rather than mixtures and wastes
	- Integrated systems (multi-feed-multi-product) are not (yet) considered
	- Rudimental knowledge on enzyme/ microorganism tolerance against ILs
	- Reduced biomass solubility due to water present; drying time and energy intensity
	- Availability of chemical engineering data on IL-bio-based chemicals mixtures: modelling hampered
	- Availability of material compatibility data
	- Availability of occupational safety and health (HSE), biodegradability and toxicity data of specific ILs

biorefinery, *i.e.* on the pre-treatment (dissolution and fractionation) of biomass, and the hydrolysis of biopolymers to (fermentable) oligo- and monomeric carbohydrates. Somewhat smaller volumes of ILs will be necessary for the selective extraction of valuable biomolecules from various feeds (Chapters 5 and 8) or as solvent for biocatalytic processes (Chapters 6 and 9). Chapter 7 concentrates on one platform chemical, *i.e.* 5-hydroxymethylfurfural, showing that although extensive research has being conducted, it has not yet achieved market entry. (There have been 612 publications with 'hydroxymethylfurfural' in title since 2000, using ILs or otherwise, according to Web of Science, 08.04.2015.)

Table 10.2　SWOT analysis (opportunities and threats): application potential of ionic liquids in the biorefinery.

Opportunities	Threats
- Similar development status as other solvent-based processes, *i.e.* little threat posed by solvent alternatives - Networking between industry and academia - Economic advantages for agrarian societies, in particular if value addition (downstream processing) is accomplished locally	- Alternative processing technologies for native biomass (no information how technologies compare) - Lack of communication and information exchange (research groups – growers – manufacturers – purchasers) - Lack of standardized procedures and data for comparison of process alternatives - Insecurity regarding future developments of the bio-economy in general - Large investments required - Dependencies and uncertainties (provision of feed independent of season and weather, logistics, economies of scale) - Technical and bio-based product resilience (purchaser industries, consumers) - Lack of proof of principle for integrated biorefineries - Strong global competition (both in research and in industry)

A SWOT analysis[1] is a tool of strategic management and the basis of several market strategies. Conventionally, a company or product is rationally analysed regarding (intrinsic) 'strengths' and 'weaknesses' in view of the market environment's (extrinsic) 'opportunities' and 'threats'. In this context, we define ILs as technology platform which competes with other technologies in biorefinery processes.

Applying a SWOT analysis to the state-of-the-art of ILs in biorefinery applications aims at identifying strategies which make use of opportunities which arise from the properties inherent to ILs. Furthermore, weaknesses are acknowledged, and can therefore be tackled by additional research, which may shift that aspect from the 'weaknesses' to the 'strengths' category. In the extrinsic part, describing 'opportunities' and 'threats' of the application of ILs in the biorefinery context, some more general aspects are considered, without trying to be comprehensive. In particular, this SWOT analysis does not in detail consider the pros and cons of the biorefinery compared to fossil-based refineries, for example, such as greenhouse gas emission, land and water depletion, socioeconomic factors, *etc.*

10.2　Strength and Weaknesses

ILs have only been investigated in the biorefinery context since about 2007,[2] although certain aspects have been investigated much earlier.[3] (2007 was the date of the first mention of 'ILs' and 'biorefinery' in topic, according to Web of Science.) Due to their unusual properties, including the ability to fine tune the structure, the variety of solvent–solute interactions they may display, and their unique physicochemical properties (low vapour pressure, low melting points, often high thermal stability),[4] they have already demonstrated enormous potential for the conversion of biomass.[5–7] The ability of some ILs to dissolve biopolymers while maintaining their high degree of polymerization (little hydrolysis) is quite unique, and only few, yet difficult to handle, solvent systems exist that can provide homogeneous solutions without the need to derivatize the biopolymer.[8] Dissolution is a prerequisite for regeneration, and regenerates with distinctive properties (*e.g.* lignin, cellulose, keratin, suberin, *etc.*)[5,9–11] can result, which may be useful for specific applications (composite materials, textiles, medical tissues, *etc.*).[12–14] Furthermore, homogeneous phase chemistry can be conducted,[15] such as the derivatization of cellulose, yielding high-performance specialty chemicals.

Besides regeneration and derivatization, the extraction of specialties from native materials by means of selective interaction of the IL and hence removal from the (solid) biomass matrix should be mentioned. Chapters 5 and 8 gave detailed examples on the extraction of nutraceuticals, food additives, cosmetics, essential oils, or pharmaceutical precursors, such as astaxanthin[16] and chitin[16,17] from shrimp waste (antioxidant for animal feed and cosmetic industries, and bio-compatible polymers, respectively), or keratin[18] from poultry feathers (for biomaterial applications). With this approach, the IL volume is low, as only the value-added product is selectively removed, leaving residual biomass behind for further downstream processing.

Due to the limited size of the targeted markets, and the prices that can be achieved for highly specialized products, it appears likely that ILs can and will be used in the future. For these particular materials, the most obvious obstacle for commercialization lies in insufficient communication and networking between academic R&D, growers, manufacturers and purchaser industries, and we suspect that is the reason that potentially interesting materials and compounds often do not find suitable applications.

For larger-scale processes, such as biomass fractionation and hydrolysis to fermentable sugars, IL technologies stand in direct competition with other novel technologies (steam or ammonia fibre explosion, microwave, ultrasound, *etc.*),[19–22] or established processes (being already in operation, and running under highly optimized conditions).[23] In these latter cases, a unique selling point is much more difficult to identify. A novel technology will only be introduced if the advantages compensate the switching costs by far.

The advantages of completely liquefying the biomass as one of the first steps in a biorefinery may lie in a lower waste fraction, a reduction in the number of required processing steps, a higher selectivity, lower energy demand, or greater flexibility in feed variety for the IL-based dissolution compared to heterogeneous processes. The flexibility regarding the feed may substantially contribute to the viability of a biorefinery at a given location. Although most publications currently investigate the dissolution, fractionation and hydrolysis only of specific feeds (*dedicated feed*, *e.g.* wheat straw, grass, spruce, *etc.*), it can be predicted that ILs will be able to perform in a similar fashion for mixed feeds (including various residues of other industries), or feeds with varying compositions, in order to lessen the biorefinery's seasonal, weather and logistic dependencies, and so profit from the economy of scale. The same is obviously true for *dedicated product* biorefineries, where a single (or limited number of) product(s) are produced, such as in pulp-/paper or sugar cane mills, where a strong dependency on market prices, and a limited flexibility to balance fluctuations of these, prevails.

The advantages resulting from the homogenization of the biomass by dissolution must outweigh the enormous IL inventory required for such an approach, taking into account the moderate loading of 5–15 wt.%.[24] For example, the required solvent–solute interactions might also be exhibited by binary IL–organic solvent mixtures, hence reducing the overall solvent cost. An example for such mixture is (1-ethyl-3-methylimidazolium acetate + DMSO),[25,26] which has been demonstrated to reduce the necessary IL content in cellulose dissolution. Further research will show if other, less problematic, co-solvents could be used.

In most instances, comparative studies over the whole lifecycle or at least gate-to-gate have not been carried out. In fact, as pointed out in Chapter 3 on the example of biomass pre-treatment and hydrolysis, there are no standardized methods in place (nor is sufficient process data available) that allow for the comparison of process alternatives, including type of feed, concentrations, temperature, time, auxiliary loading (*e.g.* enzymes), fermentation efficiency in terms of yield, and sugar fermentability. Furthermore, the tolerance of enzymes or microorganisms involved in downstream processing regarding the presence of ILs (irrespective of being present as solvent or residual impurities in the intermediates) must be assessed.

An often cited aspect is the unarguably high IL cost when compared to conventional organic solvents. As pointed out above, the acceptable cost of a solvent will depend on whether there are any alternative cheaper solvents or processes. If alternatives exist, the structural variability of ILs may be employed to seek for less costly ion combinations with the same or similar process performance. In many cases, the cation, which is mostly responsible for the physical properties (melting point, viscosity)[27] and not for the chemical activity (*e.g.* acidity) can be substituted, resulting in a decrease of cost. An example has been presented by Chen *et al.*, where the pre-treatment of biomass could be affected using less costly triethylammonium hydrogensulfate rather than 1-ethyl-3-methylimidazolium acetate.[28]

Further weaknesses of IL technology include their high affinity to polar surfaces (loss of IL due to adsorption or even chemisorption[29]), and potential issues arise regarding long-term thermal stability (affecting recyclability) and viscosity (affecting *e.g.* mass transport, but also energy costs). While the unusual physicochemical properties may be mentioned as the unique strength of ILs, this aspect also needs to be pointed out as a potential weakness, in particular regarding their recyclability: ILs are composed of ions only, and hence feature a low vapour pressure. Therefore, distillation is not in general an adequate means of removing impurities from ILs (as is conventionally done for the purification of organic solvents). Furthermore, their high dissolution capacity for polar molecules, an advantage for homogenizing biomass, can also turn into a disadvantage, as many polar compounds, such as water, inorganic salts, acids, monosaccharides and their decomposition products, accumulate and in turn affect the properties.[30]

It is common knowledge that water, which may accumulate in a process due to moisture in the feed, but also due to absorption from the air (in particular in those ILs that possess strong hydrogen bond accepting anions),[31,32] is difficult to remove from ILs. However, to do so is essential to regain a medium capable of dissolving biomass, as shown in Chapter 2; drying of ILs is time and energy consuming.

Additionally, ionic decomposition products or salts introduced with the feed may lead to ion exchange and 'ion leaching', an aspect not yet investigated in detail. The regeneration efficiency under process conditions must be determined, and possibly it will be necessary to develop specific unit operations for ILs. For example, in order to compare with an enzyme-based hydrolysis process, Binder *et al.*[30] allow replacement costs of $0.13 IL or enzyme per litre ethanol product. For the IL-based process, with a yield of 0.38 L ethanol per kg biomass, a biomass : IL ratio of 1 and IL costs of $2.5 per kg IL, a tolerable loss of 2.0% results. It is clear that this benchmark will be difficult to achieve, both in terms of IL costs and regeneration efficiency. In order to improve regeneration and recycling, interesting approaches of alternative ILs have been discussed in the literature, including switchable,[33] or distillable[34,35] ILs.

Finally, it can be stated that in the past 20 years, tremendous advances in IL technology have been achieved. These solvents are available on a large scale, and expertise exists regarding the tuneability of their properties, comprehending their particularities and handling. Nevertheless, due to the sheer number of different ILs, there is a lack of data for specific ILs, including chemical engineering data on biorefinery-relevant mixtures of ILs and bio-based chemicals (required for modelling purposes), material compatibility, occupational safety and health (HSE), biodegradability and toxicity data.

10.3 Opportunities and Threats

The opportunities and threats for the application of ILs in the biorefinery are the same as for any other technology in this emerging field. Threats are related to the overall insecurity regarding political drivers, legal implementations, market development and the general assessment of the future of the competing fossil-based refineries, *etc.*, hampering developments of the biorefinery.

As pointed out above, where processing alternatives of bio-based feed exist, little information is available on how these compare. Where processes for unique materials, chemicals or bio-fuels have been developed, markets have not yet been established. Although the most likely approach to integrated biorefineries is to add further feed streams and product streams onto a dedicated biorefinery (*e.g.* sugar or pulp/paper mill), rather than building an integrated biorefinery from scratch, large investments are required, and the technical as well as bio-based product resilience needs to be overcome. Clearly, the risk currently is very high as no example for an integrated biorefinery, let alone one involving IL processes, is in operation as proof of principle.

Although a high degree of global competition exists, both in R&D and in the production industry, the shift to a worldwide largely bio-based economy carries the opportunity to strengthening the agrarian sector, in particular in developing countries, and to reduce their dependence on (imported) fossil-based raw materials. Of course, beneficiation and value addition by downstream processing must be accomplished close to the producers to ascertain sustainability.

10.4 Conclusions

In conclusion, whatever technology is employed for their production, biorefinery products must compete with fossil-based products, and in many cases issues of market entry are not solved. Hence, research on the production of fuels, chemicals and materials from renewable resources is often curiosity and technology driven, lacking a business plan for market entry. The dialogue with purchaser industries and consumers must come to the fore to be able to specify market sectors, required quantities and qualities, and hence acceptable production costs in comparison with competitor processes. How far the provision of state-financed incentives of substituting for bio-based products, and the conditions for obtaining these, will be necessary to catalyse the paradigm shift to biorefinery products is debatable, but must be discussed, in particular in the context of a global economy. Possibly a new approach to attenuate risks and make them calculable is called for to aid innovation transfer and adoption.

As opposed to other areas, where IL-based process developments compete with established processes and other advanced technologies, there is a definite chance for ILs to be used in some applications in certain processes of the biorefinery, if they are added as new production lines and no switching costs are anticipated. This is especially the case for products where the unique properties of ILs lead to specific process or product characteristics, *i.e.* featuring an exclusive selling point.

Acknowledgements

This work was supported in part by the South African sugarcane processing industry through the Sugar Milling Research Institute NPC.

References

1. W. Skinner, *Harvard Business Review*, 1969, **47**, 136–145.
2. A. Stark, *Energy Environ. Sci.*, 2011, **4**, 19–32.
3. C. Graenacher, US Pat. 1943176, 1934.
4. P. Wasserscheid and T. Welton, *Ionic Liquids in Synthesis*, Wiley-VCH, Weinheim, Germany, 2nd edn, 2008.
5. R. P. Swatloski, S. K. Spear, J. D. Holbrey and R. D. Rogers, *J. Am. Chem. Soc.*, 2002, **124**, 4974–4975.
6. D. A. Fort, R. C. Remsing, R. P. Swatloski, P. Moyna, G. Moyna and R. D. Rogers, *Green Chem.*, 2007, **9**, 63–69.
7. H. Xie and T. Shi, *Holzforschung*, 2006, **60**, 509–512.
8. T. F. Liebert, in *Cellulose Solvents: For Analysis, Shaping and Chemical Modification*, ed. T. F. Liebert, T. J. Heinze and K. J. Edgar, American Chemical Society, 2010, ch. 1, vol. 1033, pp. 3–54.
9. Y. Pu, N. Jiang and A. J. Ragauskas, *J. Wood Chem. Technol.*, 2007, **27**, 23–33.
10. H. Xie, S. Li and S. Zhang, *Green Chem.*, 2005, **7**, 606–608.
11. H. Garcia, R. Ferreira, M. Petkovic, J. L. Ferguson, M. C. Leitao, H. Q. N. Gunaratne, K. R. Seddon, L. P. N. Rebelo and C. S. Pereira, *Green Chem.*, 2010, **12**, 367–369.
12. J. Lee, R. M. Broughton, S. D. Worley and T. S. Huang, *J. Eng. Fibers Fabr.*, 2007, **2**, 25–32.
13. M. B. Turner, S. K. Spear, J. D. Holbrey, D. T. Daly and R. D. Rogers, *Biomacromolecules*, 2005, **6**, 2497–2502.
14. C. Tsioptsias and C. Panayiotou, *Carbohydr. Polym.*, 2008, **74**, 99–105.
15. S. Barthel and T. Heinze, *Green Chem.*, 2006, **8**, 301–306.
16. R. Praveenkumar, K. Lee, J. Lee and Y.-K. Oh, *Green Chem.*, 2015, **17**, 1226–1234.
17. S. K. Spear, W. M. Reichert, R. P. Swatloski and R. D. Rogers, *Abstr. Pap., Am. Chem. Soc.*, 2001, **221**, U630.
18. Y. X. Wang and X. J. Cao, *Process Biochem.*, 2012, **47**, 896–899.
19. T. P. Schultz, C. J. Blermann and G. D. McGinnis, *Ind. Eng. Chem. Prod. Res. Dev.*, 1983, **22**, 344–348.
20. B. E. Dale, J. Weaver and F. M. Byers, *Appl. Biochem. Biotechnol.*, 1999, **77**, 35–45.
21. A. R. Goncalves and U. Schuchardt, *Appl. Biochem. Biotechnol.*, 2002, **98**, 1213–1219.
22. Z. Hu and Z. Wen, *Biochem. Eng. J.*, 2008, **38**, 369–378.
23. A. V. Someshar and J. E. Pinkerton, in *Air Pollution Engineering Manual*, ed. A. D. Buonicore and W. T. Davis, Reinhold Van Nostrand, New York, 1992.
24. I. A. Kilpeläinen, H. Xie, A. King, M. Granstrom, S. Heikkinen and D. S. Argyropoulos, *J. Agric. Food Chem.*, 2007, **55**, 9142–9148.
25. N. L. Mai, S. H. Ha and Y.-M. Koo, *Process Biochem.*, 2014, **49**, 1144–1151.
26. M. Kostag, T. Liebert and T. Heinze, *Macromol. Rapid Commun.*, 2014, **35**, 1419–1422.

27. A. Stark, *Top. Curr. Chem.*, 2009, **290**, 41–81.
28. L. Chen, M. Sharifzadeh, N. MacDowell, T. Welton, N. Shah and J. P. Hallett, *Green Chem.*, 2014, **16**, 3098–3106.
29. G. Ebner, S. Schiehser, A. Potthast and T. Rosenau, *Tetrahedron Lett.*, 2008, **49**, 7322–7324.
30. J. B. Binder and R. T. Raines, *Proc. Natl. Acad. Sci. U. S. A.*, 2010, **107**, 4516–4521.
31. K. R. Seddon, A. Stark and M. J. Torres, *Pure Appl. Chem.*, 2000, **72**, 2275.
32. A. Stark, M. Sellin, B. Ondruschka and K. Massonne, *Sci. China: Chem.*, 2012, **55**, 1663–1670.
33. P. G. Jessop, D. J. Heldebrant, X. W. Li, C. A. Eckert and C. L. Liotta, *Nature*, 2005, **436**, 1102.
34. A. W. King, J. Asikkala, I. Mutikainen, P. Järvi and I. Kilpeläinen, *Angew. Chem.*, 2011, **123**, 6425–6429.
35. U. P. Kreher, A. E. Rosamilia, C. L. Raston, J. L. Scott and C. R. Strauss, *Molecules*, 2004, **9**, 387–393.